Modern Birkhäuser Classics

Many of the original research and survey monographs, as well as textbooks, in pure and applied mathematics published by Birkhäuser in recent decades have been groundbreaking and have come to be regarded as foundational to the subject. Through the MBC Series, a select number of these modern classics, entirely uncorrected, are being re-released in paperback (and as eBooks) to ensure that these treasures remain accessible to new generations of students, scholars, and researchers.

A Probability Path

Sidney I. Resnick

Reprint of the 2005 Edition

Birkhäuser

Sidney I. Resnick
School of Operations Research
and Information Engineering
Cornell University
Ithaca, NY, USA

ISSN 2197-1803 ISSN 2197-1811 (electronic)
ISBN 978-0-8176-8408-2 ISBN 978-0-8176-8409-9 (eBook)
DOI 10.1007/978-0-8176-8409-9
Springer New York Heidelberg Dordrecht London

Library of Congress Control Number: 2013953550

Printed on acid-free paper

Springer is part of Springer Science+Business Media (www.birkhauser-science.com)

Sidney I. Resnick

A Probability Path

Birkhäuser
Boston • Basel • Berlin

Sidney I. Resnick
School of Operations Research and
Industrial Engineering
Ithaca, NY 14853
USA

AMS Subject Classifications: 60-XX, 60-01, 60A10, 60Exx, 60E10, 60E15, 60Fxx, 60G40, 60G48

Library of Congress Cataloging-in-Publication Data
Resnick, Sidney I.
A probability path / Sidney Resnick.
p. cm.
Includes bibliographical references and index.
ISBN 0-8176-4055-X (hardcover : alk. paper).
1. Probabilities. I. Title.
QA273.R437 1998
519'.24–dc21 98-21749
CIP

ISBN 0-8176-4055-X printed on acid-free paper.

Birkhäuser

Cover design by Minna Resnick.
Typeset by the author in LaTeX.
Printed in the United States of America. (HP)

9 8 7 6 5 SPIN 11361374

www.birkhauser.com

Contents

Preface

There are several good current probability books — Billingsley (1995), Durrett (1991), Port (1994), Fristedt and Gray (1997), and I still have great affection for the books I was weaned on — Breiman (1992), Chung (1974), Feller (1968, 1971) and even Loève (1977). The books by Neveu (1965, 1975) are educational and models of good organization. So why publish another? Many of the existing books are encyclopedic in scope and seem intended as reference works, with navigation problems for the beginner. Some neglect to teach any measure theory, assuming students have already learned all the foundations elsewhere. Most are written by mathematicians and have the built in bias that the reader is assumed to be a mathematician who is coming to the material for its beauty. Most books do not clearly indicate a one-semester syllabus which will offer the essentials.

I and my students have consequently found difficulties using currently available probability texts. There is a large market for measure theoretic probability by students whose primary focus is not mathematics for its own sake. Rather, such students are motivated by examples and problems in statistics, engineering, biology and finance to study probability with the expectation that it will be useful to them in their research work. Sometimes it is not clear where their work will take them, but it is obvious they need a deep understanding of advanced probability in order to read the literature, understand current methodology, and prove that the new technique or method they are dreaming up is superior to standard practice.

So the clientele for an advanced or measure theoretic probability course that is primarily motivated by applications outnumbers the clientele deeply embedded in pure mathematics. Thus, I have tried to show links to statistics and operations research. The pace is quick and disciplined. The course is designed for one semester with an overstuffed curriculum that leaves little time for interesting excursions or personal favorites. A successful book needs to cover the basics clearly. Equally important, the exposition must be efficient, allowing for time to cover the next important topic.

Chapters 1, 2 and 3 cover enough measure theory to give a student access to advanced material. Independence is covered carefully in Chapter 4 and expectation and Lebesgue integration in Chapter 5. There is some attention to comparing the Lebesgue vs the Riemann integral, which is usually an area that concerns students. Chapter 6 surveys and compares different modes of convergence and must be carefully studied since limit theorems are a central topic in classical probability and form the core results. This chapter naturally leads into laws of large numbers (Chapter 7), convergence in distribution, and the central limit theorem (Chapters 8 and 9). Chapter 10 offers a careful discussion of conditional expectation and martingales, including a short survey of the relevance of martingales to mathematical finance.

Suggested syllabi: If you have one semester, you have the following options: You could cover Chapters 1–8 plus 9, or Chapters 1–8 plus 10. You would have to move along at unacceptable speed to cover both Chapters 9 and 10. If you have two quarters, do Chapters 1–10. If you have two semesters, you could do Chapters 1–10, and then do the random walk Chapter 7 and the Brownian Motion Chapter 6 from Resnick (1992), or continue with stochastic calculus from one of many fine sources.

Exercises are included and students should be encouraged or even forced to do many of them.

Harry is on vacation.

Acknowledgements. Cornell University continues to provide a fine, stimulating environment. NSF and NSA have provided research support which, among other things, provides good computing equipment. I am pleased that AMS-TEXand LATEX merged into AMS-LATEX, which is a marvelous tool for writers. Rachel, who has grown into a terrific adult, no longer needs to share her mechanical pencils with me. Nathan has stopped attacking my manuscripts with a hole puncher and gives ample evidence of the fine adult he will soon be. Minna is the ideal companion on the random path of life. Ann Kostant of Birkhäuser continues to be a pleasure to deal with.

Sidney I. Resnick
School of Operations Research and Industrial Engineering
Cornell University

1

Sets and Events

1.1 Introduction

The core classical theorems in probability and statistics are the following:

- *The law of large numbers (LLN)*: Suppose $\{X_n, n \geq 1\}$ are independent, identically distributed (iid) random variables with common mean $E(X_n) = \mu$. The LLN says the sample average is approximately equal to the mean, so that

$$\frac{1}{n}\sum_{i=1}^{n} X_i \to \mu.$$

  An immediate concern is what does the convergence arrow "$\to$" mean? This result has far-reaching consequences since, if

$$X_i = \begin{cases} 1, & \text{if event } A \text{ occurs,} \\ 0, & \text{otherwise} \end{cases}$$

  then the average $\sum_{i=1}^{n} X_i/n$ is the relative frequency of occurrence of A in n repetitions of the experiment and $\mu = P(A)$. The LLN justifies the frequency interpretation of probabilities and much statistical estimation theory where it underlies the notion of *consistency* of an estimator.

- *Central limit theorem (CLT)*: The central limit theorem assures us that sample averages when centered and scaled to have mean 0 and variance 1 have a distribution that is approximately normal. If $\{X_n, n \geq 1\}$ are iid with common mean $E(X_n) = \mu$ and variance $\text{Var}(X_n) = \sigma^2$, then

$$P\left[\frac{\sum_{i=1}^n X_i - n\mu}{\sigma\sqrt{n}} \le x\right] \to N(x) := \int_{-\infty}^{x} \frac{e^{-u^2/2}}{\sqrt{2\pi}}\, du.$$

This result is arguably the most important and most frequently applied result of probability and statistics. How is this result and its variants proved?

- *Martingale convergence theorems and optional stopping*: A martingale is a stochastic process $\{X_n, n \ge 0\}$ used to model a fair sequence of gambles (or, as we say today, investments). The conditional expectation of your wealth X_{n+1} after the next gamble or investment given the past equals the current wealth X_n. The martingale results on convergence and optimal stopping underlie the modern theory of stochastic processes and are essential tools in application areas such as mathematical finance. What are the basic results and why do they have such far reaching applicability?

Historical references to the CLT and LLN can be found in such texts as Breiman (1968), Chapter I; Feller, volume I (1968) (see the background on coin tossing and the de Moivre-Laplace CLT); Billingsley (1995), Chapter 1; Port (1994), Chapter 17.

1.2 Basic Set Theory

Here we review some basic set theory which is necessary before we can proceed to carve a path through classical probability theory. We start by listing some basic notation.

- Ω: An abstract set representing the sample space of some experiment. The points of Ω correspond to the outcomes of an experiment (possibly only a thought experiment) that we want to consider.

- $\mathcal{P}(\Omega)$: The power set of Ω, that is, the set of all subsets of Ω.

- Subsets $A, B, \ldots$ of Ω which will usually be written with roman letters at the beginning of the alphabet. Most (but maybe not all) subsets will be thought of as *events*, that is, collections of simple events (points of Ω).

  The necessity of restricting the class of subsets which will have probabilities assigned to them to something perhaps smaller than $\mathcal{P}(\Omega)$ is one of the sophistications of modern probability which separates it from a treatment of discrete sample spaces.

- Collections of subsets $\mathcal{A}, \mathcal{B}, \ldots$ which will usually be written by calligraphic letters from the beginning of the alphabet.

- An individual element of Ω: $\omega \in \Omega$.

- The empty set $\emptyset$, not to be confused with the Greek letter ϕ.

$\mathcal{P}(\Omega)$ has the structure of a Boolean algebra. This is an abstract way of saying that the usual set operations perform in the usual way. We will proceed using naive set theory rather than by axioms. The *set operations which you should know and will be commonly used* are listed next. These are often used to manipulate sets in a way that parallels the construction of complex events from simple ones.

1. *Complementation*: The complement of a subset $A \subset \Omega$ is
$$A^c := \{\omega : \omega \notin A\}.$$

2. *Intersection over arbitrary index sets*: Suppose T is some index set and for each $t \in T$ we are given $A_t \subset \Omega$. We define
$$\bigcap_{t \in T} A_t := \{\omega : \omega \in A_t, \quad \forall\, t \in T\}.$$
The collection of subsets $\{A_t, t \in T\}$ is *pairwise disjoint* if whenever $t, t' \in T$, but $t \neq t'$, we have
$$A_t \cap A_{t'} = \emptyset.$$
A synonym for pairwise disjoint is *mutually disjoint*. Notation: When we have a small number of subsets, perhaps two, we write for the intersection of subsets A and B
$$AB = A \cap B,$$
using a "multiplication" notation as shorthand.

3. *Union over arbitrary index sets*: As above, let T be an index set and suppose $A_t \subset \Omega$. Define the union as
$$\bigcup_{t \in T} A_t := \{\omega : \omega \in A_t, \quad \text{for some } \ t \in T\}.$$
When sets $A_1, A_2, \ldots$ are mutually disjoint, we sometimes write
$$A_1 + A_2 + \ldots$$
or even $\sum_{i=1}^{\infty} A_i$ to indicate $\cup_{i=1}^{\infty} A_i$, the union of mutually disjoint sets.

4. *Set difference* Given two sets A, B, the part that is in A but not in B is
$$A \setminus B := AB^c.$$
This is most often used when $B \subset A$; that is, when $AB = B$.

5. *Symmetric difference*: If A, B are two subsets, the points that are in one but not in both are called the symmetric difference
$$A \,\Delta\, B = (A \setminus B) \cup (B \setminus A).$$

You may wonder why we are interested in arbitrary index sets. Sometimes the natural indexing of sets can be rather exotic. Here is one example. Consider the space $\mathrm{USC}_+([0,\infty))$, the space of non-negative *upper semi-continuous* functions with domain $[0,\infty)$. For $f \in \mathrm{USC}_+([0,\infty))$, define the hypograph $\mathrm{hypo}(f)$ by

$$\mathrm{hypo}(f) = \{(s,x) : 0 \le x \le f(s)\},$$

so that $\mathrm{hypo}(f)$ is the portion of the plane between the horizontal axis and the graph of f. Thus we have a family of sets indexed by the upper semi-continuous functions, which is a somewhat more exotic index set than the usual subsets of the integers or real line.

The previous list described common ways of constructing new sets from old. Now we list ways sets can be compared. Here are some simple *relations between sets.*

1. *Containment*: A is a subset of B, written $A \subset B$ or $B \supset A$, iff $AB = A$ or equivalently iff $\omega \in A$ implies $\omega \in B$.
2. *Equality*: Two subsets A, B are equal, written $A = B$, iff $A \subset B$ and $B \subset A$. This means $\omega \in A$ iff $\omega \in B$.

Example 1.2.1 Here are two simple examples of set equality on the real line for you to verify.

(i) $\bigcup_{n=1}^{\infty}[0, n/(n+1)) = [0,1)$.

(ii) $\bigcap_{n=1}^{\infty}(0, 1/n) = \emptyset$. □

Here are some straightforward *properties of set containment* that are easy to verify:

$$\begin{aligned}
&A \subset A,\\
&A \subset B \text{ and } B \subset C \text{ implies } A \subset C,\\
&A \subset C \text{ and } B \subset C \text{ implies } A \cup B \subset C,\\
&A \supset C \text{ and } B \supset C \text{ implies } AB \supset C,\\
&A \subset B \text{ iff } B^c \subset A^c
\end{aligned}$$

Here is a list of *simple connections* between the set operations:

1. *Complementation*:

$$(A^c)^c = A, \quad \emptyset^c = \Omega, \quad \Omega^c = \emptyset.$$

2. *Commutativity* of set union and intersection:

$$A \cup B = B \cup A, \quad A \cap B = B \cap A.$$

Note as a consequence of the definitions, we have

$$\begin{aligned} A \cup A &= A, & A \cap A &= A, \\ A \cup \emptyset &= A, & A \cap \emptyset &= \emptyset \\ A \cup \Omega &= \Omega, & A \cap \Omega &= A, \\ A \cup A^c &= \Omega, & A \cap A^c &= \emptyset. \end{aligned}$$

3. *Associativity* of union and intersection:

$$(A \cup B) \cup C = A \cup (B \cup C), \quad (A \cap B) \cap C = A \cap (B \cap C).$$

4. *De Morgan's laws*, a relation between union, intersection and complementation: Suppose as usual that T is an index set and $A_t \subset \Omega$. Then we have

$$(\bigcup_{t \in T} A_t)^c = \bigcap_{t \in T} (A_t^c), \quad (\bigcap_{t \in T} A_t)^c = \bigcup_{t \in T} (A_t^c).$$

The two De Morgan's laws given are equivalent.

5. *Distributivity* laws providing connections between union and intersection:

$$\begin{aligned} B \cap \left(\bigcup_{t \in T} A_t \right) &= \bigcup_{t \in T} (BA_t), \\ B \cup \left(\bigcap_{t \in T} A_t \right) &= \bigcap_{t \in T} (B \cup A_t). \end{aligned}$$

1.2.1 *Indicator functions*

There is a very nice and useful duality between sets and functions which emphasizes the algebraic properties of sets. It has a powerful expression when we see later that taking the expectation of a random variable is theoretically equivalent to computing the probability of an event. If $A \subset \Omega$, we define the *indicator function of A* as

$$1_A(\omega) = \begin{cases} 1, & \text{if } \omega \in A, \\ 0, & \text{if } \omega \in A^c. \end{cases}$$

This definition quickly yields the simple properties:

$$1_A \le 1_B \text{ iff } A \subset B,$$

and

$$1_{A^c} = 1 - 1_A.$$

Note here that we use the convention that for two functions f, g with domain Ω and range $\mathbb{R}$, we have

$$f \le g \text{ iff } f(\omega) \le g(\omega) \text{ for all } \omega \in \Omega$$

and

$$f = g \text{ if } f \le g \text{ and } g \le f.$$

1.3 Limits of Sets

The definition of convergence concepts for random variables rests on manipulations of sequences of events which require limits of sets. Let $A_n \subset \Omega$. We define

$$\inf_{k \geq n} A_k := \bigcap_{k=n}^{\infty} A_k, \quad \sup_{k \geq n} A_k := \bigcup_{k=n}^{\infty} A_k$$

$$\liminf_{n\to\infty} A_n = \bigcup_{n=1}^{\infty}\bigcap_{k=n}^{\infty} A_k,$$

$$\limsup_{n\to\infty} A_n = \bigcap_{n=1}^{\infty}\bigcup_{k=n}^{\infty} A_k.$$

The *limit* of a sequence of sets is defined as follows: If for some sequence $\{B_n\}$ of subsets

$$\limsup_{n\to\infty} B_n = \liminf_{n\to\infty} B_n = B,$$

then B is called the limit of B_n and we write $\lim_{n\to\infty} B_n = B$ or $B_n \to B$. It will be demonstrated soon that

$$\liminf_{n\to\infty} A_n = \lim_{n\to\infty}\left(\inf_{k\geq n} A_k\right)$$

and

$$\limsup_{n\to\infty} A_n = \lim_{n\to\infty}\left(\sup_{k\geq n} A_k\right).$$

To make sure you understand the definitions, you should check the following example as an exercise.

Example 1.3.1 Check

$$\liminf_{n\to\infty}[0, n/(n+1)) = \limsup_{n\to\infty}[0, n/(n+1)) = [0, 1). \qquad \square$$

We can now give an interpretation of $\liminf_{n\to\infty} A_n$ and $\limsup_{n\to\infty} A_n$.

Lemma 1.3.1 *Let $\{A_n\}$ be a sequence of subsets of Ω.*

(a) For lim sup *we have the interpretation*

$$\limsup_{n\to\infty} A_n = \left\{\omega : \sum_{n=1}^{\infty} 1_{A_n}(\omega) = \infty\right\}$$
$$= \{\omega : \omega \in A_{n_k}, k = 1, 2 \dots\}$$

for some subsequence n_k depending on ω. Consequently, we write

$$\limsup_{n\to\infty} A_n = [A_n \ i.o.\ ]$$

where i.o. stands for infinitely often.

(b) For lim inf *we have the interpretation*

$$\begin{aligned}\liminf_{n\to\infty} A_n &= \{\omega : \omega \in A_n \text{ for all } n \text{ except a finite number }\}\\ &= \{\omega : \sum_n 1_{A_n^c}(\omega) < \infty\}\\ &= \{\omega : \omega \in A_n, \forall n \geq n_0(\omega)\}.\end{aligned}$$

Proof. (a) If

$$\omega \in \limsup_{n\to\infty} A_n = \bigcap_{n=1}^{\infty}\bigcup_{k=n}^{\infty} A_k,$$

then for every n, $\omega \in \cup_{k\geq n} A_k$ and so for all n, there exists some $k_n \geq n$ such that $\omega \in A_{k_n}$, and therefore

$$\sum_{j=1}^{\infty} 1_{A_j}(\omega) \geq \sum_n 1_{A_{k_n}}(\omega) = \infty,$$

which implies

$$\omega \in \left\{\omega : \sum_{n=1}^{\infty} 1_{A_n}(\omega) = \infty\right\};$$

thus

$$\limsup_{n\to\infty} A_n \subset \{\omega : \sum_{j=1}^{\infty} 1_{A_j}(\omega) = \infty\}.$$

Conversely, if

$$\omega \in \{\omega : \sum_{j=1}^{\infty} 1_{A_j}(\omega) = \infty\},$$

then there exists $k_n \to \infty$ such that $\omega \in A_{k_n}$, and therefore for all n, $\omega \in \cup_{j\geq n} A_j$ so that $\omega \in \limsup_{n\to\infty} A_n$. By defininition

$$\{\omega : \sum_{j=1}^{\infty} 1_{A_j}(\omega) = \infty\} \subset \limsup_{n\to\infty} A_n.$$

This proves the set inclusion in both directions and shows equality.

The proof of (b) is similar. □

The properties of lim sup and lim inf are analogous to what we expect with real numbers. The link is through the indicator functions and will be made explicit shortly. Here are two simple connections:

1. The relationship between lim sup and lim inf is

$$\liminf_{n\to\infty} A_n \subset \limsup_{n\to\infty} A_n$$

since

$$\{\omega : \omega \in A_n, \text{ for all } n \geq n_0(\omega)\} \subset \{\omega : \omega \in A_n \text{ infinitely often}\}$$
$$= \limsup_{n\to\infty} A_n.$$

2. Connections via de Morgan's laws:

$$(\liminf_{n\to\infty} A_n)^c = \limsup_{n\to\infty} A_n^c$$

since applying de Morgan's laws twice yields

$$\left(\bigcup_{n=1}^{\infty}\bigcap_{k\geq n} A_k\right)^c = \bigcap_{n=1}^{\infty}\left(\bigcap_{k\geq n} A_k\right)^c$$
$$= \bigcap_{n=1}^{\infty}\left(\bigcup_{k\geq n} A_k^c\right)$$
$$= \limsup_{n\to\infty} A_n^c.$$

For a sequence of random variables $\{X_n, n \geq 0\}$, suppose we need to show $X_n \to X_0$ almost surely. As we will see, this means that we need to show

$$P\{\omega : \lim_{n\to\infty} X_n(\omega) = X_0(\omega)\} = 1.$$

We will show later that a criterion for this is that for all $\varepsilon > 0$

$$P\{[|X_n - X_0| > \varepsilon] \text{ i.o.}\} = 0.$$

That is, with $A_n = [|X_n - X_0| > \varepsilon]$, we need to check

$$P\left(\limsup_{n\to\infty} A_n\right) = 0.$$

1.4 Monotone Sequences

A sequence of sets $\{A_n\}$ is monotone non-decreasing if $A_1 \subset A_2 \subset \cdots$. The sequence $\{A_n\}$ is monotone non-increasing if $A_1 \supset A_2 \supset A_3 \cdots$. To indicate a monotone sequence we will use the notation $A_n \nearrow$ or $A_n \uparrow$ for non-decreasing sets and $A_n \searrow$ or $A_n \downarrow$ for non-increasing sets. For a monotone sequence of sets, the limit always exists.

Proposition 1.4.1 *Suppose $\{A_n\}$ is a monotone sequence of subsets.*

(1) If $A_n \nearrow$, then $\lim_{n\to\infty} A_n = \cup_{n=1}^{\infty} A_n$.

(2) If $A_n \searrow$, then $\lim_{n\to\infty} A_n = \cap_{n=1}^{\infty} A_n$.

Consequently, since for any sequences B_n, we have

$$\inf_{k\geq n} B_k \nearrow, \quad \sup_{k\geq n} B_k \searrow,$$

it follows that

$$\liminf_{n\to\infty} B_n = \lim_{n\to\infty}\left(\inf_{k\geq n} B_k\right), \quad \limsup_{n\to\infty} B_n = \lim_{n\to\infty}\left(\sup_{k\geq n} B_k\right).$$

Proof. (1) We need to show

$$\liminf_{n\to\infty} A_n = \limsup_{n\to\infty} A_n = \bigcup_{n=1}^{\infty} A_n.$$

Since $A_j \subset A_{j+1}$,

$$\bigcap_{k\geq n} A_k = A_n,$$

and therefore

$$\liminf_{n\to\infty} A_n = \bigcup_{n=1}^{\infty}\left(\bigcap_{k\geq n} A_k\right) = \bigcup_{n=1}^{\infty} A_n. \tag{1.1}$$

Likewise

$$\begin{aligned}\limsup_{n\to\infty} A_n &= \bigcap_{n=1}^{\infty}\bigcup_{k\geq n} A_k \subset \bigcup_{k\geq 1} A_k \\ &= \liminf_{n\to\infty} A_n \qquad \text{(from (1.1))} \\ &\subset \limsup_{n\to\infty} A_n.\end{aligned}$$

Thus equality prevails and

$$\limsup_{n\to\infty} A_n \subset \bigcup_{k\geq 1} A_k \subset \limsup_{n\to\infty} A_n;$$

therefore

$$\limsup_{n\to\infty} A_n = \bigcup_{k=1}^{\infty} A_k.$$

This coupled with (1.1) yields (1).

The proof of (2) is similar. □

Example 1.4.1 As an easy exercise, check that you believe that

$$\lim_{n\to\infty}[0, 1-1/n] = [0,1)$$
$$\lim_{n\to\infty}[0, 1-1/n) = [0,1)$$
$$\lim_{n\to\infty}[0, 1+1/n] = [0,1]$$
$$\lim_{n\to\infty}[0, 1+1/n) = [0,1].$$

□

Here are relations that provide additional parallels between sets and functions and further illustrate the algebraic properties of sets. As usual let $\{A_n\}$ be a sequence of subsets of Ω.

1. We have

$$1_{\inf_{n\geq k} A_n} = \inf_{n\geq k} 1_{A_n}, \quad 1_{\sup_{n\geq k} A_n} = \sup_{n\geq k} 1_{A_n}.$$

2. The following inequality holds:

$$1_{\bigcup_n A_n} \leq \sum_n 1_{A_n}$$

and if the sequence $\{A_i\}$ is mutually disjoint, then equality holds.

3. We have

$$1_{\limsup_{n\to\infty} A_n} = \limsup_{n\to\infty} 1_{A_n}, \quad 1_{\liminf_{n\to\infty} A_n} = \liminf_{n\to\infty} 1_{A_n}.$$

4. Symmetric difference satisfies the relation

$$1_{A\Delta B} = 1_A + 1_B \pmod 2\,.$$

Note (3) follows from (1) since

$$1_{\limsup_{n\to\infty} A_n} = 1_{\inf_{n\geq 1}\sup_{k\geq n} A_k},$$

and from (1) this is

$$\inf_{n\geq 1} 1_{\sup_{k\geq n} A_k}.$$

Again using (1) we get

$$\inf_{n\geq 1}\sup_{k\geq n} 1_{A_k} = \limsup_{n\to\infty} 1_{A_n},$$

from the definition of the lim sup of a sequence of numbers.

To prove (1), we must prove two functions are equal. But $1_{\inf_{n\geq k} A_n}(\omega) = 1$ iff $\omega \in \inf_{n\geq k} A_n = \bigcap_{n=k}^{\infty} A_n$ iff $\omega \in A_n$ for all $n \geq k$ iff $1_{A_n}(\omega) = 1$ for all $n \geq k$ iff $\inf_{n\geq k} 1_{A_n}(\omega) = 1$. □

1.5 Set Operations and Closure

In order to think about what it means for a class to be closed under certain set operations, let us first consider some typical set operations. Suppose $\mathcal{C} \subset \mathcal{P}(\Omega)$ is a collection of subsets of Ω.

(1) Arbitrary union: Let T be any arbitrary index set and assume for each $t \in T$ that $A_t \in \mathcal{C}$. The word arbitrary here reminds us that T is not necessarily finite, countable or a subset of the real line. The arbitrary union is

$$\bigcup_{t \in T} A_t.$$

(2) Countable union: Let $A_n, n \geq 1$ be any sequence of subsets in $\mathcal{C}$. The countable union is

$$\bigcup_{j=1}^{\infty} A_j.$$

(3) Finite union: Let $A_1, \ldots, A_n$ be any finite collection of subsets in $\mathcal{C}$. The finite union is

$$\bigcup_{j=1}^{n} A_j.$$

(4) Arbitrary intersection: As in (1) the arbitrary intersection is

$$\bigcap_{t \in T} A_t.$$

(5) Countable intersection: As in (2), the countable intersection is

$$\bigcap_{j=1}^{\infty} A_j.$$

(6) Finite intersection: As in (3), the finite intersection is

$$\bigcap_{j=1}^{n} A_j.$$

(7) Complementation: If $A \in \mathcal{C}$, then A^c is the set of points not in A.

(8) Monotone limits: If $\{A_n\}$ is a monotone sequence of sets in $\mathcal{C}$, the monotone limit

$$\lim_{n \to \infty} A_n$$

is $\cup_{j=1}^{\infty} A_j$ in case $\{A_n\}$ is non-decreasing and is $\cap_{j=1}^{\infty} A_j$ if $\{A_n\}$ is non-increasing.

Definition 1.5.1 (Closure.) Let $\mathcal{C}$ be a collection of subsets of Ω. $\mathcal{C}$ is *closed* under one of the set operations 1–8 listed above if the set obtained by performing the set operation on sets in $\mathcal{C}$ yields a set in $\mathcal{C}$.

For example, $\mathcal{C}$ is closed under (3) if for any finite collection $A_1, \ldots, A_n$ of sets in $\mathcal{C}$, $\cup_{j=1}^n A_j \in \mathcal{C}$.

Example 1.5.1 1. Suppose $\Omega = \mathbb{R}$, and

$$\begin{aligned} \mathcal{C} &= \text{finite intervals} \\ &= \{(a, b], -\infty < a \leq b < \infty\}. \end{aligned}$$

$\mathcal{C}$ is *not* closed under finite unions since $(1, 2] \cup (36, 37]$ is not a finite interval. $\mathcal{C}$ is closed under finite intersections since $(a, b] \cap (c, d] = (a \vee c, d \wedge b]$. Here we use the notation $a \vee b = \max\{a, b\}$ and $a \wedge b = \min\{a, b\}$.

2. Suppose $\Omega = \mathbb{R}$ and $\mathcal{C}$ consists of the open subsets of $\mathbb{R}$. Then $\mathcal{C}$ is not closed under complementation since the complement of an open set is not open.

Why do we need the notion of closure? A probability model has an event space. This is the class of subsets of Ω to which we know how to assign probabilities. In general, we cannot assign probabilities to all subsets, so we need to distinguish a class of subsets that have assigned probabilities. The subsets of this class are called *events*. We combine and manipulate events to make more complex events via set operations. We need to be sure we can still assign probabilities to the results of the set operations. We do this by postulating that allowable set operations applied to events yield events; that is, we require that certain set operations do not carry events outside the event space. This is the idea behind *closure*.

Definition 1.5.2 A *field* is a non-empty class of subsets of Ω closed under finite union, finite intersection and complements. A synonym for field is algebra.

A minimal set of postulates for $\mathcal{A}$ to be a field is

(i) $\Omega \in \mathcal{A}$.

(ii) $A \in \mathcal{A}$ implies $A^c \in \mathcal{A}$.

(iii) $A, B \in \mathcal{A}$ implies $A \cup B \in \mathcal{A}$.

Note if $A_1, A_2, A_3 \in \mathcal{A}$, then from (iii)

$$A_1 \cup A_2 \cup A_3 = (A_1 \cup A_2) \cup A_3 \in \mathcal{A}$$

and similarly if $A_1, \ldots, A_n \in \mathcal{A}$, then $\cup_{i=1}^n A_i \in \mathcal{A}$. Also if $A_i \in \mathcal{A}$, $i = 1, \ldots, n$, then $\cap_{i=1}^n A_i \in \mathcal{A}$ since

$$A_i \in \mathcal{A} \quad \text{implies} \quad A_i^c \in \mathcal{A} \qquad \text{(from (ii))}$$

$$A_i^c \in \mathcal{A} \quad \text{implies} \quad \bigcup_{i=1}^n A_i^c \in \mathcal{A} \qquad \text{(from (iii))}$$

$$\bigcup_{i=1}^n A_i^c \quad \text{implies} \quad \left(\bigcup_{i=1}^n A_i^c\right)^c \in \mathcal{A} \qquad \text{(from (ii))}$$

and finally

$$\left(\bigcup_1^n A_i^c\right)^c = \bigcap_1^n A_i$$

by de Morgan's laws so $\mathcal{A}$ is closed under finite intersections.

Definition 1.5.3 A σ-field $\mathcal{B}$ is a non-empty class of subsets of Ω closed under countable union, countable intersection and complements. A synonym for σ-field is σ-algebra.

A mimimal set of postulates for $\mathcal{B}$ to be a σ-field is

(i) $\Omega \in \mathcal{B}$.

(ii) $B \in \mathcal{B}$ implies $B^c \in \mathcal{B}$.

(iii) $B_i \in \mathcal{B}$, $i \geq 1$ implies $\cup_{i=1}^\infty B_i \in \mathcal{B}$.

As in the case of the postulates for a field, if $B_i \in \mathcal{B}$, for $i \geq 1$, then $\bigcap_{i=1}^\infty B_i \in \mathcal{B}$.

In probability theory, the event space is a σ-field. This allows us enough flexibility constructing new-events from old ones (closure) but not so much flexibility that we have trouble assigning probabilities to the elements of the σ-field.

1.5.1 Examples

The definitions are amplified by some examples of fields and σ-fields.

(1) The power set. Let $\mathcal{B} = \mathcal{P}(\Omega)$, the power set of Ω so that $\mathcal{P}(\Omega)$ is the class of all subsets of Ω. This is obviously a σ-field since it satisfies all closure postulates.

(2) The trivial σ-field. Let $\mathcal{B} = \{\emptyset, \Omega\}$. This is also a σ-field since it is easy to verify the postulates hold.

(3) The countable/co-countable σ-field. Let $\Omega = \mathbb{R}$, and

$$\mathcal{B} = \{A \subset \mathbb{R} : A \text{ is countable }\} \cup \{A \subset \mathbb{R} : A^c \text{ is countable }\},$$

so $\mathcal{B}$ consists of the subsets of $\mathbb{R}$ that are either countable or have countable complements. $\mathcal{B}$ is a σ-field since

(i) $\Omega \in \mathcal{B}$ (since $\Omega^c = \emptyset$ is countable).

(ii) $A \in \mathcal{B}$ implies $A^c \in \mathcal{B}$.

(iii) $A_i \in \mathcal{B}$ implies $\cap_{i=1}^{\infty} A_i \in \mathcal{B}$.

To check this last statement, there are 2 cases. Either

(a) at least one A_i is countable so that $\cap_{i=1}^{\infty} A_i$ is countable and hence in $\mathcal{B}$, or

(b) no A_i is countable, which means A_i^c is countable for every i. So $\cup_{i=1}^{\infty} A_i^c$ is countable and therefore

$$(\bigcup_{i=1}^{\infty}(A_i^c))^c = \bigcap_{i=1}^{\infty} A_i \in \mathcal{B}.$$

Note two points for this example:

- If $A = (-\infty, 0]$, then $A^c = (0, \infty)$ and neither A nor A^c is countable which means $A \notin \mathcal{B}$. So $\mathcal{B} \neq \mathcal{P}(\Omega)$.
- $\mathcal{B}$ is not closed under arbitrary unions. For example, for each $t \leq 0$, the singleton set $\{t\} \in \mathcal{B}$, since it is countable. But $A = \cup_{t \leq 0}\{t\} = (-\infty, 0] \notin \mathcal{B}$.

(4) **A field that is not a σ-field.** Let $\Omega = (0, 1]$ and suppose $\mathcal{A}$ consists of the empty set $\emptyset$ and all finite unions of disjoint intervals of the form $(a, a']$, $0 \leq a \leq a' \leq 1$. A typical set in $\mathcal{A}$ is of the form $\cup_{i=1}^{m}(a_i, a_i']$ where the intervals are disjoint. We assert that $\mathcal{A}$ is a field. To see this, observe the following.

(i) $\Omega = (0, 1] \in \mathcal{A}$.

(ii) $\mathcal{A}$ is closed under complements: For example, consider the union represented by dark lines

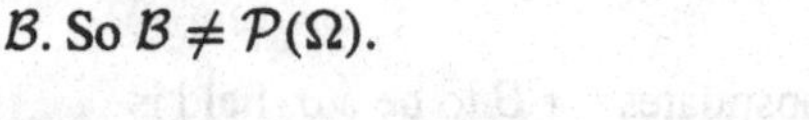

FIGURE 1.1

which has complement.

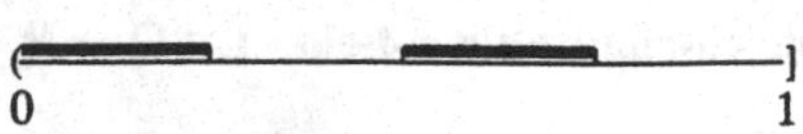

FIGURE 1.2

which is a disjoint union.

(iii) $\mathcal{A}$ is closed under finite intersections. (By the de Morgan laws, verification of this assertion is equivalent to showing closure under finite unions.) Closure under finite intersections is true because

$$(a, a'] \cap (b, b'] = (a \vee b, a' \wedge b'].$$

Note that $\mathcal{A}$ is *NOT* a σ-field. The set

$$(0, \frac{1}{2}] \cup (\frac{1}{2} + \frac{1}{2^2}, \frac{1}{2} + \frac{1}{2^2} + \frac{1}{2^3}]$$
$$\cup (\frac{1}{2} + \frac{1}{2^2} + \frac{1}{2^3} + \frac{1}{2^4}, \frac{1}{2} + \frac{1}{2^2} + \frac{1}{2^3} + \frac{1}{2^4} + \frac{1}{2^5}] \cup \cdots$$

is a countable union of members of $\mathcal{A}$ but is not in $\mathcal{A}$. □

1.6 The σ-field Generated by a Given Class $\mathcal{C}$

It is a sad fact that σ-fields cannot always be constructed by a countable set of operations on simple sets. Sometimes only abstraction works. The exception to the sad fact is if Ω is finite, in which case construction and enumeration work. However, in general, the motto of the budding measure theorist is "induction not construction".

We now discuss how to guarantee a desired σ-field exists.

Let $\mathcal{O}$ be one of the 8 set operations listed starting on page 11. For example, $\mathcal{O}$ could be "countable union". Let $\{\mathcal{C}_t, t \in T\}$ be an indexed family of subsets such that for each t, $\mathcal{C}_t$ is closed under $\mathcal{O}$. Then

$$\mathcal{C} = \bigcap_{t \in T} \mathcal{C}_t \text{ is closed under } \mathcal{O}. \tag{1.2}$$

(This is NOT true for $\bigcup_{t \in T} \mathcal{C}_t$.) Observe that the intersection can be with respect to an arbitrary index set. This will be used when we discuss the minimal σ-field generated by a class.

Here is a sample verification of (1.2) when $\mathcal{O}$ is countable union: Suppose for $i \geq 1$ that $B_i \in \mathcal{C}$. Then for any $i \geq 1$, $B_i \in \mathcal{C}_t$ for all $t \in T$. Due to the fact that $\mathcal{C}_t$ is closed under $\mathcal{O}$, we conclude $\cup_{i=1}^{\infty} B_i \in \mathcal{C}_t$ for all $t \in T$. Since $\cup_{i=1}^{\infty} B_i \in \mathcal{C}_t$ for all t, $\cup_{i=1}^{\infty} B_i \in \cap_{t \in T} \mathcal{C}_t$. Thus $\cap_{t \in T} \mathcal{C}_t$ is closed under $\mathcal{O}$.

Applying the principle in (1.2) using the set operations of complementation and countable union, we get the following result.

Corollary 1.6.1 *The intersection of σ-fields is a σ-field.*

Definition 1.6.1 Let $\mathcal{C}$ be a collection of subsets of Ω. The σ-field generated by $\mathcal{C}$, denoted $\sigma(\mathcal{C})$, is a σ-field satisfying

(a) $\sigma(\mathcal{C}) \supset \mathcal{C}$

(b) If $\mathcal{B}'$ is some other σ-field containing $\mathcal{C}$, then $\mathcal{B}' \supset \sigma(\mathcal{C})$.

Another name for $\sigma(\mathcal{C})$ is the minimal σ-field over $\mathcal{C}$. Part (b) of the definition makes the name *minimal* apt.

The next result shows why a σ-field containing a given class exists.

Proposition 1.6.1 *Given a class $\mathcal{C}$ of subsets of Ω, there is a unique minimal σ-field containing $\mathcal{C}$.*

Proof. Let

$$\aleph = \{\mathcal{B} : \mathcal{B} \text{ is a } \sigma\text{-field}, \mathcal{B} \supset \mathcal{C}\}$$

be the set of all σ-fields containing $\mathcal{C}$. Then $\aleph \neq \emptyset$ since $\mathcal{P}(\Omega) \in \aleph$. Let

$$\mathcal{B}^\sharp = \bigcap_{\mathcal{B} \in \aleph} \mathcal{B}.$$

Since each class $\mathcal{B} \in \aleph$ is a σ-field, so is $\mathcal{B}^\sharp$ by Corollary 1.6.1. Since $\mathcal{B} \in \aleph$ implies $\mathcal{B} \supset \mathcal{C}$, we have $\mathcal{B}^\sharp \supset \mathcal{C}$. We claim $\mathcal{B}^\sharp = \sigma(\mathcal{C})$. We checked $\mathcal{B}^\sharp \supset \mathcal{C}$ and, for minimality, note that if $\mathcal{B}'$ is a σ-field such that $\mathcal{B}' \supset \mathcal{C}$, then $\mathcal{B}' \in \aleph$ and hence $\mathcal{B}^\sharp \subset \mathcal{B}'$. □

Note this is abstract and completely non-constructive. If Ω is finite, we can construct $\sigma(\mathcal{C})$ but otherwise explicit construction is usually hopeless.

In a probability model, we start with $\mathcal{C}$, a restricted class of sets to which we know how to assign probabilities. For example, if $\Omega = (0, 1]$, we could take

$$\mathcal{C} = \{(a, b], 0 \leq a \leq b \leq 1\}$$

and

$$P((a, b]) = b - a.$$

Manipulations involving a countable collection of set operations may take us outside $\mathcal{C}$ but not outside $\sigma(\mathcal{C})$. Measure theory stresses that if we know how to assign probabilities to $\mathcal{C}$, we know (in principle) how to assign probabilities to $\sigma(\mathcal{C})$.

1.7 Borel Sets on the Real Line

Suppose $\Omega = \mathbb{R}$ and let

$$\mathcal{C} = \{(a, b], -\infty \leq a \leq b < \infty\}.$$

Define

$$\mathcal{B}(\mathbb{R}) := \sigma(\mathcal{C})$$

and call $\mathcal{B}(\mathbb{R})$ the Borel subsets of $\mathbb{R}$. Thus the Borel subsets of $\mathbb{R}$ are elements of the σ-field generated by intervals that are open on the left and closed on the right. A fact which is dull to prove, but which you nonetheless need to know, is that there are many equivalent ways of generating the Borel sets and the following are all valid descriptions of Borel sets:

$$\begin{aligned}\mathcal{B}(\mathbb{R}) &= \sigma((a,b), -\infty \le a \le b \le \infty)\\ &= \sigma([a,b), -\infty < a \le b \le \infty)\\ &= \sigma([a,b], -\infty < a \le b < \infty)\\ &= \sigma((-\infty, x], x \in \mathbb{R})\\ &= \sigma(\text{open subsets of } \mathbb{R}).\end{aligned}$$

Thus we can generate the Borel sets with any kind of interval: open, closed, semi-open, finite, semi-infinite, etc.

Here is a sample proof for two of the equivalences. Let

$$\mathcal{C}^{()} = \{(a,b), -\infty \le a \le b \le \infty\}$$

be the open intervals and let

$$\mathcal{C}^{(]} = \{(a,b], -\infty \le a \le b < \infty\}$$

be the semi-open intervals open on the left. We will show

$$\sigma(\mathcal{C}^{()}) = \sigma(\mathcal{C}^{(]}).$$

Observe $(a,b) = \bigcup_{n=1}^{\infty}(a, b - 1/n]$. Now $(a, b-1/n] \in \mathcal{C}^{(]} \subset \sigma(\mathcal{C}^{(]})$, for all n implies $\bigcup_{n=1}^{\infty}(a, b-1/n] \in \sigma(\mathcal{C}^{(]})$. So $(a,b) \in \sigma(\mathcal{C}^{(]})$ which implies that $\mathcal{C}^{()} \subset \sigma(\mathcal{C}^{(]})$. Now $\sigma(\mathcal{C}^{(]})$ is a σ-field containing $\mathcal{C}^{()}$ and hence contains the *minimal* σ-field over $\mathcal{C}^{()}$, that is, $\sigma(\mathcal{C}^{()}) \subset \sigma(\mathcal{C}^{(]})$.

Conversely, $(a,b] = \bigcap_{n=1}^{\infty}(a, b+1/n)$. Now $(a, b+1/n) \in \mathcal{C}^{()} \subset \sigma(\mathcal{C}^{()})$ so that $\bigcap_{n=1}^{\infty}(a, b+1/n) \in \sigma(\mathcal{C}^{()})$ which implies $(a,b] \in \sigma(\mathcal{C}^{()})$ and hence $\mathcal{C}^{(]} \subset \sigma(\mathcal{C}^{()})$. This implies $\sigma(\mathcal{C}^{(]}) \subset \sigma(\mathcal{C}^{()})$.

From the two inclusions, we conclude

$$\sigma(\mathcal{C}^{(]}) = \sigma(\mathcal{C}^{()})$$

as desired.

Here is a sample proof of the fact that

$$\mathcal{B}(\mathbb{R}) = \sigma(\text{open sets in } \mathbb{R}).$$

We need the result from real analysis that if $O \subset \mathbb{R}$ is open, $O = \bigcup_{j=1}^{\infty} I_j$, where I_j are open, disjoint intervals. This makes it relatively easy to show that

$$\sigma(\text{ open sets }) = \sigma(\mathcal{C}^{()}).$$

If O is an open set, then we can write

$$O = \bigcup_{j=1}^{\infty} I_j.$$

We have $I_j \in \mathcal{C}^{()} \subset \sigma(\mathcal{C}^{()})$ so that $O = \cup_{j=1}^{\infty} I_j \in \sigma(\mathcal{C}^{()})$ and hence any open set belongs to $\sigma(\mathcal{C}^{()})$, which implies that

$$\sigma(\text{ open sets }) \subset \sigma(\mathcal{C}^{()}).$$

Conversely, $\mathcal{C}^{()}$ is contained in the class of open sets and therefore $\sigma(\mathcal{C}^{()}) \subset \sigma(\text{ open sets })$.

Remark. If $\mathbb{E}$ is a metric space, it is usual to define $\mathcal{B}(\mathbb{E})$, the σ-field on $\mathbb{E}$, to be the σ-field generated by the open subsets of $\mathbb{E}$. Then $\mathcal{B}(\mathbb{E})$, is called the *Borel* σ-field. Examples of metric spaces $\mathbb{E}$ that are useful to consider are

- $\mathbb{R}$, the real numbers,
- $\mathbb{R}^d$, d-dimensional Euclidean space,
- $\mathbb{R}^\infty$, sequence space; that is, the space of all real sequences.
- $C[0, \infty)$, the space of continuous functions on $[0, \infty)$.

1.8 Comparing Borel Sets

We have seen that the Borel subsets of $\mathbb{R}$ is the σ-field generated by the intervals of $\mathbb{R}$. A natural definition of Borel sets on $(0, 1]$, denoted $\mathcal{B}((0, 1])$ is to take $\mathcal{C}(0, 1]$ to be the subintervals of $(0, 1]$ and to define

$$\mathcal{B}((0, 1]) := \sigma(\mathcal{C}(0, 1]).$$

If a Borel set $A \in \mathcal{B}(\mathbb{R})$ has the property $A \subset (0, 1]$, we would hope $A \in \mathcal{B}((0, 1])$. The next result assures us this is true.

Theorem 1.8.1 *Let $\Omega_0 \subset \Omega$.*

(1) If $\mathcal{B}$ is a σ-field of subsets of Ω, then $\mathcal{B}_0 := \{A\Omega_0 : A \in \mathcal{B}\}$ is a σ-field of subsets of Ω_0. (Notation: $\mathcal{B}_0 =: \mathcal{B} \cap \Omega_0$. We hope to verify $\mathcal{B}((0, 1]) = \mathcal{B}(\mathbb{R}) \cap (0, 1]$.)

(2) Suppose $\mathcal{C}$ is a class of subsets of Ω and $\mathcal{B} = \sigma(\mathcal{C})$. Set

$$\mathcal{C} \cap \Omega_0 =: \mathcal{C}_0 = \{A\Omega_0 : A \in \mathcal{C}\}.$$

Then

$$\sigma(\mathcal{C}_0) = \sigma(\mathcal{C}) \cap \Omega_0$$

in Ω_0.

In symbols (2) can be expressed as

$$\sigma(\mathcal{C} \cap \Omega_0) = \sigma(\mathcal{C}) \cap \Omega_0$$

so that specializing to the Borel sets on the real line we get

$$\mathcal{B}(0,1] = \mathcal{B}(\mathbb{R}) \cap (0,1].$$

Proof. (1) We proceed in a series of steps to verify the postulates defining a σ-field.

(i) First, observe that $\Omega_0 \in \mathcal{B}_0$ since $\Omega\Omega_0 = \Omega_0$ and $\Omega \in \mathcal{B}$.

(ii) Next, we have that if $B = A\Omega_0 \in \mathcal{B}_0$, then

$$\Omega_0 \setminus B = \Omega_0 \setminus A\Omega_0 = \Omega_0(\Omega \setminus A) \in \mathcal{B}_0$$

since $\Omega \setminus A \in \mathcal{B}$.

(iii) Finally, if for $n \geq 1$ we have $B_n = A_n\Omega_0$, and $A_n \in \mathcal{B}$, then

$$\bigcup_{n=1}^{\infty} B_n = \bigcup_{n=1}^{\infty} A_n\Omega_0 = \Big(\bigcup_{n=1}^{\infty} A_n\Big) \cap \Omega_0 \in \mathcal{B}_0$$

since $\bigcup_n A_n \in \mathcal{B}$.

(2) Now we show $\sigma(\mathcal{C}_0) = \sigma(\mathcal{C}) \cap \Omega_0$. We do this in two steps.
Step 1: We have that

$$\mathcal{C}_0 := \mathcal{C} \cap \Omega_0 \subset \sigma(\mathcal{C}) \cap \Omega_0$$

and since (i) assures us that $\sigma(\mathcal{C}) \cap \Omega_0$ is a σ-field, it contains the minimal σ-field generated by $\mathcal{C}_0$, and thus we conclude that

$$\sigma(\mathcal{C}_0) \subset \sigma(\mathcal{C}) \cap \Omega_0.$$

Step 2: We show the reverse inclusion. Define

$$\mathcal{G} := \{A \subset \Omega : A\Omega_0 \in \sigma(\mathcal{C}_0)\}.$$

We hope to show $\mathcal{G} \supset \sigma(\mathcal{C})$.

First of all, $\mathcal{G} \supset \mathcal{C}$, since if $A \in \mathcal{C}$ then $A\Omega_0 \in \mathcal{C}_0 \subset \sigma(\mathcal{C}_0)$. Secondly, observe that $\mathcal{G}$ is a σ-field since

(i) $\Omega \in \mathcal{G}$ since $\Omega\Omega_0 = \Omega_0 \in \sigma(\mathcal{C}_0)$).

(ii) If $A \in \mathcal{G}$ then $A^c = \Omega \setminus A$ and we have

$$A^c \cap \Omega_0 = (\Omega \setminus A)\Omega_0 = \Omega_0 \setminus A\Omega_0.$$

Since $A \in \mathcal{G}$, we have $A\Omega_0 \in \sigma(\mathcal{C}_0)$ which implies $\Omega_0 \setminus A\Omega_0 \in \sigma(\mathcal{C}_0)$, so we conclude $A^c \in \mathcal{G}$.

(iii) If $A_n \in \mathcal{G}$, for $n \geq 1$, then

$$(\bigcup_{n=1}^{\infty} A_n) \cap \Omega_0 = \bigcup_{n=1}^{\infty} A_n \Omega_0.$$

Since $A_n \Omega_0 \in \sigma(\mathcal{C}_0)$, it is also true that $\cup_{n=1}^{\infty} A_n \Omega_0 \in \sigma(\mathcal{C}_0)$ and thus $\cup_{n=1}^{\infty} A_n \in \mathcal{G}$.

So $\mathcal{G}$ is a σ-field and $\mathcal{G} \supset \mathcal{C}$ and therefore $\mathcal{G} \supset \sigma(\mathcal{C})$. From the definition of $\mathcal{G}$, if $A \in \sigma(\mathcal{C})$, then $A \in \mathcal{G}$ and so $A\Omega_0 \in \sigma(\mathcal{C}_0)$. This means

$$\sigma(\mathcal{C}) \cap \Omega_0 \subset \sigma(\mathcal{C}_0)$$

as required. □

Corollary 1.8.1 *If $\Omega_0 \in \sigma(\mathcal{C})$, then*

$$\sigma(\mathcal{C}_0) = \{A : A \subset \Omega_0, A \in \sigma(\mathcal{C})\}.$$

Proof. We have that

$$\begin{aligned}\sigma(\mathcal{C}_0) =&\sigma(\mathcal{C}) \cap \Omega_0 = \{A\Omega_0 : A \in \sigma(\mathcal{C})\}\\ =&\{B : B \in \sigma(\mathcal{C}), B \subset \Omega_0\}\end{aligned}$$

if $\Omega_0 \in \sigma(\mathcal{C})$. □

This shows how Borel sets on $(0, 1]$ compare with those on $\mathbb{R}$.

1.9 Exercises

1. Suppose $\Omega = \{0, 1\}$ and $\mathcal{C} = \{\{0\}\}$. Enumerate $\aleph$, the class of all σ-fields containing $\mathcal{C}$.

2. Suppose $\Omega = \{0, 1, 2\}$ and $\mathcal{C} = \{\{0\}\}$. Enumerate $\aleph$, the class of all σ-fields containing $\mathcal{C}$ and give $\sigma(\mathcal{C})$.

3. Let A_n, A, B_n, B be subsets of Ω. Show

$$\limsup_{n\to\infty} A_n \cup B_n = \limsup_{n\to\infty} A_n \cup \limsup_{n\to\infty} B_n.$$

If $A_n \to A$ and $B_n \to B$, is it true that

$$A_n \cup B_n \to A \cup B, \quad A_n \cap B_n \to A \cap B?$$

4. Suppose

$$A_n = \{\frac{m}{n} : m \in \mathbb{N}\}, \quad n \in \mathbb{N},$$

where $\mathbb{N}$ are non-negative integers. What is

$$\liminf_{n\to\infty} A_n \text{ and } \limsup_{n\to\infty} A_n?$$

5. Let f_n, f be real functions on Ω. Show

$$\{\omega : f_n(\omega) \not\to f(\omega)\} = \bigcup_{k=1}^{\infty} \bigcap_{N=1}^{\infty} \bigcup_{n=N}^{\infty} \{\omega : |f_n(\omega) - f(\omega)| \geq \frac{1}{k}\}.$$

6. Suppose $a_n > 0$, $b_n > 1$ and

$$\lim_{n\to\infty} a_n = 0, \quad \lim_{n\to\infty} b_n = 1.$$

Define

$$A_n = \{x : a_n \leq x < b_n\}.$$

Find

$$\limsup_{n\to\infty} A_n \text{ and } \liminf_{n\to\infty} A_n.$$

7. Let

$$I = \{(x, y) : |x| \leq 1, \ |y| \leq 1\}$$

be the square with sides of length 2. Let I_n be the square pinned at (0, 0) rotated through an angle $2\pi n\theta$. Describe $\limsup_{n\to\infty} I_n$ and $\liminf_{n\to\infty} I_n$ when

(a) $\theta = 1/8$,
(b) θ is rational.
(c) θ is irrational. (Hint: A theorem of Weyl asserts that $\{e^{2\pi i n\theta}, n \geq 1\}$ is dense in the unit circle when θ is irrational.)

8. Let

$$B \subset \Omega, \quad C \subset \Omega$$

and define

$$A_n = \begin{cases} B, & \text{if } n \text{ is odd,} \\ C, & \text{if } n \text{ is even.} \end{cases}$$

What is

$$\liminf_{n\to\infty} A_n \text{ and } \limsup_{n\to\infty} A_n?$$

9. Check that

$$A \triangle B = A^c \triangle B^c.$$

10. Check that

$$A_n \to A$$

iff

$$1_{A_n} \to 1_A$$

pointwise.

11. Let $0 \le a_n < \infty$ be a sequence of numbers. Prove that

$$\sup_{n\ge 1}[0, a_n) = [0, \sup_{n\ge 1} a_n)$$

$$\sup_{n\ge 1}[0, \frac{n}{n+1}] \neq [0, \sup_{n\ge 1} \frac{n}{n+1}].$$

12. Let $\Omega = \{1,2,3,4,5,6\}$ and let $\mathcal{C} = \{\{2,4\},\{6\}\}$. What is the field generated by $\mathcal{C}$ and what is the σ-field?

13. Suppose $\Omega = \cup_{t\in T} C_t$, where $C_s \cap C_t = \emptyset$ for all $s, t \in T$ and $s \neq t$. Suppose $\widehat{\mathcal{F}}$ is a σ-field on $\widehat{\Omega} = \{C_t, t \in T\}$. Show

$$\mathcal{F} := \{A = \bigcup_{C_t \in \widehat{A}} C_t : \widehat{A} \in \widehat{\mathcal{F}}\}$$

is a σ-field and show that

$$f : \widehat{A} \mapsto \bigcup_{C_t \in \widehat{A}} C_t$$

is a 1-1 mapping from $\widehat{\mathcal{F}}$ to $\mathcal{F}$.

14. Suppose that $\mathcal{A}_n$ are fields satisfying $\mathcal{A}_n \subset \mathcal{A}_{n+1}$. Show that $\cup_n \mathcal{A}_n$ is a field. (But see also the next problem.)

15. Check that the union of a countable collection of σ-fields $\mathcal{B}_j, j \ge 1$ need not be a σ-field even if $\mathcal{B}_j \subset \mathcal{B}_{j+1}$. Is a countable union of σ-fields whether monotone or not a field?

    Hint: Try setting Ω equal to the set of positive integers and let $\mathcal{C}_j$ be all subsets of $\{1, \dots, j\}$ and $\mathcal{B}_j = \sigma(\mathcal{C}_j)$.

    If $\mathcal{B}_i, i = 1, 2$ are two σ-fields, $\mathcal{B}_1 \cup \mathcal{B}_2$ need not be a field.

16. Suppose $\mathcal{A}$ is a class of subsets of Ω such that

    - $\Omega \in \mathcal{A}$
    - $A \in \mathcal{A}$ implies $A^c \in \mathcal{A}$.
    - $\mathcal{A}$ is closed under finite *disjoint* unions.

Show $\mathcal{A}$ does not have to be a field.

Hint: Try $\Omega = \{1, 2, 3, 4\}$ and let $\mathcal{A}$ be the field generated by two point subsets of Ω.

17. Prove

$$\liminf_{n\to\infty} A_n = \{\omega : \lim_{n\to\infty} 1_{A_n}(\omega) = 1\}.$$

18. Suppose $\mathcal{A}$ is a class of sets containing Ω and satisfying

$$A, B \in \mathcal{A} \text{ implies } A \setminus B = AB^c \in \mathcal{A}.$$

Show $\mathcal{A}$ is a field.

19. For sets A, B show

$$1_{A\cup B} = 1_A \vee 1_B,$$

and

$$1_{A\cap B} = 1_A \wedge 1_B.$$

20. Suppose $\mathcal{C}$ is a non-empty class of subsets of Ω. Let $\mathcal{A}(\mathcal{C})$ be the minimal field over $\mathcal{C}$. Show that $\mathcal{A}(\mathcal{C})$ consists of sets of the form

$$\bigcup_{i=1}^{m} \bigcap_{j=1}^{n_i} A_{ij},$$

where for each i, j either $A_{ij} \in \mathcal{C}$ or $A_{ij}^c \in \mathcal{C}$ and where the m sets $\cap_{j=1}^{n_i} A_{ij}$, $1 \le i \le m$, are disjoint. Thus, we can explicitly represent the sets in $\mathcal{A}(\mathcal{C})$ even though this is impossible for the σ-field over $\mathcal{C}$.

21. Suppose $\mathcal{A}$ is a field and suppose also that $\mathcal{A}$ has the property that it is closed under countable disjoint unions. Show $\mathcal{A}$ is a σ-field.

22. Let Ω be a non-empty set and let $\mathcal{C}$ be all one point subsets. Show that

$$\sigma(\mathcal{C}) = \{A \subset \Omega : A \text{ is countable }\} \bigcup \{A \subset \Omega : A^c \text{ is countable }\}.$$

23. (a) Suppose on $\mathbb{R}$ that $t_n \downarrow t$. Show

$$(-\infty, t_n] \downarrow (-\infty, t].$$

(b) Suppose

$$t_n \uparrow t, \quad t_n < t.$$

Show

$$(-\infty, t_n] \uparrow (-\infty, t).$$

24. Let $\Omega = \mathbb{N}$, the integers. Define

$$\mathcal{A} = \{A \subset \mathbb{N} : A \text{ or } A^c \text{ is finite.}\}$$

Show $\mathcal{A}$ is a field, but not a σ-field.

25. Suppose $\Omega = \{e^{i2\pi\theta}, 0 \le \theta < 1\}$ is the unit circle. Let $\mathcal{A}$ be the collection of arcs on the unit circle with rational endpoints. Show $\mathcal{A}$ is a field but not a σ-field.

26. (a) Suppose $\mathcal{C}$ is a finite partition of Ω; that is

$$\mathcal{C} = \{A_1, \dots, A_k\}, \quad \Omega = \sum_{i=1}^{k} A_i, \quad A_i A_j = \emptyset, i \neq j.$$

Show that the minimal algebra (synonym: field) $\mathcal{A}(\mathcal{C})$ generated by $\mathcal{C}$ is the class of unions of subfamilies of $\mathcal{C}$; that is

$$\mathcal{A}(\mathcal{C}) = \{\cup_I A_j : I \subset \{1, \dots, k\}\}.$$

(This includes the empty set.)

(b) What is the σ-field generated by the partition $A_1, \dots, A_n$?

(c) If $A_1, A_2, \dots$ is a countable partition of Ω, what is the induced σ-field?

(d) If $\mathcal{A}$ is a field of subsets of Ω, we say $A \in \mathcal{A}$ is an *atom* of $\mathcal{A}$; if $A \neq \emptyset$ and if $\emptyset \neq B \subset A$ and $B \in \mathcal{A}$, then $B = A$. (So A cannot be split into smaller sets that are nonempty and still in $\mathcal{A}$.) Example: If $\Omega = \mathbb{R}$ and $\mathcal{A}$ is the field generated by intervals with integer endpoints of the form $(a, b]$ (a, b are integers) what are the atoms?

As a converse to (a), prove that if $\mathcal{A}$ is a finite field of subsets of Ω, then the atoms of $\mathcal{A}$ constitute a finite partition of Ω that generates $\mathcal{A}$.

27. Show that $\mathcal{B}(\mathbb{R})$ is countably generated; that is, show the Borel sets are generated by a countable class $\mathcal{C}$.

28. Show that the periodic sets of $\mathbb{R}$ form a σ-field; that is, let $\mathcal{B}$ be the class of sets A with the property that $x \in A$ implies $x \pm n \in A$ for all natural numbers n. Then show $\mathcal{B}$ is a σ-field.

29. Suppose $\mathcal{C}$ is a class of subsets of $\mathbb{R}$ with the property that $A \in \mathcal{C}$ implies A^c is a countable union of elements of $\mathcal{C}$. For instance, the finite intervals in $\mathbb{R}$ have this property.

Show that $\sigma(\mathcal{C})$ is the smallest class containing $\mathcal{C}$ which is closed under the formation of countable unions and intersections.

30. Let $\mathcal{B}_i$ be σ-fields of subsets of Ω for $i = 1, 2$. Show that the σ-field $\mathcal{B}_1 \vee \mathcal{B}_2$ defined to be the smallest σ-field containing both $\mathcal{B}_1$ and $\mathcal{B}_2$ is generated by sets of the form $B_1 \cap B_2$ where $B_i \in \mathcal{B}_i$ for $i = 1, 2$.

31. Suppose Ω is uncountable and let $\mathcal{G}$ be the σ-field consisting of sets A such that either A is countable or A^c is countable. Show $\mathcal{G}$ is NOT countably generated. (Hint: If $\mathcal{G}$ were countably generated, it would be generated by a countable collection of one point sets.)

    In fact, if $\mathcal{G}$ is the σ-field of subsets of Ω consisting of the countable and co-countable sets, $\mathcal{G}$ is countably generated iff Ω is countable.

32. Suppose $\mathcal{B}_1, \mathcal{B}_2$ are σ-fields of subsets of Ω such that $\mathcal{B}_1 \subset \mathcal{B}_2$ and $\mathcal{B}_2$ is countably generated. Show by example that it is not necessarily true that $\mathcal{B}_1$ is countably generated.

33. **The extended real line.** Let $\bar{\mathbb{R}} = \mathbb{R} \cup \{-\infty\} \cup \{\infty\}$ be the *extended* or *closed* real line with the points $-\infty$ and ∞ added. The Borel sets $\mathcal{B}(\bar{\mathbb{R}})$ is the σ-field generated by the sets $[-\infty, x], x \in \mathbb{R}$, where $[-\infty, x] = \{-\infty\} \cup (-\infty, x]$. Show $\mathcal{B}(\bar{\mathbb{R}})$ is also generated by the following collections of sets:

    (i) $[-\infty, x), x \in \mathbb{R}$,

    (ii) $(x, \infty], x \in \mathbb{R}$,

    (ii) all finite intervals and $\{-\infty\}$ and $\{\infty\}$.

    Now think of $\bar{\mathbb{R}} = [-\infty, \infty]$ as homeomorphic in the topological sense to $[-1, 1]$ under the transformation

$$x \mapsto \frac{x}{1 - |x|}$$

    from $[-1, 1]$ to $[-\infty, \infty]$. (This transformation is designed to stretch the finite interval onto the infinite interval.) Consider the usual topology on $[-1, 1]$ and map it onto a topology on $[-\infty, \infty]$. This defines a collection of open sets on $[-\infty, \infty]$ and these open sets can be used to generate a Borel σ-field. How does this σ-field compare with $\mathcal{B}(\bar{\mathbb{R}})$ described above?

34. Suppose $\mathcal{B}$ is a σ-field of subsets of Ω and suppose $A \notin \mathcal{B}$. Show that $\sigma(\mathcal{B} \cup \{A\})$, the smallest σ-field containing both $\mathcal{B}$ and A consists of sets of the form

$$AB \cup A^c B', \quad B, B' \in \mathcal{B}.$$

35. A σ-field cannot be countably infinite. Its cardinality is either finite or at least that of the continuum.

36. Let $\Omega = \{f, a, n, g\}$, and $\mathcal{C} = \{\{f, a, n\}, \{a, n\}\}$. Find $\sigma(\mathcal{C})$.

37. Suppose $\Omega = \mathbb{Z}$, the natural numbers. Define for integer k

$$k\mathbb{Z} = \{kz : z \in \mathbb{Z}\}.$$

    Find $\mathcal{B}(\mathcal{C})$ when $\mathcal{C}$ is

(i) $\{3\mathbb{Z}\}$.

(ii) $\{3\mathbb{Z}, 4\mathbb{Z}\}$.

(iii) $\{3\mathbb{Z}, 4\mathbb{Z}, 5\mathbb{Z}\}$.

(iv) $\{3\mathbb{Z}, 4\mathbb{Z}, 5\mathbb{Z}, 6\mathbb{Z}\}$.

38. Let $\Omega = \mathbb{R}^\infty$, the space of all sequences of the form

$$\omega = (x_1, x_2, \dots) \tag{**}$$

where $x_i \in \mathbb{R}$. Let σ be a permutation of $1, \dots, n$; that is, σ is a 1-1 and onto map of $\{1, \dots, n\} \mapsto \{1, \dots, n\}$. If ω is the sequence defined in (**), define $\sigma\omega$ to be the new sequence

$$(\sigma\omega)_j = \begin{cases} x_{\sigma(j)}, & \text{if } j \le n, \\ x_j, & \text{if } j > n. \end{cases}$$

A *finite permutation* is of the form σ for *some* n; that is, it juggles a finite initial segment of all positive integers. A set $\Lambda \subset \Omega$ is *permutable* if

$$\Lambda = \sigma\Lambda := \{\sigma\omega : \omega \in \Lambda\}$$

for all finite permutations σ.

(i) Let $B_n, n \ge 1$ be a sequence of subsets of $\mathbb{R}$. Show that

$$\{\omega = (x_1, x_2, \dots) : \sum_{i=1}^{n} x_i \in B_n \text{ i.o.}\}$$

and

$$\{\omega = (x_1, x_2, \dots) : \bigvee_{i=1}^{n} x_i \in B_n \text{ i.o.}\}$$

are permutable.

(ii) Show the permutable sets form a σ-field.

39. For a subset $A \subset \mathbb{N}$ of non-negative integers, write $\text{card}(A)$ for the number of elements in A. A set $A \subset \mathbb{N}$ has *asymptotic density* d if

$$\lim_{n\to\infty} \frac{\text{card}(A \cap \{1, 2, \dots, n\})}{n} = d.$$

Let $\mathcal{A}$ be the collection of subsets that have an asymptotic density. Is $\mathcal{A}$ a field? Is it a σ-field?

Hint: $\mathcal{A}$ is closed under complements, proper differences and finite disjoint unions but is not closed under formation of countable disjoint unions or finite unions that are not disjoint.

40. Show that $\mathcal{B}((0,1])$ is generated by the following countable collection: For an integer r,

$$\{[kr^{-n},(k+1)r^{-n}), 0 \le k < r^n,\ n = 1, 2, \ldots.\}.$$

41. A *monotone class* $\mathcal{M}$ is a non-empty collection of subsets of Ω closed under monotone limits; that is, if $A_n \nearrow$ and $A_n \in \mathcal{M}$, then $\lim_{n\to\infty} A_n = \cup_n A_n \in \mathcal{M}$ and if $A_n \searrow$ and $A_n \in \mathcal{M}$, then $\lim_{n\to\infty} A_n = \cap_n A_n \in \mathcal{M}$. Show that a σ-field is a field that is also a monotone class and conversely, a field that is a monotone class is a σ-field.

42. Assume $\mathcal{P}$ is a π-system (that is, $\mathcal{P}$ is closed under finite intersections) and $\mathcal{M}$ is a monotone class. (Cf. Exercise 41.) Show $\mathcal{P} \subset \mathcal{M}$ does not imply $\sigma(\mathcal{P}) \subset \mathcal{M}$.

43. **Symmetric differences.** For subsets A, B, C, D show

$$1_{A\Delta B} = 1_A + 1_B \pmod 2,$$

and hence

(a) $(A\Delta B)\Delta C = A\Delta(B\Delta C)$,
(b) $(A\Delta B)\Delta(B\Delta C) = (A\Delta C)$,
(c) $(A\Delta B)\Delta(C\Delta D) = (A\Delta C)\Delta(B\Delta D)$,
(d) $A\Delta B = C$ iff $A = B\Delta C$,
(e) $A\Delta B = C\Delta D$ iff $A\Delta C = B\Delta D$.

44. Let $\mathcal{A}$ be a field of subsets of Ω and define

$$\bar{\mathcal{A}} = \{A \subset \Omega : \exists A_n \in \mathcal{A} \text{ and } A_n \to A\}.$$

Show $\mathcal{A} \subset \bar{\mathcal{A}}$ and $\bar{\mathcal{A}}$ is a field.

2

Probability Spaces

This chapter discusses the basic properties of probability spaces, and in particular, probability measures. It also introduces the important ideas of set induction.

2.1 Basic Definitions and Properties

A *probability space* is a triple $(\Omega, \mathcal{B}, P)$ where

- Ω is the sample space corresponding to outcomes of some (perhaps hypothetical) experiment.
- $\mathcal{B}$ is the σ-algebra of subsets of Ω. These subsets are called events.
- P is a probability measure; that is, P is a function with domain $\mathcal{B}$ and range $[0, 1]$ such that

  (i) $P(A) \geq 0$ for all $A \in \mathcal{B}$.

  (ii) P is σ-additive: If $\{A_n, n \geq 1\}$ are events in $\mathcal{B}$ that are disjoint, then

  $$P(\bigcup_{n=1}^{\infty} A_n) = \sum_{n=1}^{\infty} P(A_n).$$

  (iii) $P(\Omega) = 1$.

Here are some simple **consequences** of the definition of a probability measure P.

1. We have
$$P(A^c) = 1 - P(A)$$
since from (iii)
$$1 = P(\Omega) = P(A \cup A^c) = P(A) + P(A^c),$$
the last step following from (ii).

2. We have
$$P(\emptyset) = 0$$
since $P(\emptyset) = P(\Omega^c) = 1 - P(\Omega) = 1 - 1$.

3. For events A, B we have
$$P(A \cup B) = PA + PB - P(AB). \tag{2.1}$$
To see this note
$$\begin{aligned} P(A) &= P(AB^c) + P(AB) \\ P(B) &= P(BA^c) + P(AB) \end{aligned}$$
and therefore
$$\begin{aligned} P(A \cup B) &= P(AB^c \cup BA^c \cup AB) \\ &= P(AB^c) + P(BA^c) + P(AB) \\ &= P(A) - P(AB) + P(B) - P(AB) + P(AB) \\ &= P(A) + P(B) - P(AB). \end{aligned}$$

4. The *inclusion–exclusion formula*: If $A_1, \dots, A_n$ are events, then
$$\begin{aligned} P(\bigcup_{j=1}^{n} A_j) &= \sum_{j=1}^{n} P(A_j) - \sum_{1 \le i < j \le n} P(A_i A_j) \\ &\quad + \sum_{1 \le i < j < k \le n} P(A_i A_j A_k) - \cdots \\ &\quad (-1)^{n+1} P(A_1 \cdots A_n). \end{aligned} \tag{2.2}$$
We may prove (2.2) by induction using (2.1) for $n = 2$. The terms on the right side of (2.2) alternate in sign and give inequalities called Bonferroni inequalities when we neglect remainders. Here are two examples:
$$P\left(\bigcup_{j=1}^{n} A_j\right) \le \sum_{j=1}^{n} PA_j$$
$$P\left(\bigcup_{j=1}^{n} A_j\right) \ge \sum_{j=1}^{n} PA_j - \sum_{1 \le i < j \le n} P(A_i A_j).$$

5. *The monotonicity property*: The measure P is non-decreasing: For events A, B

$$\text{If } A \subset B \text{ then } P(A) \leq P(B),$$

since

$$P(B) = P(A) + P(B \setminus A) \geq P(A).$$

6. *Subadditivity*: The measure P is σ-subadditive: For events $A_n, n \geq 1$,

$$P\left(\bigcup_{n=1}^{\infty} A_n\right) \leq \sum_{n=1}^{\infty} P(A_n).$$

To verify this we write

$$\bigcup_{n=1}^{\infty} A_n = A_1 + A_1^c A_2 + A_3 A_1^c A_2^c + \cdots,$$

and since P is σ-additive,

$$\begin{aligned} P(\bigcup_{n=1}^{\infty} A_n) &= P(A_1) + P(A_1^c A_2) + P(A_3 A_1^c A_2^c) + \cdots \\ &\leq P(A_1) + P(A_2) + P(A_3) + \cdots \end{aligned}$$

by the non-decreasing property of P.

7. *Continuity*: The measure P is continuous for monotone sequences in the sense that

   (i) If $A_n \uparrow A$, where $A_n \in \mathcal{B}$, then $P(A_n) \uparrow P(A)$.

   (ii) If $A_n \downarrow A$, where $A_n \in \mathcal{B}$, then $P(A_n) \downarrow P(A)$.

   To **prove** (i), assume

$$A_1 \subset A_2 \subset A_3 \subset \cdots \subset A_n \subset \cdots$$

and define

$$B_1 = A_1, B_2 = A_2 \setminus A_1, \ldots, B_n = A_n \setminus A_{n-1}, \ldots.$$

Then $\{B_i\}$ is a disjoint sequence of events and

$$\bigcup_{i=1}^{n} B_i = A_n, \quad \bigcup_{i=1}^{\infty} B_i = \bigcup_i A_i = A.$$

By σ-additivity

$$P(A) = P(\bigcup_{i=1}^{\infty} B_i) = \sum_{i=1}^{\infty} P(B_i) = \lim_{n\to\infty} \uparrow \sum_{i=1}^{n} P(B_i)$$
$$= \lim_{n\to\infty} \uparrow P(\bigcup_{i=1}^{n} B_i) = \lim_{n\to\infty} \uparrow P(A_n).$$

To prove (ii), note if $A_n \downarrow A$, then $A_n^c \uparrow A^c$ and by part (i)

$$P(A_n^c) = 1 - P(A_n) \uparrow P(A^c) = 1 - P(A)$$

so that $PA_n \downarrow PA$. □

8. *More continuity and Fatou's lemma*: Suppose $A_n \in \mathcal{B}$, for $n \geq 1$.

   (i) Fatou Lemma: We have the following inequalities

$$\begin{aligned} P(\liminf_{n\to\infty} A_n) &\leq \liminf_{n\to\infty} P(A_n) \\ &\leq \limsup_{n\to\infty} P(A_n) \leq P(\limsup_{n\to\infty} A_n). \end{aligned}$$

   (ii) If $A_n \to A$, then $P(A_n) \to P(A)$.

Proof of 8. (ii) follows from (i) since, if $A_n \to A$, then

$$\limsup_{n\to\infty} A_n = \liminf_{n\to\infty} A_n = A.$$

Suppose (i) is true. Then we get

$$\begin{aligned} P(A) = P(\liminf_{n\to\infty} A_n) &\leq \liminf_{n\to\infty} P(A_n) \\ &\leq \limsup_{n\to\infty} P(A_n) \leq P(\limsup_{n\to\infty} A_n) = P(A), \end{aligned}$$

so equality pertains throughout.

Now consider the proof of (i): We have

$$\begin{aligned} P(\liminf_{n\to\infty} A_n) &= P(\lim_{n\to\infty} \uparrow (\bigcap_{k\geq n} A_k)) \\ &= \lim_{n\to\infty} \uparrow P(\bigcap_{k\geq n} A_k) \end{aligned}$$

(from the monotone continuity property 7)

$$\leq \liminf_{n\to\infty} P(A_n)$$

since $P(\cap_{k\geq n}A_k) \leq P(A_n)$. Likewise

$$P(\limsup_{n\to\infty} A_n) = P(\lim_{n\to\infty} \downarrow (\bigcup_{k\geq n} A_k))$$
$$= \lim_{n\to\infty} \downarrow P(\bigcup_{k\geq n} A_k)$$

(from continuity property 7)

$$\geq \limsup_{n\to\infty} P(A_n),$$

completing the proof. □

Example 2.1.1 Let $\Omega = \mathbb{R}$, and suppose P is a probability measure on $\mathbb{R}$. Define $F(x)$ by

$$F(x) = P((-\infty, x]), \quad x \in \mathbb{R}. \tag{2.3}$$

Then

(i) F is right continuous,

(ii) F is monotone non-decreasing,

(iii) F has limits at $\pm\infty$

$$F(\infty) := \lim_{x\uparrow\infty} F(x) = 1$$
$$F(-\infty) := \lim_{x\downarrow-\infty} F(x) = 0.$$

Definition 2.1.1 A function $F : \mathbb{R} \mapsto [0, 1]$ satisfying (i), (ii), (iii) is called a (probability) distribution function. We abbreviate distribution function by df.

Thus, starting from P, we get F from (2.3). In practice we need to go in the other direction: we start with a known df and wish to construct a probability space $(\Omega, \mathcal{B}, P)$ such that (2.3) holds. See Section 2.5.

Proof of (i), (ii), (iii). For (ii), note that if $x < y$, then

$$(-\infty, x] \subset (-\infty, y]$$

so by monotonicity of P

$$F(x) = P((-\infty, x]) \leq P((-\infty, y]) \leq F(y).$$

Now consider (iii). We have

$$\begin{aligned}F(\infty) &= \lim_{x_n \uparrow \infty} F(x_n) \qquad \text{(for any sequence } x_n \uparrow \infty)\\ &= \lim_{x_n \uparrow \infty} \uparrow P((-\infty, x_n])\\ &= P(\lim_{x_n \uparrow \infty} \uparrow (-\infty, x_n]) \qquad \text{(from property 7)}\\ &= P(\bigcup_n (-\infty, x_n]) = P((-\infty, \infty))\\ &= P(\mathbb{R}) = P(\Omega) = 1.\end{aligned}$$

Likewise,

$$\begin{aligned}F(-\infty) &= \lim_{x_n \downarrow -\infty} F(x_n) = \lim_{x_n \downarrow -\infty} \downarrow P((-\infty, x_n])\\ &= P(\lim_{x_n \downarrow -\infty} (-\infty, x_n]) \qquad \text{(from property 7)}\\ &= P(\bigcap_n (-\infty, x_n]) = P(\emptyset) = 0.\end{aligned}$$

For the proof of (i), we may show F is right continuous as follows: Let $x_n \downarrow x$. We need to prove $F(x_n) \downarrow F(x)$. This is immediate from the continuity property 7 of P and

$$(-\infty, x_n] \downarrow (-\infty, x]. \qquad \square$$

Example 2.1.2 (Coincidences) The inclusion-exclusion formula (2.2) can be used to compute the probability of a coincidence. Suppose the integers $1, 2, \dots, n$ are randomly permuted. What is the probability that there is an integer left unchanged by the permutation?

To formalize the question, we construct a probability space. Let Ω be the set of all permutations of $1, 2, \dots, n$ so that

$$\Omega = \{(x_1, \dots, x_n) : x_i \in \{1, \dots, n\}; i = 1, \dots, n; x_i \neq x_j\}.$$

Thus Ω is the set of outcomes from the experiment of sampling n times without replacement from the population $1, \dots, n$. We let $\mathcal{B} = \mathcal{P}(\Omega)$ be the power set of Ω and define for $(x_1, \dots, x_n) \in \Omega$

$$P((x_1, \dots, x_n)) = \frac{1}{n!},$$

and for $B \in \mathcal{B}$

$$P(B) = \frac{1}{n!} \#\text{elements in } B.$$

For $i = 1, \dots, n$, let A_i be the set of all elements of Ω with i in the ith spot. Thus, for instance,

$$\begin{aligned}A_1 &= \{(1, x_2, \dots, x_n) : (1, x_2, \dots, x_n) \in \Omega\},\\ A_2 &= \{(x_1, 2, \dots, x_n) : (x_1, 2, \dots, x_n) \in \Omega\}.\end{aligned}$$

and so on. We need to compute $P(\cup_{i=1}^n A_i)$. From the inclusion-exclusion formula (2.2) we have

$$P(\bigcup_{i=1}^n A_i) = \sum_{i=1}^n P(A_i) - \sum_{1\le i<j\le n} P(A_iA_j) + \sum_{1\le i<j<k\le n} P(A_iA_jA_k) - \dots (-1)^{n+1} P(A_1A_2\dots A_n).$$

To compute $P(A_i)$, we fix integer i in the ith spot and count the number of ways to distribute $n-1$ objects in $n-1$ spots, which is $(n-1)!$ and then divide by $n!$. To compute $P(A_iA_j)$ we fix i and j and count the number of ways to distribute $n-2$ integers into $n-2$ spots, and so on. Thus

$$P(\bigcup_{i=1}^n A_i) = n\frac{(n-1)!}{n!} - \binom{n}{2}\frac{(n-2)!}{n!} + \binom{n}{3}\frac{(n-3)!}{n!} - \dots (-1)^n \frac{1}{n!}$$
$$= 1 - \frac{1}{2!} + \frac{1}{3!} - \dots (-1)^n \frac{1}{n!}.$$

Taking into account the expansion of e^x for $x = -1$ we see that for large n, the probability of a coincidence is approximately

$$P(\bigcup_{i=1}^n A_i) \approx 1 - e^{-1} \approx 0.632.$$

□

2.2 More on Closure

A σ-field is a collection of subsets of Ω satisfying certain closure properties, namely closure under complementation and countable union. We will have need of collections of sets satisfying different closure axioms. We define a *structure* $\mathcal{G}$ to be a collection of subsets of Ω satisfying certain specified closure axioms. Here are some other structures. Some have been discussed, some will be discussed and some are listed but will not be discussed or used here.

- field
- σ-field
- semialgebra
- semiring
- ring
- σ-ring
- monotone class (closed under monotone limits)

- π-system ($\mathcal{P}$ is a π-system, if it is closed under finite intersections: $A, B \in \mathcal{P}$ implies $A \cap B \in \mathcal{P}$).
- λ-system (synonyms: σ-additive class, Dynkin class); this will be used extensively as the basis of our most widely used induction technique.

Fix a structure in mind. Call it $\mathcal{S}$. As with σ-algebras, we can make the following definition.

Definition 2.2.1 *The minimal structure $\mathcal{S}$ generated by a class $\mathcal{C}$ is a non-empty structure satisfying*

(i) $\mathcal{S} \supset \mathcal{C}$,

(ii) If $\mathcal{S}'$ is some other structure containing $\mathcal{C}$, then $\mathcal{S}' \supset \mathcal{S}$.

Denote the minimal structure by $\mathcal{S}(\mathcal{C})$.

Proposition 2.2.1 *The minimal structure $\mathcal{S}$ exists and is unique.*

As we did with generating a minimal σ-field, let

$$\aleph = \{\mathcal{G} : \mathcal{G} \text{ is a structure }, \mathcal{G} \supset \mathcal{C}\}$$

and

$$\mathcal{S}(\mathcal{C}) = \cap_{\mathcal{G} \in \aleph} \mathcal{G}.$$

2.2.1 *Dynkin's theorem*

Dynkin's theorem is a remarkably flexible device for performing set inductions which is ideally suited to probability theory.

A class of subsets $\mathcal{L}$ of Ω is a called a λ-system if it satisfies either the *new* postulates $\lambda_1, \lambda_2, \lambda_3$ or the old postulates $\lambda'_1, \lambda'_2, \lambda'_3$ given in the following table.

λ-system postulates			
old		new	
λ'_1	$\Omega \in \mathcal{L}$	λ_1	$\Omega \in \mathcal{L}$
λ'_2	$A, B \in \mathcal{L}, A \subset B \Rightarrow B \setminus A \in \mathcal{L}$	λ_2	$A \in \mathcal{L} \Rightarrow A^c \in \mathcal{L}$
λ'_3	$A_n \uparrow, A_n \in \mathcal{L} \Rightarrow \cup_n A_n \in \mathcal{L}$	λ_3	$n \neq m, A_n A_m = \emptyset,$ $A_n \in \mathcal{L} \Rightarrow \cup_n A_n \in \mathcal{L}.$

The *old* postulates are equivalent to the *new* ones. Here we only check that *old* implies *new*. Suppose $\lambda'_1, \lambda'_2, \lambda'_3$ are true. Then λ_1 is true. Since $\Omega \in \mathcal{L}$, if $A \in \mathcal{L}$, then $A \subset \Omega$ and by λ'_2, $\Omega \setminus A = A^c \in \mathcal{L}$, which shows that λ_2 is true. If $A, B \in \mathcal{L}$ are disjoint, we show that $A \cup B \in \mathcal{L}$. Now $\Omega \setminus A \in \mathcal{L}$ and $B \subset \Omega \setminus A$ (since $\omega \in B$ implies $\omega \notin A$ which means $\omega \in A^c = \Omega \setminus A$) so by λ'_2 we have $(\Omega \setminus A) \setminus B = A^c B^c \in \mathcal{L}$ and by λ_2 we have $(A^c B^c)^c = A \cup B \in \mathcal{L}$ which is λ_3 for finitely many sets. Now if $A_j \in \mathcal{L}$ are mutually disjoint for $j = 1, 2, \ldots,$

define $B_n = \cup_{j=1}^n A_j$. Then $B_n \in \mathcal{L}$ by the prior argument for 2 sets and by λ_3' we have $\cup_n B_n = \lim_{n\to\infty} \uparrow B_n \in \mathcal{L}$. Since $\cup_n B_n = \cup_n A_n$ we have $\cup_n A_n \in \mathcal{L}$ which is λ_3. □

Remark. It is clear that a σ-field is always a λ-system since the *new* postulates obviously hold.

Recall that a π-system is a class of sets closed under finite intersections; that is, $\mathcal{P}$ is a π-system if whenever $A, B \in \mathcal{P}$ we have $AB \in \mathcal{P}$.

We are now in a position to state Dynkin's theorem.

Theorem 2.2.2 (Dynkin's theorem) *(a) If $\mathcal{P}$ is a π-system and $\mathcal{L}$ is a λ-system such that $\mathcal{P} \subset \mathcal{L}$, then $\sigma(\mathcal{P}) \subset \mathcal{L}$.*

(b) If $\mathcal{P}$ is a π-system

$$\sigma(\mathcal{P}) = \mathcal{L}(\mathcal{P}),$$

that is, the minimal σ-field over $\mathcal{P}$ equals the minimal λ-system over $\mathcal{P}$.

Note (b) follows from (a). To see this assume (a) is true. Since $\mathcal{P} \subset \mathcal{L}(\mathcal{P})$, we have from (a) that $\sigma(\mathcal{P}) \subset \mathcal{L}(\mathcal{P})$. On the other hand, $\sigma(\mathcal{P})$, being a σ-field, is a λ-system containing $\mathcal{P}$ and hence contains the minimal λ-system over $\mathcal{P}$, so that $\sigma(\mathcal{P}) \supset \mathcal{L}(\mathcal{P})$.

Before the proof of (a), here is a significant application of Dynkin's theorem.

Proposition 2.2.3 *Let P_1, P_2 be two probability measures on $(\Omega, \mathcal{B})$. The class*

$$\mathcal{L} := \{A \in \mathcal{B} : P_1(A) = P_2(A)\}$$

is a λ-system.

Proof of Proposition 2.2.3. We show the *new* postulates hold:

(λ_1) $\Omega \in \mathcal{L}$ since $P_1(\Omega) = P_2(\Omega) = 1$.

(λ_2) $A \in \mathcal{L}$ implies $A^c \in \mathcal{L}$, since $A \in \mathcal{L}$ means $P_1(A) = P_2(A)$, from which

$$P_1(A^c) = 1 - P_1(A) = 1 - P_2(A) = P_2(A^c).$$

(λ_3) If $\{A_j\}$ is a mutually disjoint sequence of events in $\mathcal{L}$, then $P_1(A_j) = P_2(A_j)$ for all j, and hence

$$P_1(\bigcup_j A_j) = \sum_j P_1(A_j) = \sum_j P_2(A_j) = P_2(\bigcup_j A_j)$$

so that

$$\bigcup_j A_j \in \mathcal{L}.$$

□

Corollary 2.2.1 *If P_1, P_2 are two probability measures on $(\Omega, \mathcal{B})$ and if $\mathcal{P}$ is a π-system such that*

$$\forall A \in \mathcal{P}: \quad P_1(A) = P_2(A),$$

then

$$\forall B \in \sigma(\mathcal{P}): \quad P_1(B) = P_2(B).$$

Proof of Corollary 2.2.1. We have

$$\mathcal{L} = \{A \in \mathcal{B} : P_1(A) = P_2(A)\}$$

is a λ-system. But $\mathcal{L} \supset \mathcal{P}$ and hence by Dynkin's theorem $\mathcal{L} \supset \sigma(\mathcal{P})$. □

Corollary 2.2.2 *Let $\Omega = \mathbb{R}$. Let P_1, P_2 be two probability measures on $(\mathbb{R}, \mathcal{B}(\mathbb{R}))$ such that their distribution functions are equal:*

$$\forall x \in \mathbb{R}: \quad F_1(x) = P_1((-\infty, x]) = F_2(x) = P_2((-\infty, x]).$$

Then

$$P_1 \equiv P_2$$

on $\mathcal{B}(\mathbb{R})$.

So a probability measure on $\mathbb{R}$ is uniquely determined by its distribution function.

Proof of Corollary 2.2.2. Let

$$\mathcal{P} = \{(-\infty, x] : x \in \mathbb{R}\}.$$

Then $\mathcal{P}$ is a π-system since

$$(-\infty, x] \cap (-\infty, y] = (-\infty, x \wedge y] \in \mathcal{P}.$$

Furthermore $\sigma(\mathcal{P}) = \mathcal{B}(\mathbb{R})$ since the Borel sets can be generated by the semi-infinite intervals (see Section 1.7). So $F_1(x) = F_2(x)$ for all $x \in \mathbb{R}$, means $P_1 = P_2$ on $\mathcal{P}$ and hence $P_1 = P_2$ on $\sigma(\mathcal{P}) = \mathcal{B}(\mathbb{R})$. □

2.2.2 *Proof of Dynkin's theorem*

Recall that we only need to prove: If $\mathcal{P}$ is a π-system and $\mathcal{L}$ is a λ-system then $\mathcal{P} \subset \mathcal{L}$ implies $\sigma(\mathcal{P}) \subset \mathcal{L}$.

We begin by proving the following proposition.

Proposition 2.2.4 *If a class $\mathcal{C}$ is both a π-system and a λ-system, then it is a σ-field.*

Proof of Proposition 2.2.4. First we show $\mathcal{C}$ is a field: We check the field postulates.

(i) $\Omega \in \mathcal{C}$ since $\mathcal{C}$ is a λ-system.

(ii) $A \in \mathcal{C}$ implies $A^c \in \mathcal{C}$ since $\mathcal{C}$ is a λ-system.

(iii) If $A_j \in \mathcal{C}$, for $j = 1, \ldots, n$, then $\cap_{j=1}^n A_j \in \mathcal{C}$ since $\mathcal{C}$ is a π-system.

Knowing that $\mathcal{C}$ is a field, in order to show that it is a σ-field we need to show that if $A_j \in \mathcal{C}$, for $j \geq 1$, then $\cup_{j=1}^\infty A_j \in \mathcal{C}$. Since

$$\bigcup_{j=1}^{\infty} A_j = \lim_{n\to\infty} \uparrow \bigcup_{j=1}^{n} A_j$$

and $\cup_{j=1}^n A_j \in \mathcal{C}$ (since $\mathcal{C}$ is a field) it suffices to show $\mathcal{C}$ is closed under monotone non-decreasing limits. This follows from the *old* postulate λ_3'. □

We can now prove Dynkin's theorem.

Proof of Dynkin's Theorem 2.2.2. It suffices to show $\mathcal{L}(\mathcal{P})$ is a π-system since $\mathcal{L}(\mathcal{P})$ is both a π-system and a λ-system, and thus by Proposition 2.2.4 also a σ-field. This means that

$$\mathcal{L} \supset \mathcal{L}(\mathcal{P}) \supset \mathcal{P}.$$

Since $\mathcal{L}(\mathcal{P})$ is a σ-field containing $\mathcal{P}$,

$$\mathcal{L}(\mathcal{P}) \supset \sigma(\mathcal{P})$$

from which

$$\mathcal{L} \supset \mathcal{L}(\mathcal{P}) \supset \sigma(\mathcal{P}),$$

and therefore we get the desired conclusion that

$$\mathcal{L} \supset \sigma(\mathcal{P}).$$

We now concentrate on showing that $\mathcal{L}(\mathcal{P})$ is a π-system. Fix a set $A \in \sigma(\mathcal{P})$ and relative to this A, define

$$\mathcal{G}_A = \{B \in \sigma(\mathcal{P}) : AB \in \mathcal{L}(\mathcal{P})\}.$$

We proceed in a series of steps.

[A] If $A \in \mathcal{L}(\mathcal{P})$, we claim that $\mathcal{G}_A$ is a λ-system.

To prove [A] we check the *new* λ-system postulates.

(i) We have

$$\Omega \in \mathcal{G}_A$$

since $A\Omega = A \in \mathcal{L}(\mathcal{P})$ by assumption.

(ii) Suppose $B \in \mathcal{G}_A$. We have that $B^c A = A \setminus AB$. But $B \in \mathcal{G}_A$ means $AB \in \mathcal{L}(\mathcal{P})$ and since by assumption $A \in \mathcal{L}(\mathcal{P})$, we have $A \setminus AB = B^c A \in \mathcal{L}(\mathcal{P})$ since λ-systems are closed under proper differences. Since $B^c A \in \mathcal{L}(\mathcal{P})$, it follows that $B^c \in \mathcal{G}_A$ by definition.

(iii) Suppose $\{B_j\}$ is a mutually disjoint sequence and $B_j \in \mathcal{G}_A$. Then

$$A \cap (\bigcup_{j=1}^{\infty} B_j) = \bigcup_{j=1}^{\infty} AB_j$$

is a disjoint union of sets in $\mathcal{L}(\mathcal{P})$, and hence in $\mathcal{L}(\mathcal{P})$.

[B] Next, we claim that if $A \in \mathcal{P}$, then $\mathcal{L}(\mathcal{P}) \subset \mathcal{G}_A$.

To prove this claim, observe that since $A \in \mathcal{P} \subset \mathcal{L}(\mathcal{P})$, we have from [A] that $\mathcal{G}_A$ is a λ-system.

For $B \in \mathcal{P}$, we have $AB \in \mathcal{P}$ since by assumption $A \in \mathcal{P}$ and $\mathcal{P}$ is a π-system. So if $B \in \mathcal{P}$, then $AB \in \mathcal{P} \subset \mathcal{L}(\mathcal{P})$ implies $B \in \mathcal{G}_A$; that is

$$\mathcal{P} \subset \mathcal{G}_A. \tag{2.4}$$

Since $\mathcal{G}_A$ is a λ-system, $\mathcal{G}_A \supset \mathcal{L}(\mathcal{P})$.

[B′] We may rephrase [B] using the definition of $\mathcal{G}_A$ to get the following statement. If $A \in \mathcal{P}$, and $B \in \mathcal{L}(\mathcal{P})$, then $AB \in \mathcal{L}(\mathcal{P})$. (So we are making progress toward our goal of showing $\mathcal{L}(\mathcal{P})$ is a π-system.)

[C] We now claim that if $A \in \mathcal{L}(\mathcal{P})$, then $\mathcal{L}(\mathcal{P}) \subset \mathcal{G}_A$.

To prove [C]: If $B \in \mathcal{P}$ and $A \in \mathcal{L}(\mathcal{P})$, then from [B′] (interchange the roles of the sets A and B) we have $AB \in \mathcal{L}(\mathcal{P})$. So when $A \in \mathcal{L}(\mathcal{P})$,

$$\mathcal{P} \subset \mathcal{G}_A.$$

From [A], $\mathcal{G}_A$ is a λ-system so $\mathcal{L}(\mathcal{P}) \subset \mathcal{G}_A$.

[C′] To finish, we rephrase [C]: If $A \in \mathcal{L}(\mathcal{P})$, then for any $B \in \mathcal{L}(\mathcal{P})$, $B \in \mathcal{G}_A$. This says that

$$AB \in \mathcal{L}(\mathcal{P})$$

as desired. □

2.3 Two Constructions

Here we give two simple examples of how to construct probability spaces. These examples will be familiar from earlier probability studies and from Example 2.1.2, but can now be viewed from a more mature perspective. The task of constructing more general probability models will be considered in the next Section 2.4

(i) Discrete models: Suppose $\Omega = \{\omega_1, \omega_2, \dots\}$ is countable. For each i, associate to ω_i the number p_i where

$$\forall i \geq 1, \; p_i \geq 0 \text{ and } \sum_{i=1}^{\infty} p_i = 1.$$

Define $\mathcal{B} = \mathcal{P}(\Omega)$, and for $A \in \mathcal{B}$, set

$$P(A) = \sum_{\omega_i \in A} p_i.$$

Then we have the following properties of P:

(i) $P(A) \geq 0$ for all $A \in \mathcal{B}$.

(ii) $P(\Omega) = \sum_{i=1}^{\infty} p_i = 1$.

(iii) P is σ-additive: If $A_j, j \geq 1$ are mutually disjoint subsets, then

$$P(\bigcup_{j=1}^{\infty} A_j) = \sum_{\omega_i \in \cup_j A_j} p_i = \sum_j \sum_{\omega_i \in A_j} p_i$$
$$= \sum_j P(A_j).$$

Note this last step is justified because the series, being positive, can be added in any order.

This gives the general construction of probabilities when Ω is countable. Next comes a time honored specific example of countable state space model.

(ii) Coin tossing N times: What is an appropriate probability space for the experiment "toss a weighted coin N times"? Set

$$\Omega = \{0, 1\}^N = \{(\omega_1, \dots, \omega_N) : \omega_i = 0 \text{ or } 1\}.$$

For $p \geq 0, q \geq 0, p + q = 1$, define

$$p_{(\omega_1, \dots, \omega_N)} = p^{\sum_{j=1}^N \omega_j} q^{N - \sum_{j=1}^N \omega_j} = p^{\#1's} q^{\#0's}.$$

Construct a probability measure P as in (i) above: Let $\mathcal{B} = \mathcal{P}(\Omega)$ and for $A \subset \Omega$ define

$$P(A) = \sum_{\omega \in A} p_\omega.$$

As in (i) above, this gives a probability model provided $\sum_{\omega\in\Omega} p_\omega = 1$. Note the product form

$$p_{(\omega_1,\dots,\omega_N)} = \prod_{i=1}^{N} p^{\omega_i} q^{1-\omega_i}$$

so

$$\begin{aligned}\sum_{\omega_1,\dots,\omega_N} p_{\omega_1,\dots,\omega_N} &= \sum_{\omega_1,\dots,\omega_N} \prod_{i=1}^{n} p^{\omega_i} q^{1-\omega_i} \\ &= \sum_{\omega_1,\dots,\omega_{N-1}} \prod_{i=1}^{N-1} p^{\omega_i} q^{1-\omega_i} \underbrace{(p^1q^0 + p^0q^1)}_{1} = \cdots = 1.\end{aligned}$$

□

2.4 Constructions of Probability Spaces

The previous section described how to construct a probability space when the sample space Ω is countable. A more complex case but very useful in applications is when Ω is uncountable, for example, when $\Omega = \mathbb{R}, \mathbb{R}^k, \mathbb{R}^\infty$, and so on. For these and similar cases, how do we construct a probability space which will have given desirable properties? For instance, consider the following questions.

(i) Given a distribution function $F(x)$, let $\Omega = \mathbb{R}$. How do we construct a probability measure P on $\mathcal{B}(\mathbb{R})$ such that the distribution function corresponding to P is F:

$$P((-\infty, x]) = F(x).$$

(ii) How do you construct a probability space containing an iid sequence of random variables or a sequence of random variables with given finite dimensional distributions.

A simple case of this question: How do we build a model of an infinite sequence of coin tosses so we can answer questions such as:

(a) What is the probability that heads occurs infinitely often in an infinite sequence of coin tosses; that is, how do we compute

$$P[\text{ heads occurs i.o. }]?$$

(b) How do we compute the probability that ultimately the excess of heads over tails is at least 17?

(c) In a gambling game where a coin is tossed repeatedly and a heads results in a gain of one dollar and a tail results in a loss of one dollar, what is the probability that starting with a fortune of x, ruin eventually occurs; that is, eventually my stake is wiped out?

For these and similar questions, we need uncountable spaces. For the coin tossing problems we need the sample space

$$\begin{aligned}\Omega =&\{0,1\}^{\mathbb{N}}\\ =&\{(\omega_1,\omega_2,\dots):\omega_i\in\{0,1\},\ i\geq 1\}.\end{aligned}$$

2.4.1 General Construction of a Probability Model

The general method is to start with a sample space Ω and a restricted, simple class of subsets $\mathcal{S}$ of Ω to which the assignment of probabilities is obvious or natural. Then this assignment of probabilities is extended to $\sigma(\mathcal{S})$. For example, if $\Omega = \mathbb{R}$, the real line, and we are given a distribution function F, we could take $\mathcal{S}$ to be

$$\mathcal{S} = \{(a,b] : -\infty \leq a \leq b \leq \infty\}$$

and then define P on $\mathcal{S}$ to be

$$P((a,b]) = F(b) - F(a).$$

The problem is to extend the definition of P from $\mathcal{S}$ to $\mathcal{B}(\mathbb{R})$, the Borel sets.

For what follows, recall the notational convention that $\sum_{i=1}^{n} A_i$ means a disjoint union; that is, that $A_1,\dots,A_n$ are mutually disjoint and

$$\sum_{i=1}^{n} A_i = \bigcup_{i=1}^{n} A_i.$$

The following definitions help clarify language and proceedings. Given two structures $\mathcal{G}_1, \mathcal{G}_2$ of subsets of Ω such that $\mathcal{G}_1 \subset \mathcal{G}_2$ and two set functions

$$P_i : \mathcal{G}_i \mapsto [0,1], \quad i = 1,2,$$

we say P_2 is an *extension* of P_1 (or P_1 extends to P_2) if P_2 restricted to $\mathcal{G}_1$ equals P_1. This is written

$$P_2|_{\mathcal{G}_1} = P_1$$

and means $P_2(A_1) = P_1(A_1)$ for all $A_1 \in \mathcal{G}_1$. A set function P with structure $\mathcal{G}$ as domain and range $[0,1]$,

$$P : \mathcal{G} \mapsto [0,1],$$

is *additive* if for any $n \geq 1$ and any disjoint $A_1,\dots,A_n \in \mathcal{G}$ such that $\sum_{i=1}^{n} A_i \in \mathcal{G}$ we have

$$P\Big(\sum_{i=1}^{n} A_i\Big) = \sum_{i=1}^{n} P(A_i). \tag{2.5}$$

Call P *σ-additive* if the index n can be replaced by ∞; that is, (2.5) holds for mutually disjoint $\{A_n, n \geq 1\}$ with $A_j \in \mathcal{G}$, $j \geq 1$ and $\sum_{j=1}^{\infty} A_j \in \mathcal{G}$.

We now define a primitive structure called a *semialgebra*.

Definition 2.4.1 A class $\mathcal{S}$ of subsets of Ω is a *semialgebra* if the following postulates hold:

(i) $\emptyset, \Omega \in \mathcal{S}$.

(ii) $\mathcal{S}$ is a π-system; that is, it is closed under finite intersections.

(iii) If $A \in \mathcal{S}$, then there exist some finite n and disjoint sets $C_1, \dots, C_n$, with each $C_i \in \mathcal{S}$ such that $A^c = \sum_{i=1}^n C_i$.

The plan is to start with a probability measure on the primitive structure $\mathcal{S}$, show there is a unique extension to $\mathcal{A}(\mathcal{S})$, the algebra (field) generated by $\mathcal{S}$ (first extension theorem) and then show there is a unique extension from $\mathcal{A}(\mathcal{S})$ to $\sigma(\mathcal{A}(\mathcal{S})) = \sigma(\mathcal{S})$, the σ-field generated by $\mathcal{S}$ (second extension theorem).

Before proceeding, here are standard examples of semialgebras.

Examples:

(a) Let $\Omega = \mathbb{R}$, and suppose $\mathcal{S}_1$ consists of intervals including $\emptyset$, the empty set:

$$\mathcal{S}_1 = \{(a, b] : -\infty \le a \le b \le \infty\}.$$

If $I_1, I_2 \in \mathcal{S}_1$, then $I_1 I_2$ is an interval and in $\mathcal{S}_1$ and if $I \in \mathcal{S}_1$, then I^c is a union of disjoint intervals.

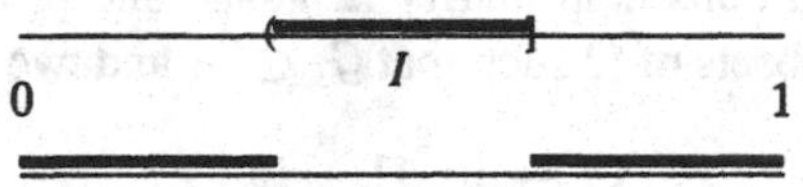

FIGURE 2.1 Intervals

(b) Let

$$\Omega = \mathbb{R}^k = \{(x_1, \dots, x_k) : x_i \in \mathbb{R}, i = 1, \dots, k\}$$
$$\mathcal{S}_k = \text{ all rectangles (including } \emptyset, \text{ the empty set).}$$

Note that we call A a rectangle if it is of the form

$$A = I_1 \times \cdots \times I_k$$

where $I_j \in \mathcal{S}_1$ is an interval, $j = 1, \dots, k$ as in item (a) above. Obviously $\emptyset, \Omega$ are rectangles and intersections of rectangles are rectangles. When $k = 2$ and A is a rectangle, the picture of A^c appears in Figure 2.2, showing A^c can be written as a disjoint union of rectangles.

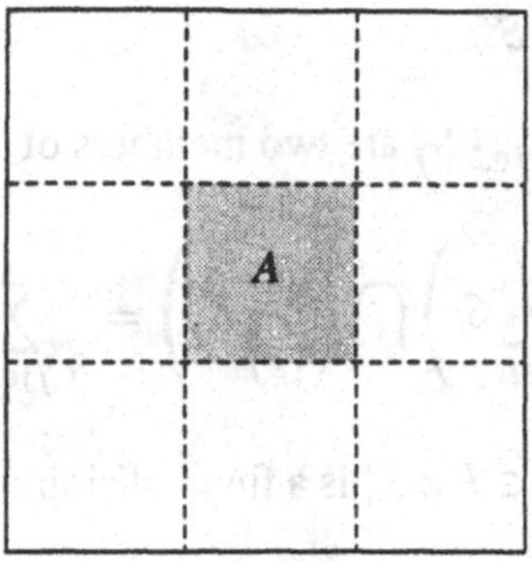

FIGURE 2.2 Rectangles

For general k, let

$$A = I_1 \times \cdots \times I_k, = \bigcap_{i=1}^{k} \{(x_1, \ldots, x_k) : x_i \in I_i\}$$

so that

$$A^c = \left(\bigcap_{i=1}^{k} \{(x_1, \ldots, x_k) : x_i \in I_i\}\right)^c = \bigcup_{i=1}^{k} \{(x_1, \ldots, x_k) : x_i \in I_i^c\}.$$

Since $I_i \in \mathcal{S}_1$, we have $I_i^c = I_i' + I_i''$, where $I_i', I_i'' \in \mathcal{S}_1$ are intervals. Consider

$$\mathcal{D} := \{U_1 \times \cdots \times U_k : U_\alpha = I_\alpha \text{ or } I_\alpha' \text{ or } I_\alpha'',\ \alpha = 1, \ldots, k\}.$$

When $U_\alpha = I_\alpha,\ \alpha = 1, \ldots, k$, then $U_1 \times \ldots \times U_k = A$. So

$$A^c = \sum_{\substack{U_1 \times \cdots \times U_k \in \mathcal{D} \\ \text{Not all } U_\alpha = I_\alpha,\ \alpha = 1, \ldots, k}} U_1 \times \cdots \times U_k.$$

This shows that $\mathcal{S}_k$ is a semialgebra. □

Starting with a semialgebra $\mathcal{S}$, we first discuss the structure of $\mathcal{A}(\mathcal{S})$, the smallest algebra or field containing $\mathcal{S}$.

Lemma 2.4.1 (The field generated by a semialgebra) *Suppose $\mathcal{S}$ is a semialgebra of subsets of Ω. Then*

$$\mathcal{A}(\mathcal{S}) = \{\sum_{i \in I} S_i : I \text{ finite}, \{S_i, i \in I\} \text{ disjoint}, S_i \in \mathcal{S}\}, \tag{2.6}$$

is the family of all sums of finite families of mutually disjoint subsets of Ω in $\mathcal{S}$.

Proof. Let Λ be the collection on the right side of (2.6). It is clear that $\Lambda \supset \mathcal{S}$ (take I to be a singleton set) and we claim Λ is a field. We check the field postulates in Definition 1.5.2, Chapter 1 on page 12:

(i) $\Omega \in \Lambda$ since $\Omega \in \mathcal{S}$.

(iii) If $\sum_{i \in I} S_i$ and $\sum_{j \in J} S'_j$ are two members of Λ, then

$$\left(\sum_{i \in I} S_i\right) \bigcap \left(\sum_{j \in J} S'_j\right) = \sum_{(i,j) \in I \times J} S_i S'_j \in \Lambda$$

since $\{S_i S'_j, (i, j) \in I \times J\}$ is a finite, disjoint collection of members of the π-system $\mathcal{S}$.

(ii) To check closure under complementation, let $\sum_{i \in I} S_i \in \Lambda$ and observe

$$\left(\sum_{i \in I} S_i\right)^c = \bigcap_{i \in I} S_i^c.$$

But from the axioms defining a semialgebra, $S_i \in \mathcal{S}$ implies

$$S_i^c = \sum_{j \in J_i} S_{ij}$$

for a finite index set J_i and disjoint sets $\{S_{ij}, j \in J_i\}$ in $\mathcal{S}$. Now observe that $\bigcap_{i \in I} S_i^c \in \Lambda$ by the previously proven (iii).

So Λ is a field, $\Lambda \supset \mathcal{S}$ and hence $\Lambda \supset \mathcal{A}(\mathcal{S})$. Since also

$$\sum_{i \in I} S_i \in \Lambda \text{ implies } \sum_{i \in I} S_i \in \mathcal{A}(\mathcal{S}),$$

we get $\Lambda \subset \mathcal{A}(\mathcal{S})$ and thus, as desired, $\Lambda = \mathcal{A}(\mathcal{S})$. □

It is now relatively easy to extend a probability measure from $\mathcal{S}$ to $\mathcal{A}(\mathcal{S})$.

Theorem 2.4.1 (First Extension Theorem) *Suppose $\mathcal{S}$ is a semialgebra of subsets of Ω and $P : \mathcal{S} \mapsto [0, 1]$ is σ-additive on $\mathcal{S}$ and satisfies $P(\Omega) = 1$. There is a unique extension P' of P to $\mathcal{A}(\mathcal{S})$, defined by*

$$P'(\sum_{i \in I} S_i) = \sum_{i \in I} P(S_i), \tag{2.7}$$

which is a probability measure on $\mathcal{A}(\mathcal{S})$; that is $P'(\Omega) = 1$ and P' is σ-additive on $\mathcal{A}(\mathcal{S})$.

Proof. We must first check that (2.7) defines P' unambiguously and to do this, suppose $A \in \mathcal{A}(\mathcal{S})$ has two distinct representations

$$A = \sum_{i \in I} S_i = \sum_{j \in J} S'_j.$$

We need to verify that

$$\sum_{i\in I} P(S_i) = \sum_{j\in J} P(S'_j) \tag{2.8}$$

so that P' has a unique value at A. Confirming (2.8) is easy since $S_i \subset A$ and therefore

$$\begin{aligned}\sum_{i\in I} P(S_i) &= \sum_{i\in I} P(S_i A) = \sum_{i\in I} P(S_i \cap \sum_{j\in J} S'_j)\\ &= \sum_{i\in I} P(\sum_{j\in J} S_i S'_j)\end{aligned}$$

and using the fact that $S_i = \sum_{j\in J} S_i S'_j \in \mathcal{S}$ and P is additive on $\mathcal{S}$, we get the above equal to

$$= \sum_{i\in I}\sum_{j\in J} P(S_i S'_j) = \sum_{j\in J}\sum_{i\in I} P(S_i S'_j).$$

Reversing the logic, this equals

$$= \sum_{j\in J} P(S'_j)$$

as required.

Now we check that P' is σ-additive on $\mathcal{A}(\mathcal{S})$. Thus suppose for $i \ge 1$,

$$A_i = \sum_{j\in J_i} S_{ij} \in \mathcal{A}(\mathcal{S}), \quad S_{ij} \in \mathcal{S},$$

and $\{A_i, i \ge 1\}$ are mutually disjoint and

$$A = \sum_{i=1}^{\infty} A_i \in \mathcal{A}(\mathcal{S}).$$

Since $A \in \mathcal{A}(\mathcal{S})$, A also has a representation

$$A = \sum_{k\in K} S_k, \quad S_k \in \mathcal{S},\ k \in K,$$

where K is a finite index set. From the definition of P', we have

$$P'(A) = \sum_{k\in K} P(S_k).$$

Write

$$S_k = S_k A = \sum_{i=1}^{\infty} S_k A_i = \sum_{i=1}^{\infty}\sum_{j\in J_i} S_k S_{ij}.$$

Now $S_k S_{ij} \in \mathcal{S}$ and $\sum_{i=1}^{\infty} \sum_{j \in J_i} S_k S_{ij} = S_k \in \mathcal{S}$, and since P is σ-additive on $\mathcal{S}$, we have

$$\sum_{k \in K} P(S_k) = \sum_{k \in K} \sum_{i=1}^{\infty} \sum_{j \in J_i} P(S_k S_{ij}) = \sum_{i=1}^{\infty} \sum_{j \in J_i} \sum_{k \in K} P(S_k S_{ij}).$$

Again observe

$$\sum_{k \in K} S_k S_{ij} = A S_{ij} = S_{ij} \in \mathcal{S}$$

and by additivity of P on $\mathcal{S}$

$$\sum_{i=1}^{\infty} \sum_{j \in J_i} \sum_{k \in K} P(S_k S_{ij}) = \sum_{i=1}^{\infty} \sum_{j \in J_i} P(S_{ij}),$$

and continuing in the same way, we get this equal to

$$= \sum_{i=1}^{\infty} P(\sum_{j \in J_i} S_{ij}) = \sum_{i=1}^{\infty} P'(A_i)$$

as desired.

Finally, it is clear that P has a unique extension from $\mathcal{S}$ to $\mathcal{A}(\mathcal{S})$, since if P_1' and P_2' are two additive extensions, then for any

$$A = \sum_{i \in I} S_i \in \mathcal{A}(\mathcal{S})$$

we have

$$P_1'(A) = \sum_{i \in I} P(S_i) = P_2'(A).$$

□

Now we know how to extend a probability measure from $\mathcal{S}$ to $\mathcal{A}(\mathcal{S})$. The next step is to extend the probability measure from the algebra to the σ-algebra.

Theorem 2.4.2 (Second Extension Theorem) *A probability measure P defined on a field $\mathcal{A}$ of subsets has a unique extension to a probability measure on $\sigma(\mathcal{A})$, the σ-field generated by $\mathcal{A}$.*

Combining the First and Second Extension Theorems 2.4.1 and 2.4.2 yields the final result.

Theorem 2.4.3 (Combo Extension Theorem) *Suppose $\mathcal{S}$ is a semialgebra of subsets of Ω and that P is a σ-additive set function mapping $\mathcal{S}$ into $[0, 1]$ such that $P(\Omega) = 1$. There is a unique probability measure on $\sigma(\mathcal{S})$ that extends P.*

The ease with which this result can be applied depends largely on how easily one can check that a set function P defined on $\mathcal{S}$ is σ-additive (as opposed to just being additive). Sometimes some sort of compactness argument is needed.

The proof of the Second Extension Theorem 2.4.2 is somewhat longer than the proof of the First Extension Theorem and is deferred to the next Subsection 2.4.2.

2.4.2 Proof of the Second Extension Theorem

We now prove the Second Extension Theorem. We start with a field $\mathcal{A}$ and a probability measure P on $\mathcal{A}$ so that $P(\Omega) = 1$, and for all $A \in \mathcal{A}$, $P(A) \geq 0$ and for $\{A_i\}$ disjoint, $A_i \in \mathcal{A}$, $\sum_{i=1}^{\infty} A_i \in \mathcal{A}$, we have $P(\sum_{i=1}^{\infty} A_i) = \sum_{i=1}^{\infty} P(A_i)$.

The proof is broken into 3 parts. In Part I, we extend P to a set function Π on a class $\mathcal{G} \supset \mathcal{A}$. In Part II we extend Π to a set function Π^* on a class $\mathcal{D} \supset \sigma(\mathcal{A})$ and in Part III we restrict Π^* to $\sigma(\mathcal{A})$ yielding the desired extension.

PART I. We begin by defining the class $\mathcal{G}$:

$$\begin{aligned}\mathcal{G} :=& \{\bigcup_{j=1}^{\infty} A_j : A_j \in \mathcal{A}\}\\ =& \{\lim_{n\to\infty} \uparrow B_n : B_n \in \mathcal{A}, B_n \subset B_{n+1}, \forall n\}.\end{aligned}$$

So $\mathcal{G}$ is the class of unions of countable collections of sets in $\mathcal{A}$, or equivalently, since $\mathcal{A}$ is a field, $\mathcal{G}$ is the class of non-decreasing limits of elements of $\mathcal{A}$.

We also define a set function $\Pi : \mathcal{G} \mapsto [0, 1]$ via the following definition: If $G = \lim_{n\to\infty} \uparrow B_n \in \mathcal{G}$, where $B_n \in \mathcal{A}$, define

$$\Pi(G) = \lim_{n\to\infty} \uparrow P(B_n). \tag{2.9}$$

Since P is σ-additive on $\mathcal{A}$, P is monotone on $\mathcal{A}$, so the monotone convergence indicated in (2.9) is justified. Call the sequence $\{B_n\}$ the *approximating* sequence to G. To verify that Π is well defined, we need to check that if G has two approximating sequences $\{B_n\}$ and $\{B'_n\}$,

$$G = \lim_{n\to\infty} \uparrow B_n = \lim_{n\to\infty} \uparrow B'_n$$

then

$$\lim_{n\to\infty} \uparrow P(B_n) = \lim_{n\to\infty} \uparrow P(B'_n).$$

This is verified in the next lemma whose proof is typical of this sort of uniqueness proof in that some sort of merging of two approximating sequences takes place.

Lemma 2.4.2 *If $\{B_n\}$ and $\{B'_n\}$ are two non-decreasing sequences of sets in $\mathcal{A}$ and*

$$\bigcup_{n=1}^{\infty} B_n \subset \bigcup_{n=1}^{\infty} B'_n,$$

then

$$\lim_{n\to\infty} \uparrow P(B_n) \leq \lim_{n\to\infty} \uparrow P(B'_n).$$

Proof. For fixed m

$$\lim_{n\to\infty} \uparrow B_m B'_n = B_m. \tag{2.10}$$

Since also

$$B_m B'_n \subset B'_n$$

and P is continuous with respect to monotonely converging sequences as a consequence of being σ-additive (see Item 7 on page 31), we have

$$\lim_{n\to\infty} \uparrow P(B'_n) \geq \lim_{n\to\infty} \uparrow P(B_m B'_n) = P(B_m),$$

where the last equality results from (2.10) and P being continuous. The inequality holds for all m, so we conclude that

$$\lim_{n\to\infty} \uparrow P(B'_n) \geq \lim_{m\to\infty} \uparrow P(B_m)$$

as desired. □

Now we list some properties of Π and $\mathcal{G}$:

Property 1. We have

$$\emptyset \in \mathcal{G}, \quad \Pi(\emptyset) = 0,$$
$$\Omega \in \mathcal{G}, \quad \Pi(\Omega) = 1,$$

and for $G \in \mathcal{G}$

$$0 \leq \Pi(G) \leq 1. \tag{2.11}$$

More generally, we have $\mathcal{A} \subset \mathcal{G}$ and

$$\Pi|_{\mathcal{A}} = P;$$

that is, $\Pi(A) = P(A)$, for $A \in \mathcal{A}$.

The first statements are clear since, for example, if we set $B_n = \Omega$ for all n, then

$$\mathcal{A} \ni B_n = \Omega \uparrow \Omega,$$

and

$$\Pi(\Omega) = \lim_{n\to\infty} \uparrow P(\Omega) = 1$$

and a similar argument holds for $\emptyset$. The statement (2.11) follows from $0 \leq P(B_n) \leq 1$ for approximating sets $\{B_n\}$ in $\mathcal{A}$. To show $\Pi(A) = P(A)$ for $A_n \in \mathcal{A}$, take the approximating sequence to be identically equal to A.

Property 2. If $G_i \in \mathcal{G}$ for $i = 1, 2$ then

$$G_1 \cup G_2 \in \mathcal{G}, \quad G_1 \cap G_2 \in \mathcal{G},$$

and

$$\Pi(G_1 \cup G_2) + \Pi(G_1 \cap G_2) = \Pi(G_1) + \Pi(G_2). \tag{2.12}$$

This implies Π is additive on $\mathcal{G}$.

To see this, pick approximating sets $B_{n1}, B_{n2} \in \mathcal{A}$ such that $B_{ni} \uparrow G_i$ for $i = 1, 2$ as $n \to \infty$ and then, since $\mathcal{A}$ is a field, it follows that

$$\mathcal{A} \ni B_{n1} \cup B_{n2} \uparrow G_1 \cup G_2,$$
$$\mathcal{A} \ni B_{n1} \cap B_{n2} \uparrow G_1 \cap G_2,$$

showing that $G_1 \cup G_2$ and $G_1 \cap G_2$ are in $\mathcal{G}$. Further

$$P(B_{n1} \cup B_{n2}) + P(B_{n1} \cap B_{n2}) = P(B_{n1}) + P(B_{n2}), \tag{2.13}$$

from (2.1) on page 30. If we let $n \to \infty$ in (2.13), we get (2.12).

Property 3. Π is monotone on $\mathcal{G}$: If $G_i \in \mathcal{G}$, $i = 1, 2$ and $G_1 \subset G_2$, then $\Pi(G_1) \le \Pi(G_2)$. This follows directly from Lemma 2.4.2.

Property 4. If $G_n \in \mathcal{G}$ and $G_n \uparrow G$, then $G \in \mathcal{G}$ and

$$\Pi(G) = \lim_{n\to\infty} \Pi(G_n).$$

So $\mathcal{G}$ is closed under non-decreasing limits and Π is sequentially monotonely continuous. Combining this with Property 2, we get that if $\{A_i, i \ge 1\}$ is a disjoint sequence of sets in $\mathcal{G}$, $\sum_{i=1}^{\infty} A_i \in \mathcal{G}$ and

$$\begin{aligned}\Pi(\sum_{i=1}^{\infty} A_i) &= \Pi(\lim_{n\to\infty} \uparrow \sum_{i=1}^{n} A_i) = \lim_{n\to\infty} \uparrow \Pi(\sum_{i=1}^{n} A_i)\\ &= \lim_{n\to\infty} \uparrow \sum_{i=1}^{n} \Pi(A_i) = \sum_{i=1}^{\infty} P(A_i).\end{aligned}$$

So Π is σ-additive on $\mathcal{G}$.

For each n, G_n has an approximating sequence $B_{m,n} \in \mathcal{A}$ such that

$$\lim_{m\to\infty} \uparrow B_{m,n} = G_n. \tag{2.14}$$

Define $D_m = \cup_{n=1}^{m} B_{m,n}$. Since $\mathcal{A}$ is closed under finite unions, $D_m \in \mathcal{A}$. We show

$$\lim_{m\to\infty} \uparrow D_m = G, \tag{2.15}$$

and if (2.15) is true, then G has a monotone approximating sequence of sets in $\mathcal{A}$, and hence $G \in \mathcal{G}$.

To show (2.15), we first verify $\{D_m\}$ is monotone:

$$D_m = \bigcup_{n=1}^{m} B_{m,n} \subset \bigcup_{n=1}^{m} B_{m+1,n}$$

(from (2.14))

$$\subset \bigcup_{n=1}^{m+1} B_{m+1,n} = D_{m+1}.$$

Now we show $\{D_m\}$ has the correct limit. If $n \le m$, we have from the definition of D_m and (2.14)

$$B_{m,n} \subset D_m = \bigcup_{j=1}^{m} B_{m,j} \subset \bigcup_{j=1}^{m} G_j = G_m;$$

that is,

$$B_{m,n} \subset D_m \subset G_m. \tag{2.16}$$

Taking limits on m, we have for any $n \ge 1$,

$$G_n = \lim_{m\to\infty} \uparrow B_{m,n} \subset \lim_{m\to\infty} \uparrow D_m \subset \lim_{m\to\infty} \uparrow G_m = G$$

and now taking limits on n yields

$$G = \lim_{n\to\infty} \uparrow G_n \subset \lim_{m\to\infty} \uparrow D_m \subset \lim_{m\to\infty} \uparrow G_m = G \tag{2.17}$$

which shows $D_m \uparrow G$ and proves $G \in \mathcal{G}$. Furthermore, from the definition of Π, we know $\Pi(G) = \lim_{m\to\infty} \uparrow \Pi(D_m)$.

It remains to show $\Pi(G_n) \uparrow \Pi(G)$. From Property 2, all sets appearing in (2.16) are in $\mathcal{G}$ and from monotonicity property 3, we get

$$\Pi(B_{m,n}) \le \Pi(D_m) \le \Pi(G_m).$$

Let $m \to \infty$ and since $G_n = \lim_{m\to\infty} \uparrow B_{m,n}$ we get

$$\Pi(G_n) \le \lim_{m\to\infty} \uparrow \Pi(D_m) \le \lim_{m\to\infty} \uparrow \Pi(G_m)$$

which is true for all n. Thus letting $n \to \infty$ gives

$$\lim_{n\to\infty} \uparrow \Pi(G_n) \le \lim_{m\to\infty} \Pi(D_m) \le \lim_{m\to\infty} \uparrow \Pi(G_m),$$

and therefore

$$\lim_{n\to\infty} \uparrow \Pi(G_n) = \lim_{m\to\infty} \Pi(D_m).$$

The desired result follows from recalling

$$\lim_{m\to\infty} \Pi(D_m) = \Pi(G).$$

This extends P on $\mathcal{A}$ to a σ-additive set function Π on $\mathcal{G}$. □

PART 2. We next extend Π to a set function Π^* on the power set $\mathcal{P}(\Omega)$ and finally show the restriction of Π^* to a certain subclass $\mathcal{D}$ of $\mathcal{P}(\Omega)$ can yield the desired extension of P.

We define $\Pi^* : \mathcal{P}(\Omega) \mapsto [0, 1]$ by

$$\forall A \in \mathcal{P}(\Omega) : \quad \Pi^*(A) = \inf\{\Pi(G) : A \subset G \in \mathcal{G}\}, \tag{2.18}$$

so $\Pi^*(A)$ is the least upper bound of values of Π on sets $G \in \mathcal{G}$ containing A.

We now consider properties of Π^*:

Property 1. We have on $\mathcal{G}$:

$$\Pi^*|_{\mathcal{G}} = \Pi \tag{2.19}$$

and $0 \leq \Pi^*(A) \leq 1$ for any $A \in \mathcal{P}(\Omega)$.

It is clear that if $A \in \mathcal{G}$, then

$$A \in \{G : A \subset G \in \mathcal{G}\}$$

and hence the infimum in (2.18) is achieved at A.

In particular, from (2.19) we get

$$\Pi^*(\Omega) = \Pi(\Omega) = 1, \quad \Pi^*(\emptyset) = \Pi(\emptyset) = 0.$$

Property 2. We have for $A_1, A_2 \in \mathcal{P}(\Omega)$

$$\Pi^*(A_1 \cup A_2) + \Pi^*(A_1 \cap A_2) \leq \Pi^*(A_1) + \Pi^*(A_2) \tag{2.20}$$

and taking $A_1 = A$, $A_2 = A^c$ in (2.20) we get

$$1 = \Pi^*(\Omega) \leq \Pi^*(A) + \Pi^*(A^c), \tag{2.21}$$

where we used the fact that $\Pi^*(\Omega) = 1$.

To verify (2.20), fix $\epsilon > 0$ and find $G_i \in \mathcal{G}$ such that $G_i \supset A_i$, and for $i = 1, 2$,

$$\Pi^*(A_i) + \frac{\epsilon}{2} \geq \Pi(G_i).$$

Adding over $i = 1, 2$ yields

$$\Pi^*(A_1) + \Pi^*(A_2) + \epsilon \geq \Pi(G_1) + \Pi(G_2).$$

By Property 2 for Π (see (2.12)), the right side equals

$$= \Pi(G_1 \cup G_2) + \Pi(G_1 \cap G_2).$$

Since $G_1 \cup G_2 \supset A_1 \cup A_2$, $G_1 \cap G_2 \supset A_1 \cap A_2$, we get from the definition of Π^* that the above is bounded below by

$$\geq \Pi^*(A_1 \cup A_2) + \Pi^*(A_1 \cap A_2).$$

Property 3. Π^* is monotone on $\mathcal{P}(\Omega)$. This follows from the fact that Π is monotone on $\mathcal{G}$.

Property 4. Π^* is sequentially monotone continuous on $\mathcal{P}(\Omega)$ in the sense that if $A_n \uparrow A$, then $\Pi^*(A_n) \uparrow \Pi^*(A)$.

To prove this, fix $\epsilon > 0$. for each $n \geq 1$, find $G_n \in \mathcal{G}$ such that $G_n \supset A_n$ and

$$\Pi^*(A_n) + \frac{\epsilon}{2^n} \geq \Pi(G_n). \tag{2.22}$$

Define $G'_n = \cup_{m=1}^n G_m$. Since $\mathcal{G}$ is closed under finite unions, $G'_n \in \mathcal{G}$ and $\{G'_n\}$ is obviously non-decreasing. We claim for all $n \geq 1$,

$$\Pi^*(A_n) + \epsilon \sum_{i=1}^{n} 2^{-i} \geq \Pi(G'_n). \tag{2.23}$$

We prove the claim by induction. For $n = 1$, the claim follows from (2.22) and the fact that $G'_1 = G_1$. Make the induction hypothesis that (2.23) holds for n and we verify (2.23) for $n + 1$. We have

$$A_n \subset G_n \subset G'_n \text{ and } A_n \subset A_{n+1} \subset G_{n+1}$$

and therefore $A_n \subset G'_n$ and $A_n \subset G_{n+1}$, so

$$A_n \subset G'_n \cap G_{n+1} \in \mathcal{G}. \tag{2.24}$$

Thus

$$\begin{aligned}\Pi(G'_{n+1}) =& \Pi(G'_n \cup G_{n+1})\\ =& \Pi(G'_n) + \Pi(G_{n+1}) - \Pi(G'_n \cap G_{n+1})\end{aligned}$$

from (2.12) for Π on $\mathcal{G}$ and using the induction hypothesis, (2.22) and the monotonicity of Π^*, we get the upper bound

$$\begin{aligned}&\leq \left(\Pi^*(A_n) + \epsilon \sum_{i=1}^{n} 2^{-i}\right) + \left(\Pi^*(A_{n+1}) + \frac{\epsilon}{2^n}\right)\\ &\quad - \Pi^*(A_n)\\ &= \epsilon \sum_{i=1}^{n+1} 2^{-i} + \Pi^*(A_{n+1})\end{aligned}$$

which is (2.23) with n replaced by $n + 1$.

Let $n \to \infty$ in (2.23). Recalling Π^* is monotone on $\mathcal{P}(\Omega)$, Π is monotone on $\mathcal{G}$ and $\mathcal{G}$ is closed under non-decreasing limits, we get

$$\lim_{n\to\infty} \uparrow \Pi^*(A_n) + \epsilon \geq \lim_{n\to\infty} \uparrow \Pi(G'_n) = \Pi(\bigcup_{j=1}^{\infty} G'_j).$$

Since

$$A = \lim_{n\to\infty} \uparrow A_n \subset \bigcup_{j=1}^{\infty} G'_j \in \mathcal{G},$$

we conclude

$$\lim_{n\to\infty} \uparrow \Pi^*(A_n) \geq \Pi^*(A).$$

For a reverse inequality, note that monotonicity gives

$$\Pi^*(A_n) \leq \Pi^*(A)$$

and thus

$$\lim_{n\to\infty} \uparrow \Pi^*(A_n) \leq \Pi^*(A).$$

□

PART 3. We now retract Π^* to a certain subclass $\mathcal{D}$ of $\mathcal{P}(\Omega)$ and show $\Pi^*|_{\mathcal{D}}$ is the desired extension.

We define

$$\mathcal{D} := \{D \in \mathcal{P}(\Omega) : \Pi^*(D) + \Pi^*(D^c) = 1.\}$$

Lemma 2.4.3 *The class $\mathcal{D}$ has the following properties:*

1. *$\mathcal{D}$ is a σ-field.*
2. *$\Pi^*|_{\mathcal{D}}$ is a probability measure on $(\Omega, \mathcal{D})$.*

Proof. We first show $\mathcal{D}$ is a field. Obviously $\Omega \in \mathcal{D}$ since $\Pi^*(\Omega) = 1$ and $\Pi^*(\emptyset) = 0$. To see $\mathcal{D}$ is closed under complementation is easy: If $D \in \mathcal{D}$, then

$$\Pi^*(D) + \Pi^*(D^c) = 1$$

and the same holds for D^c.

Next, we show $\mathcal{D}$ is closed under finite unions and finite intersections. If $D_1, D_2 \in \mathcal{D}$, then from (2.20)

$$\Pi^*(D_1 \cup D_2) + \Pi^*(D_1 \cap D_2) \leq \Pi^*(D_1) + \Pi^*(D_2) \tag{2.25}$$

$$\Pi^*((D_1 \cup D_2)^c) + \Pi^*((D_1 \cap D_2)^c) \leq \Pi^*(D_1^c) + \Pi^*(D_2^c). \tag{2.26}$$

Add the two inequalities (2.25) and (2.26) to get

$$\Pi^*(D_1 \cup D_2) + \Pi^*((D_1 \cup D_2)^c) + \Pi^*(D_1 \cap D_2) + \Pi^*((D_1 \cap D_2)^c) \leq 2 \tag{2.27}$$

where we used $D_i \in \mathcal{D}, i = 1, 2$ on the right side. From (2.21), the left side of (2.27) is ≥ 2, so equality prevails in (2.27). Again using (2.21), we see

$$\Pi^*(D_1 \cup D_2) + \Pi^*((D_1 \cup D_2)^c) = 1$$

$$\Pi^*(D_1 \cap D_2) + \Pi^*((D_1 \cap D_2)^c) = 1.$$

Thus $D_1 \cup D_2$, $D_1 \cap D_2 \in \mathcal{D}$ and $\mathcal{D}$ is a field. Also, equality must prevail in (2.25) and (2.26) (else it would fail in (2.27)). This shows that Π^* is finitely additive on $\mathcal{D}$.

Now it remains to show that $\mathcal{D}$ is a σ-field and Π^* is σ-additive on $\mathcal{D}$. Since $\mathcal{D}$ is a field, to show it is a σ-field, it suffices by Exercise 41 of Chapter 1 to show that $\mathcal{D}$ is a monotone class. Since $\mathcal{D}$ is closed under complementation, it is enough to show that $D_n \in \mathcal{D}$, $D_n \uparrow D$ implies $D \in \mathcal{D}$. However, $D_n \uparrow D$ implies, since Π^* is monotone and sequentially monotone continuous, that

$$\lim_{n\to\infty} \uparrow \Pi^*(D_n) = \Pi^*(\bigcup_{n=1}^{\infty} D_n) = \Pi^*(D).$$

Also, for any $m \geq 1$,

$$\Pi^*((\bigcup_{n=1}^{\infty} D_n)^c) = \Pi^*(\bigcap_{n=1}^{\infty} D_n^c) \leq \Pi^*(D_m^c)$$

and therefore, from (2.21)

$$1 \leq \Pi^*(\bigcup_{n=1}^{\infty} D_n) + \Pi^*((\bigcup_{n=1}^{\infty} D_n)^c) \leq \lim_{n\to\infty} \Pi^*(D_n) + \Pi^*(D_m^c) \tag{2.28}$$

and letting $m \to \infty$, we get using $D_n \in \mathcal{D}$

$$\begin{aligned} 1 &\leq \lim_{n\to\infty} \Pi^*(D_n) + \lim_{m\to\infty} \Pi^*(D_m^c) \\ &= \lim_{n\to\infty} \left(\Pi^*(D_n) + \Pi^*(D_n^c)\right) = 1, \end{aligned}$$

and so equality prevails in (2.28). Thus, $D_n \uparrow D$ and $D_n \in \mathcal{D}$ imply $D \in \mathcal{D}$ and $\mathcal{D}$ is both an algebra and a monotone class and hence is a σ-algebra.

Finally, we show $\Pi^*|_{\mathcal{D}}$ is σ-additive. If $\{D_n\}$ is a sequence of disjoint sets in $\mathcal{D}$, then because Π^* is continuous with respect to non-decreasing sequences and $\mathcal{D}$ is a field

$$\begin{aligned} \Pi^*(\sum_{i=1}^{\infty} D_i) &= \Pi^*(\lim_{n\to\infty} \sum_{i=1}^{n} D_i) \\ &= \lim_{n\to\infty} \Pi^*(\sum_{i=1}^{n} D_i) \end{aligned}$$

and because Π^* is finitely additive on $\mathcal{D}$, this is

$$= \lim_{n\to\infty} \sum_{i=1}^{n} \Pi^*(D_i) = \sum_{i=1}^{\infty} \Pi^*(D_i),$$

as desired.

Since $\mathcal{D}$ is a σ-field and $\mathcal{D} \supset \mathcal{A}$, $\mathcal{D} \supset \sigma(\mathcal{A})$. The restriction $\Pi^*|_{\sigma(\mathcal{A})}$ is the desired extension of P on $\mathcal{A}$ to a probability measure on $\sigma(\mathcal{A})$. The extension from $\mathcal{A}$ to $\sigma(\mathcal{A})$ must be unique because of Corollary 2.2.1 to Dynkin's theorem. □

2.5 Measure Constructions

In this section we give two related constructions of probability spaces. The first discussion shows how to construct Lebesgue measure on (0, 1] and the second shows how to construct a probability on $\mathbb{R}$ with given distribution function F.

2.5.1 *Lebesgue Measure on* (0, 1]

Suppose

$$\begin{aligned}\Omega &= (0, 1],\\ \mathcal{B} &= \mathcal{B}((0, 1]),\\ \mathcal{S} &= \{(a, b] : 0 \le a \le b \le 1\}.\end{aligned}$$

Define on $\mathcal{S}$ the function $\lambda : \mathcal{S} \mapsto [0, 1]$ by

$$\lambda(\emptyset) = 0, \quad \lambda(a, b] = b - a.$$

With a view to applying Extension Theorem 2.4.3, note that $\lambda(A) \ge 0$. To show that λ has unique extension we need to show that λ is σ-additive.

We first show that λ is finitely additive on $\mathcal{S}$. Let $(a, b] \in \mathcal{S}$ and suppose

$$(a, b] = \bigcup_{i=1}^{k} (a_i, b_i],$$

where the intervals on the right side are disjoint. Assuming the intervals have been indexed conveniently, we have

$$a_1 = a, b_k = b, b_i = a_{i+1}, \ i = 1, \ldots, k-1.$$

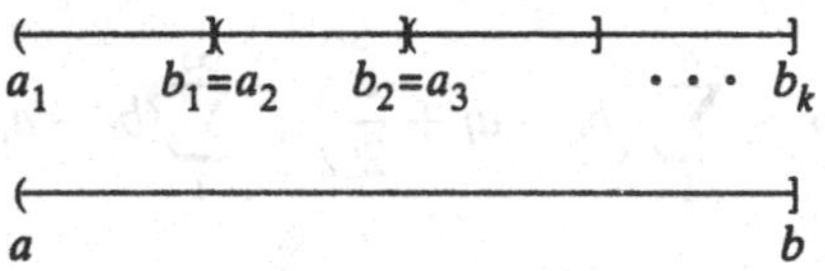

FIGURE 2.3 Abutting Intervals

Then $\lambda(a, b] = b - a$ and

$$\begin{aligned}\sum_{i=1}^{k} \lambda(a_i, b_i] &= \sum_{i=1}^{k} (b_i - a_i)\\ &= b_1 - a_1 + b_2 - a_2 + \cdots + b_k - a_k\\ &= b_k - a_1 = b - a.\end{aligned}$$

This shows λ is finitely additive.

We now show λ is σ-additive. Care must be taken since this involves an infinite number of sets and in fact a compactness argument is employed to cope with the infinities.

Let

$$(a, b] = \bigcup_{i=1}^{\infty} (a_i, b_i]$$

and we first prove that

$$b - a \leq \sum_{i=1}^{\infty} (b_i - a_i). \tag{2.29}$$

Pick $\varepsilon < b - a$ and observe

$$[a + \varepsilon, b] \subset \bigcup_{i=1}^{\infty} \left(a_i, b_i + \frac{\varepsilon}{2^i}\right). \tag{2.30}$$

The set on the left side of (2.30) is compact and the right side of (2.30) gives an open cover, so that by compactness, there is a finite subcover. Thus there exists some integer N such that

$$[a + \varepsilon, b] \subset \bigcup_{i=1}^{N} \left(a_i, b_i + \frac{\varepsilon}{2^i}\right). \tag{2.31}$$

It suffices to prove

$$b - a - \varepsilon \leq \sum_{1}^{N} \left(b_i - a_i + \frac{\varepsilon}{2^i}\right) \tag{2.32}$$

since then we would have

$$b - a - \varepsilon \leq \sum_{1}^{N} \left(b_i - a_i + \frac{\varepsilon}{2^i}\right) \leq \sum_{1}^{\infty} (b_i - a_i) + \varepsilon; \tag{2.33}$$

that is,

$$b - a \leq \sum_{1}^{\infty} (b_i - a_i) + 2\varepsilon. \tag{2.34}$$

Since ε can be arbitrarily small

$$b - a \leq \sum_{1}^{\infty} (b_i - a_i)$$

as desired.

Rephrasing relations (2.31) and (2.32) slightly, we need to prove that

$$[a,b] \subset \bigcup_{1}^{N}(a_i, b_i) \tag{2.35}$$

implies

$$b - a \leq \sum_{1}^{N}(b_i - a_i). \tag{2.36}$$

We prove this by induction. First note that the assertion that (2.35) implies (2.36) is true for $N = 1$. Now we make the induction hypothesis that whenever relation (2.35) holds for $N-1$, it follows that relation (2.36) holds for $N-1$. We now must show that (2.35) implies (2.36) for N.

Suppose $a_N = \vee_1^N a_i$, and

$$a_N < b \leq b_N, \tag{2.37}$$

with similar argument if (2.37) fails. Suppose relation (2.35) holds. We consider two cases:

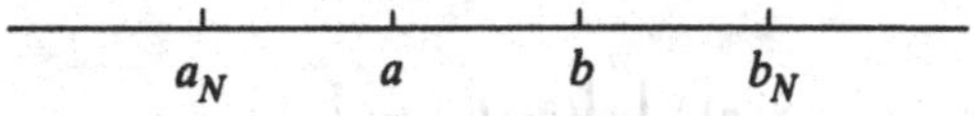

FIGURE 2.4 Case 1

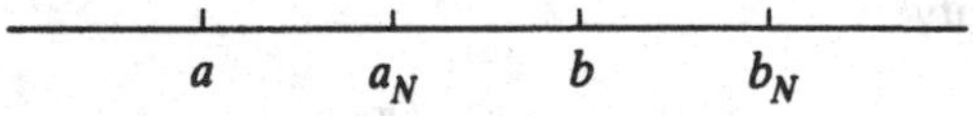

FIGURE 2.5 Case 2

CASE 1: Suppose $a_N \leq a$ Then

$$b - a \leq b_N - a_N \leq \sum_{1}^{N}(b_i - a_i).$$

CASE 2: Suppose $a_N > a$. Then if (2.35) holds

$$[a, a_N] \subset \bigcup_{1}^{N-1}(a_i, b_i)$$

so by the induction hypothesis

$$a_N - a \leq \sum_{i=1}^{N-1}(b_i - a_i)$$

so

$$\begin{aligned} b - a =& b - a_N + a_N - a \\ \le& b - a_N + \sum_{i=1}^{N-1} (b_i - a_i) \\ \le& b_N - a_N + \sum_{i=1}^{N-1} (b_i - a_i) \\ =& \sum_{i=1}^{N} (b_i - a_i) \end{aligned}$$

which is relation (2.36). This verifies (2.29).

We now obtain a reverse inequality complementary to (2.29). We claim that if $(a, b] = \sum_{i=1}^{\infty} (a_i, b_i]$, then for every n,

$$\lambda((a, b]) = b - a \ge \sum_{i=1}^{n} \lambda((a_i, b_i]) = \sum_{i=1}^{n} (b_i - a_i). \tag{2.38}$$

This is easily verified since we know λ is finitely additive on $\mathcal{S}$. For any n, $\cup_{i=1}^{n} (a_i, b_i]$ is a finite union of disjoint intervals and so is

$$(a, b] \setminus \bigcup_{i=1}^{n} (a_i, b_i] =: \bigcup_{j=1}^{m} I_j.$$

So by finite additivity

$$\lambda((a, b]) = \lambda(\bigcup_{i=1}^{n} (a_i, b_i] \cup \bigcup_{j=1}^{m} I_j),$$

which by finite additivity is

$$\begin{aligned} &= \sum_{i=1}^{n} \lambda((a_i, b_i]) + \sum_{j=1}^{m} \lambda(I_j) \\ &\ge \sum_{i=1}^{n} \lambda((a_i, b_i]). \end{aligned}$$

Let $n \to \infty$ to achieve

$$\lambda((a, b]) \ge \sum_{i=1}^{\infty} \lambda((a_i, b_i]).$$

This plus (2.29) shows λ is σ-additive on $\mathcal{S}$. □

2.5.2 Construction of a Probability Measure on $\mathbb{R}$ with Given Distribution Function $F(x)$

Given Lebesgue measure λ constructed in Section 2.5.1 and a distribution function $F(x)$, we construct a probability measure on $\mathbb{R}$, P_F, such that

$$P_F((-\infty, x]) = F(x).$$

Define the left continuous inverse of F as

$$F^{\leftarrow}(y) = \inf\{s : F(s) \geq y\}, \quad 0 < y \leq 1 \tag{2.39}$$

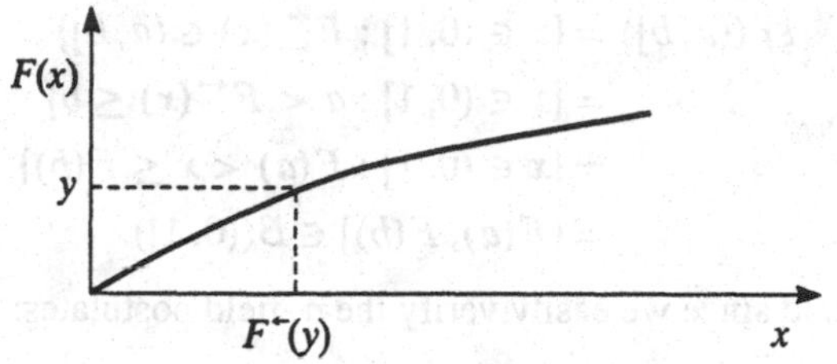

FIGURE 2.6

and define

$$A(y) := \{s : F(s) \geq y\}.$$

Here are the important properties of $A(y)$.

(a) The set $A(y)$ is closed. If $s_n \in A(y)$, and $s_n \downarrow s$, then by right continuity

$$y \leq F(s_n) \downarrow F(s),$$

so $F(s) \geq y$ and $s \in A(y)$. If $s_n \uparrow s$ and $s_n \in A(y)$, then

$$y \leq F(s_n) \uparrow F(s-) \leq F(s)$$

and $y \leq F(s)$ implies $s \in A(y)$.

(b) Since $A(y)$ closed,

$$\inf A(y) \in A(y);$$

that is,

$$F(F^{\leftarrow}(y)) \geq y.$$

(c) Consequently,

$$F^{\leftarrow}(y) > t \text{ iff } y > F(t)$$

or equivalently

$$F^{\leftarrow}(y) \leq t \text{ iff } y \leq F(t).$$

The last property is proved as follows. If $t < F^{\leftarrow}(y) = \inf A(y)$, then $t \notin A(y)$, so that $F(t) < y$. Conversely, if $F^{\leftarrow}(y) \leq t$, then $t \in A(y)$ and $F(t) \geq y$.

Now define for $A \subset \mathbb{R}$
$$\xi_F(A) = \{x \in (0,1] : F^{\leftarrow}(x) \in A\}.$$
If A is a Borel subset of $\mathbb{R}$, then $\xi_F(A)$ is a Borel subset of $(0, 1]$.

Lemma 2.5.1 *If $A \in \mathcal{B}(\mathbb{R})$, then $\xi_F(A) \in \mathcal{B}((0,1])$.*

Proof. Define
$$\mathcal{G} = \{A \subset \mathbb{R} : \xi_F(A) \in \mathcal{B}((0,1])\}.$$
$\mathcal{G}$ contains finite intervals of the form $(a, b] \subset \mathbb{R}$ since from Property (c) of $F^{\leftarrow}$
$$\begin{aligned}\xi_F((a,b]) &= \{x \in (0,1] : F^{\leftarrow}(x) \in (a,b]\}\\ &= \{x \in (0,1] : a < F^{\leftarrow}(x) \leq b\}\\ &= \{x \in (0,1] : F(a) < x \leq F(b)\}\\ &= (F(a), F(b)] \in \mathcal{B}((0,1]).\end{aligned}$$
Also $\mathcal{G}$ is a σ-field since we easily verify the σ-field postulates:

(i) We have
$$\mathbb{R} \in \mathcal{G}$$
since $\xi_F(\mathbb{R}) = (0, 1]$.

(ii) We have that $A \in \mathcal{G}$ implies $A^c \in \mathcal{G}$ since
$$\begin{aligned}\xi_F(A^c) &= \{x \in (0,1] : F^{\leftarrow}(x) \in A^c\}\\ &= \{x \in (0,1] : F^{\leftarrow}(x) \in A\}^c = (\xi_F(A))^c.\end{aligned}$$

(iii) $\mathcal{G}$ is closed under countable unions since if $A_n \in \mathcal{G}$, then
$$\xi_F(\bigcup_n A_n) = \bigcup_n \xi_F(A_n)$$
and therefore
$$\bigcup_n A_n \in \mathcal{G}.$$

So $\mathcal{G}$ contains intervals and $\mathcal{G}$ is a σ-field and therefore
$$\mathcal{G} \supset \mathcal{B}(\text{ intervals }) = \mathcal{B}(\mathbb{R}).$$
□

We now can make our definition of P_F. We define
$$P_F(A) = \lambda(\xi_F(A)),$$
where λ is Lebesgue measure on $(0, 1]$. It is easy to check that P_F is a probability measure. To compute its distribution function and check that it is F, note that
$$\begin{aligned}P_F(-\infty, x] &= \lambda(\xi_F(-\infty, x]) = \lambda\{y \in (0,1] : F^{\leftarrow}(y) \leq x\}\\ &= \lambda\{y \in (0,1] : y \leq F(x)\}\\ &= \lambda((0, F(x)]) = F(x).\end{aligned}$$
□

2.6 Exercises

1. Let Ω be a non-empty set. Let $\mathcal{F}_0$ be the collection of all subsets such that either A or A^c is finite.

   (a) Show that $\mathcal{F}_0$ is a field.

   Define for $E \in \mathcal{F}_0$ the set function P by

$$P(E) = \begin{cases} 0, & \text{if } E \text{ is finite,} \\ 1, & \text{if } E^c \text{ is finite.} \end{cases}$$

   (b) If Ω is countably infinite, show P is finitely additive but not σ-additive.

   (c) If Ω is uncountable, show P is σ-additive on $\mathcal{F}_0$.

2. Let $\mathcal{A}$ be the smallest field over the π-system $\mathcal{P}$. Use the inclusion-exclusion formula (2.2) to show that probability measures agreeing on $\mathcal{P}$ must agree also on $\mathcal{A}$.

   Hint: Use Exercise 20 of Chapter 1.

3. Let $(\Omega, \mathcal{B}, P)$ be a probability space. Show for events $B_i \subset A_i$ the following generalization of subadditivity:

$$P(\cup_i A_i) - P(\cup_i B_i) \le \sum_i (P(A_i) - P(B_i)).$$

4. Review Exercise 34 in Chapter 1 to see how to extend a σ-field. Suppose P is a probability measure on a σ-field $\mathcal{B}$ and suppose $A \notin \mathcal{B}$. Let

$$\mathcal{B}_1 = \sigma(\mathcal{B}, A)$$

   and show that P has an extension to a probability measure P_1 on $\mathcal{B}_1$. (Do this without appealing directly to the Combo Extension Theorem 2.4.3.)

5. Let P be a probability measure on $\mathcal{B}(\mathbb{R})$. For any $B \in \mathcal{B}(\mathbb{R})$ and any $\epsilon > 0$, there exists a finite union of intervals A such that

$$P(A \Delta B) < \epsilon.$$

   Hint: Define

$$\mathcal{G} := \{B \in \mathcal{B}(\mathbb{R}) : \forall \epsilon > 0, \text{ there exists a finite union of intervals } A_\epsilon \text{ such that } P(A \Delta B) < \epsilon\}.$$

6. Say events $A_1, A_2, \ldots$ are *almost disjoint* if

$$P(A_i \cap A_j) = 0, \quad i \neq j.$$

   Show for such events

$$P(\bigcup_{j=1}^{\infty} A_j) = \sum_{j=1}^{\infty} P(A_j).$$

7. **Coupon collecting.** Suppose there are N different types of coupons available when buying cereal; each box contains one coupon and the collector is seeking to collect one of each in order to win a prize. After buying n boxes, what is the probability p_n that the collector has at least one of each type? (Consider sampling with replacement from a population of N distinct elements. The sample size is $n > N$. Use inclusion–exclusion formula (2.2).)

8. We know that $P_1 = P_2$ on $\mathcal{B}$ if $P_1 = P_2$ on $\mathcal{C}$, provided that $\mathcal{C}$ generates $\mathcal{B}$ and is a π-system. Show this last property cannot be omitted. For example, consider $\Omega = \{a, b, c, d\}$ with

$$P_1(\{a\}) = P_1(\{d\}) = P_2(\{b\}) = P_2(\{c\}) = \frac{1}{6}$$

and

$$P_1(\{b\}) = P_1(\{c\}) = P_2(\{a\}) = P_2(\{d\}) = \frac{1}{3}.$$

Set

$$\mathcal{C} = \{\{a, b\}, \{d, c\}, \{a, c\}, \{b, d\}\}.$$

9. Background: Call two sets $A_1, A_2 \in \mathcal{B}$ *equivalent* if $P(A_1 \Delta A_2) = 0$. For a set $A \in \mathcal{B}$, define the equivalence class

$$A^\# = \{B \in \mathcal{B} : P(B \Delta A) = 0\}.$$

This decomposes $\mathcal{B}$ into equivalences classes. Write

$$P^\#(A^\#) = P(A), \qquad \forall A \in A^\#.$$

In practice we drop #s; that is identify the equivalence classes with the members.

An *atom* in a probability space $(\Omega, \mathcal{B}, P)$ is defined as (the equivalence class of) a set $A \in \mathcal{B}$ such that $P(A) > 0$, and if $B \subset A$ and $B \in \mathcal{B}$, then $P(B) = 0$, or $P(A \setminus B) = 0$. Furthermore the probability space is called *non-atomic* if there are no atoms; that is, $A \in \mathcal{B}$ and $P(A) > 0$ imply that there exists a $B \in \mathcal{B}$ such that $B \subset A$ and $0 < P(B) < P(A)$.

(a) If $\Omega = R$, and P is determined by a distribution function $F(x)$, show that the atoms are $\{x : F(x) - F(x-) > 0\}$.

(b) If $(\Omega, \mathcal{B}, P) = ((0, 1], \mathcal{B}((0, 1]), \lambda)$, where λ is Lebesgue measure, then the probability space is non-atomic.

(c) Show that two distinct atoms have intersection which is the empty set. (The sets A, B are distinct means $P(A \Delta B) > 0$. The exercise then requires showing $P(AB\Delta\emptyset) = 0$.)

(d) A probability space contains at most countably many atoms. (Hint: What is the maximum number of atoms that the space can contain that have probability at least $1/n$?)

(e) If a probability space $(\Omega, \mathcal{B}, P)$ contains no atoms, then for every $a \in (0, 1]$ there exists at least one set $A \in \mathcal{B}$ such that $P(A) = a$. (One way of doing this uses Zorn's lemma.)

(f) For every probability space $(\Omega, \mathcal{B}, P)$ and any $\epsilon > 0$, there exists a finite partition of Ω by $\mathcal{B}$ sets, each of whose elements either has probability $\leq \epsilon$ or is an atom with probability $> \epsilon$.

(g) **Metric space:** On the set of equivalence classes, define

$$d(A_1^\#, A_2^\#) = P(A_1 \Delta A_2)$$

where $A_i \in A_i^\#$ for $i = 1, 2$. Show d is a metric on the set of equivalence classes. Verify

$$|P(A_1) - P(A_2)| \leq P(A_1 \Delta A_2)$$

so that $P^\#$ is uniformly continuous on the set of equivalence classes. P is σ-additive is equivalent to

$$\mathcal{B} \ni A_n \downarrow \emptyset \text{ implies } d(A_n^\#, \emptyset^\#) \to 0.$$

10. Two events A, B on the probability space $(\Omega, \mathcal{B}, P)$ are equivalent (see Exercise 9) if

$$P(A \cap B) = P(A) \vee P(B).$$

11. Suppose $\{B_n, n \geq 1\}$ are events with $P(B_n) = 1$ for all n. Show

$$P(\bigcap_{n=1}^{\infty} B_n) = 1.$$

12. Suppose $\mathcal{C}$ is a class of subsets of Ω and suppose $B \subset \Omega$ satisfies $B \in \sigma(\mathcal{C})$. Show that there exists a countable class $\mathcal{C}_B \subset \mathcal{C}$ such that $B \in \sigma(\mathcal{C}_B)$.

Hint: Define

$$\mathcal{G} := \{B \subset \Omega : \exists \text{ countable } \mathcal{C}_B \subset \mathcal{C} \text{ such that } B \in \sigma(\mathcal{C}_B)\}.$$

Show that $\mathcal{G}$ is a σ-field that contains $\mathcal{C}$.

13. If $\{B_k\}$ are events such that

$$\sum_{k=1}^{n} P(B_k) > n - 1,$$

then

$$P(\bigcap_{k=1}^{n} B_k) > 0.$$

14. If F is a distribution function, then F has at most countably many discontinuities.

15. If $\mathcal{S}_1$ and $\mathcal{S}_2$ are two semialgebras of subsets of Ω, show that the class

$$\mathcal{S}_1\mathcal{S}_2 := \{A_1A_2 : A_1 \in \mathcal{S}_1, A_2 \in \mathcal{S}_2\}$$

is again a semialgebra of subsets of Ω. The field (σ-field) generated by $\mathcal{S}_1\mathcal{S}_2$ is identical with that generated by $\mathcal{S}_1 \cup \mathcal{S}_2$.

16. Suppose $\mathcal{B}$ is a σ-field of subsets of Ω and suppose $Q : \mathcal{B} \mapsto [0, 1]$ is a set function satisfying

    (a) Q is finitely additive on $\mathcal{B}$.

    (b) $0 \leq Q(A) \leq 1$ for all $A \in \mathcal{B}$ and $Q(\Omega) = 1$.

    (c) If $A_i \in \mathcal{B}$ are disjoint and $\sum_{i=1}^{\infty} A_i = \Omega$, then $\sum_{i=1}^{\infty} Q(A_i) = 1$.

    Show Q is a probability measure; that is, show Q is σ-additive.

17. For a distribution function $F(x)$, define

$$F_l^{\leftarrow}(y) = \inf\{t : F(t) \geq y\}$$
$$F_r^{\leftarrow}(y) = \inf\{t : F(t) > y\}.$$

We know $F_l^{\leftarrow}(y)$ is left-continuous. Show $F_r^{\leftarrow}(y)$ is right continuous and show

$$\lambda\{u \in (0, 1] : F_l^{\leftarrow}(u) \neq F_r^{\leftarrow}(u)\} = 0,$$

where, as usual, λ is Lebesgue measure. Does it matter which inverse we use?

18. Let A, B, C be disjoint events in a probability space with

$$P(A) = .6, \quad P(B) = .3, \quad P(C) = .1.$$

Calculate the probabilities of every event in $\sigma(A, B, C)$.

19. **Completion.** Let $(\Omega, \mathcal{B}, P)$ be a probability space. Call a set N *null* if $N \in \mathcal{B}$ and $P(N) = 0$. Call a set $B \subset \Omega$ *negligible* if there exists a null set N such that $B \subset N$. Notice that for B to be negligible, it is not required that B be measurable. Denote the set of all negligible subsets by $\mathcal{N}$. Call $\mathcal{B}$ *complete* (with respect to P) if every negligible set is null.

    What if $\mathcal{B}$ is not complete? Define

$$\mathcal{B}^* := \{A \cup M : A \in \mathcal{B}, M \in \mathcal{N}\}.$$

    (a) Show $\mathcal{B}^*$ is a σ-field.

(b) If $A_i \in \mathcal{B}$ and $M_i \in \mathcal{N}$ for $i = 1, 2$ and

$$A_1 \cup M_1 = A_2 \cup M_2,$$

then $P(A_1) = P(A_2)$.

(c) Define $P^* : \mathcal{B}^* \mapsto [0, 1]$ by

$$P^*(A \cup M) = P(A), \quad A \in \mathcal{B},\ M \in \mathcal{N}.$$

Show P^* is an extension of P to $\mathcal{B}^*$.

(d) If $B \subset \Omega$ and $A_i \in \mathcal{B}, i = 1, 2$ and $A_1 \subset B \subset A_2$ and $P(A_2 \setminus A_1) = 0$, then show $B \in \mathcal{B}^*$.

(e) Show $\mathcal{B}^*$ is complete. Thus every σ-field has a completion.

(f) Suppose $\Omega = \mathbb{R}$ and $\mathcal{B} = \mathcal{B}(\mathbb{R})$. Let $p_k \geq 0, \sum_k p_k = 1$. Let $\{a_k\}$ be any sequence in $\mathbb{R}$. Define P by

$$P(\{a_k\}) = p_k, \quad P(A) = \sum_{a_k \in A} p_k, \quad A \in \mathcal{B}.$$

What is the completion of $\mathcal{B}$?

(g) Say that the probability space $(\Omega, \mathcal{B}, P)$ has a complete extension $(\Omega, \mathcal{B}_1, P_1)$ if $\mathcal{B} \subset \mathcal{B}_1$ and $P_1|_{\mathcal{B}} = P$. The previous problem (c) showed that every probability space has a complete extension. However, this extension may not be unique. Suppose that $(\Omega, \mathcal{B}_2, P_2)$ is a second complete extension of $(\Omega, \mathcal{B}, P)$. Show P_1 and P_2 may not agree on $\mathcal{B}_1 \cap \mathcal{B}_2$. (It should be enough to suppose Ω has a small number of points.)

(h) Is there a *minimal* extension?

20. In $(0, 1]$, let $\mathcal{B}$ be the class of sets that either (a) are of the first category or (b) have complement of the first category. Show that $\mathcal{B}$ is a σ-field. For $A \in \mathcal{B}$, define $P(A)$ to be 0 in case (a) and 1 in case (b). Is P σ-additive?

21. Let $\mathcal{A}$ be a field of subsets of Ω and let μ be a *finitely* additive probability measure on $\mathcal{A}$. (This requires $\mu(\Omega) = 1$.)

    If $A_n \in \mathcal{A}$ and $A_n \downarrow \emptyset$, is it the case that $\mu(A_n) \downarrow 0$? (Hint: Review Problem 2.6.1 with $A_n = \{n, n+1, \dots\}$.)

22. Suppose $F(x)$ is a continuous distribution function on $\mathbb{R}$. Show F is *uniformly* continuous.

23. **Multidimensional distribution functions.** For $\mathbf{a}, \mathbf{b}, \mathbf{x} \in \mathcal{B}(\mathbb{R}^k)$ write

$$\mathbf{a} \le \mathbf{b} \text{ iff } a_i \le b_i,\ i = 1, \dots, k;$$
$$(-\infty, \mathbf{x}] = \{\mathbf{u} \in \mathcal{B}(\mathbb{R}^k) : \mathbf{u} \le \mathbf{x}\}$$
$$(\mathbf{a}, \mathbf{b}] = \{\mathbf{u} \in \mathcal{B}(\mathbb{R}^k) : \mathbf{a} < \mathbf{u} \le \mathbf{b}\}.$$

Let P be a probability measure on $\mathcal{B}(\mathbb{R}^k)$ and define for $\mathbf{x} \in \mathbb{R}^k$

$$F(\mathbf{x}) = P((-\infty, \mathbf{x}]).$$

Let $\mathcal{S}_k$ be the semialgebra of k-dimensional rectangles in $\mathbb{R}^k$.

(a) If $\mathbf{a} \le \mathbf{b}$, show the rectangle $I_k := (\mathbf{a}, \mathbf{b}]$ can be written as

$$I_k = (-\infty, \mathbf{b}] \setminus \Big((-\infty, (a_1, b_2, \dots, b_k)] \cup (-\infty, (b_1, a_2, \dots, b_k)] \cup \dots \cup (-\infty, (b_1, b_2, \dots, a_k)] \Big) \tag{2.40}$$

where the union is indexed by the vertices of the rectangle other than $\mathbf{b}$.

(b) Show
$$\mathcal{B}(\mathbb{R}^k) = \sigma((-\infty, \mathbf{x}], \mathbf{x} \in \mathbb{R}^k).$$

(c) Check that $\{(-\infty, \mathbf{x}], \mathbf{x} \in \mathbb{R}^k\}$ is a π-system.

(d) Show P is determined by $F(\mathbf{x}), \mathbf{x} \in \mathbb{R}^k$.

(e) Show F satisfies the following properties:

(1) If $x_i \to \infty, i = 1, \dots, k$, then $F(\mathbf{x}) \to 1$.

(2) If for some $i \in \{1, \dots, k\}$ $x_i \to -\infty$, then $F(\mathbf{x}) \to 0$.

(3) For $\mathcal{S}_k \ni I_k = (\mathbf{a}, \mathbf{b}]$, use the inclusion-exclusion formula (2.2) to show
$$P(I_k) = \Delta_{I_k} F.$$

The symbol on the right is explained as follows. Let $\mathcal{V}$ be the vertices of I_k so that

$$\mathcal{V} = \{(x_1, \dots, x_i) : x_i = a_i \text{ or } b_i,\ i = 1, \dots, k\}.$$

Define for $x \in \mathcal{V}$

$$\operatorname{sgn}(\mathbf{x}) = \begin{cases} +1, & \text{if } \operatorname{card}\{i : x_i = a_i\} \text{ is even.} \\ -1, & \text{if } \operatorname{card}\{i : x_i = a_i\} \text{ is odd.} \end{cases}$$

Then

$$\Delta_{I_k} F = \sum_{\mathbf{x} \in \mathcal{V}} \operatorname{sgn}(\mathbf{x}) F(\mathbf{x}).$$

(f) Show F is continuous from above:

$$\lim_{\mathbf{a}\le\mathbf{x}\downarrow\mathbf{a}} F(\mathbf{x}) = F(\mathbf{a}).$$

(g) Call $F : \mathbb{R}^k \mapsto [0, 1]$ a multivariate distribution function if properties (1), (2) hold as well as F is continuous from above and $\Delta_{I_k} F \geq 0$. Show any multivariate distribution function determines a unique probability measure P on $(\mathbb{R}^k, \mathcal{B}(\mathbb{R}^k))$. (Use the extension theorem.)

24. Suppose λ_2 is the uniform distribution on the unit square $[0, 1]^2$ defined by its distribution function

$$\lambda_2([0, \theta_1] \times [0, \theta_2]) = \theta_1\theta_2, \quad (\theta_1, \theta_2) \in [0, 1]^2.$$

(a) Prove that λ_2 assigns 0 probability to the boundary of $[0, 1]^2$.

(b) Calculate

$$\lambda_2\{(\theta_1, \theta_2) \in [0, 1]^2 : \theta_1 \wedge \theta_2 > \frac{2}{3}\}.$$

(c) Calculate

$$\lambda_2\{(\theta_1, \theta_2) \in [0, 1]^2 : \theta_1 \wedge \theta_2 \leq x, \theta_1 \wedge \theta_2 \leq y\}.$$

25. In the game of bridge 52 distinguishable cards constituting 4 equal suits are distributed at random among 4 players. What is the probability that at least one player has a complete suit?

26. If $A_1, \ldots, A_n$ are events, define

$$S_1 = \sum_{i=1}^{n} P(A_i)$$

$$S_2 = \sum_{1\leq i<j\leq n} P(A_iA_j)$$

$$S_3 = \sum_{1\leq i<j<k\leq n} P(A_iA_jA_k)$$

$$\vdots \qquad \vdots$$

and so on.

(a) Show the probability $(1 \leq m \leq n)$

$$p(m) = P[\sum_{i=1}^{n} 1_{A_i} = m]$$

of exactly m of the events occurring is

$$p(m) = S_m - \binom{m+1}{m} S_{m+1} + \binom{m+2}{m} S_{m+2} - + \cdots \pm \binom{n}{m} S_n. \tag{2.41}$$

Verify that the inclusion-exclusion formula (2.2) is a special case of (2.41).

(b) Referring to Example 2.1.2, compute the probability of exactly m coincidences.

27. **Regular measures.** Consider the probability space $(\mathbb{R}^k, \mathcal{B}(\mathbb{R}^k), P)$. A Borel set A is *regular* if

$$P(A) = \inf\{P(G) : G \supset A, G \text{ open,}\}$$

and

$$P(A) = \sup\{P(F) : F \subset A, F \text{ closed.}\}$$

P is *regular* if all Borel sets are regular. Define $\mathcal{C}$ to be the collection of regular sets.

(a) Show $\mathbb{R}^k \in \mathcal{C}, \emptyset \in \mathcal{C}$.

(b) Show $\mathcal{C}$ is closed under complements and countable unions.

(c) Let $\mathcal{F}(\mathbb{R}^k)$ be the closed subsets of $\mathbb{R}^k$. Show

$$\mathcal{F}(\mathbb{R}^k) \subset \mathcal{C}.$$

(d) Show $\mathcal{B}(\mathbb{R}^k) \subset \mathcal{C}$; that is, show regularity.

(e) For any Borel set A

$$P(A) = \sup\{P(K) : K \subset A, K \text{ compact.}\}$$

3
Random Variables, Elements, and Measurable Maps

In this chapter, we will precisely define a random variable. A random variable is a real valued function with domain Ω which has an extra property called measurability that allows us to make probability statements about the random variables.

Random variables are convenient tools that allow us to focus on properties of interest about the experiment being modelled. The Ω may be rich but we may want to focus on one part of the description. For example, suppose

$$\begin{aligned}\Omega &= \{0, 1\}^n \\ &= \{(\omega_1, \dots, \omega_n) : \omega_i = 0 \text{ or } 1,\ i = 1, \dots, n\}.\end{aligned}$$

We may imagine this as the sample space for n repeated trials where the outcome is 1 (success) or 0 (failure) at each trial. One example of a random variable that summarizes information and allows us to focus on an aspect of the experiment of interest is the total number of successes

$$X((\omega_1, \dots, \omega_n)) = \omega_1 + \dots + \omega_n.$$

We now proceed to the general discussion.

3.1 Inverse Maps

Suppose Ω and Ω' are two sets. Frequently $\Omega' = \mathbb{R}$. Suppose

$$X : \Omega \mapsto \Omega',$$

meaning X is a function with domain Ω and range Ω'. Then X determines a function

$$X^{-1} : \mathcal{P}(\Omega') \mapsto \mathcal{P}(\Omega)$$

defined by

$$X^{-1}(A') = \{\omega \in \Omega : X(\omega) \in A'\}$$

for $A' \subset \Omega'$. X^{-1} preserves complementation, union and intersections as the following *properties* show. For $A' \subset \Omega'$, $A'_t \subset \Omega'$, and T an arbitrary index set, the following are true.

(i) We have

$$X^{-1}(\emptyset) = \emptyset, \quad X^{-1}(\Omega') = \Omega.$$

(ii) Set inverses preserve complements:

$$X^{-1}(A'^c) = (X^{-1}(A'))^c$$

so that

$$X^{-1}(\Omega' \setminus A') = \Omega \setminus X^{-1}(A').$$

(iii) Set inverses preserve unions and intersections:

$$X^{-1}(\bigcup_{t\in T} A'_t) = \bigcup_{t\in T} X^{-1}(A'_t),$$

and

$$X^{-1}(\bigcap_{t\in T} A'_t) = \bigcap_{t\in T} X^{-1}(A'_t).$$

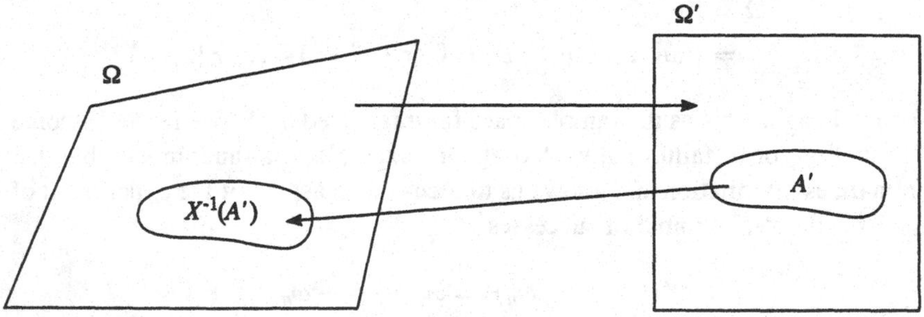

FIGURE 3.1 Inverses

Here is a **sample proof** of these properties. We verify (ii). We have $\omega \in X^{-1}(A'^c)$ iff $X(\omega) \in (A')^c$ iff $X(\omega) \notin A'$ iff $\omega \notin X^{-1}(A')$ iff $\omega \in (X^{-1}(A'))^c$.

□

Notation: If $\mathcal{C}' \subset \mathcal{P}(\Omega')$ is a class of subsets of Ω', define

$$X^{-1}(\mathcal{C}') := \{X^{-1}(C') : C' \in \mathcal{C}'\}.$$

Proposition 3.1.1 *If $\mathcal{B}'$ is a σ-field of subsets of Ω', then $X^{-1}(\mathcal{B}')$ is a σ-field of subsets of Ω.*

Proof. We verify the postulates for a σ-field.

(i) Since $\Omega' \in \mathcal{B}'$, we have

$$X^{-1}(\Omega') = \Omega \in X^{-1}(\mathcal{B}')$$

by Property (i) of the inverse map.

(ii) If $A' \in \mathcal{B}'$, then $(A')^c \in \mathcal{B}'$, and so if $X^{-1}(A') \in X^{-1}(\mathcal{B}')$, we have

$$X^{-1}((A')^c) = (X^{-1}(A'))^c \in X^{-1}(\mathcal{B}')$$

where we used Property (ii) of the inverse map.

(iii) If $X^{-1}(B'_n) \in X^{-1}(\mathcal{B}')$, then

$$\bigcup_n X^{-1}(B'_n) = X^{-1}(\bigcup_n B'_n) \in X^{-1}(\mathcal{B}')$$

since $\bigcup_n B'_n \in \mathcal{B}'$. □

A related but slightly deeper result comes next.

Proposition 3.1.2 *If $\mathcal{C}'$ is a class of subsets of Ω' then*

$$X^{-1}(\sigma(\mathcal{C}')) = \sigma(X^{-1}(\mathcal{C}')),$$

that is, the inverse image of the σ-field generated by $\mathcal{C}'$ in Ω' is the same as the σ-field generated in Ω by the inverse images.

Proof. From the previous Proposition 3.1.1, $X^{-1}(\sigma(\mathcal{C}'))$ is a σ-field, and

$$X^{-1}(\sigma(\mathcal{C}')) \supset X^{-1}(\mathcal{C}'),$$

since $\sigma(\mathcal{C}') \supset \mathcal{C}'$ and hence by minimality

$$X^{-1}(\sigma(\mathcal{C}')) \supset \sigma(X^{-1}(\mathcal{C}')).$$

Conversely, define

$$\mathcal{F}' := \{B' \in \mathcal{P}(\Omega') : X^{-1}(B') \in \sigma(X^{-1}(\mathcal{C}'))\}.$$

Then $\mathcal{F}'$ is a σ-field since

(i) $\Omega' \in \mathcal{F}'$, since $X^{-1}(\Omega') = \Omega \in \sigma(X^{-1}(\mathcal{C}'))$.

(ii) $A' \in \mathcal{F}'$ implies $(A')^c \in \mathcal{F}'$ since

$$X^{-1}(A'^c) = (X^{-1}(A'))^c \in \sigma(X^{-1}(\mathcal{C}'))$$

if $X^{-1}(A') \in \sigma(X^{-1}(\mathcal{C}'))$.

(iii) $B'_n \in \mathcal{F}'$ implies $\cup_n B'_n \in \mathcal{F}'$ since

$$X^{-1}(\cup_n B'_n) = \cup_n X^{-1}(B'_n) \in \sigma(X^{-1}(\mathcal{C}'))$$

if $X^{-1}(B'_n) \in \sigma(X^{-1}(\mathcal{C}'))$.

By definition

$$X^{-1}(\mathcal{F}') \subset \sigma(X^{-1}(\mathcal{C}')). \tag{3.1}$$

Also

$$\mathcal{C}' \subset \mathcal{F}'$$

since $X^{-1}(\mathcal{C}') \subset \sigma(X^{-1}(\mathcal{C}'))$. Since $\mathcal{F}'$ is a σ-field,

$$\sigma(\mathcal{C}') \subset \mathcal{F}' \tag{3.2}$$

and thus by (3.1) and (3.2)

$$X^{-1}(\sigma(\mathcal{C}')) \subset X^{-1}(\mathcal{F}') \subset \sigma(X^{-1}(\mathcal{C}')).$$

This suffices. □

3.2 Measurable Maps, Random Elements, Induced Probability Measures

A pair $(\Omega, \mathcal{B})$ consisting of a set and a σ-field of subsets is called a *measurable space*. (It is ready to have a measure assigned to it.) If $(\Omega, \mathcal{B})$ and $(\Omega', \mathcal{B}')$ are two measurable spaces, then a map

$$X : \Omega \to \Omega'$$

is called measurable if

$$X^{-1}(\mathcal{B}') \subset \mathcal{B}.$$

X is also called a *random element* of Ω'. We will use the notation that

$$X \in \mathcal{B}/\mathcal{B}'$$

or

$$X : (\Omega, \mathcal{B}) \mapsto (\Omega', \mathcal{B}').$$

A special case occurs when $(\Omega', \mathcal{B}') = (\mathbb{R}, \mathcal{B}(\mathbb{R}))$. In this case, X is called a *random variable*.

Let $(\Omega, \mathcal{B}, P)$ be a probability space and suppose

$$X : (\Omega, \mathcal{B}) \mapsto (\Omega', \mathcal{B}')$$

is measurable. Define for $A' \subset \Omega'$

$$[X \in A'] := X^{-1}(A') = \{\omega : X(\omega) \in A'\}.$$

Define the set function $P \circ X^{-1}$ on $\mathcal{B}'$ by

$$P \circ X^{-1}(A') = P(X^{-1}(A')).$$

$P \circ X^{-1}$ is a probability on $(\Omega', \mathcal{B}')$ called the *induced* probability or the *distribution* of X. To verify it is a probability measure on $\mathcal{B}'$, we note

(a) $P \circ X^{-1}(\Omega') = P(\Omega) = 1.$

(b) $P \circ X^{-1}(A') \geq 0$, for all $A' \in \mathcal{B}'$.

(c) If $\{A'_n, n \geq 1\}$ are disjoint,

$$\begin{aligned} P \circ X^{-1}(\bigcup_n A'_n) &= P(\bigcup_n X^{-1}(A'_n)) \\ &= \sum_n P(X^{-1}(A'_n)) \\ &= \sum_n P \circ X^{-1}(A'_n) \end{aligned}$$

since $\{X^{-1}(A'_n)\}_{n \geq 1}$ are disjoint in $\mathcal{B}$.

Usually we write

$$P \circ X^{-1}(A') = P[X \in A'].$$

If X is a random variable then $P \circ X^{-1}$ is the measure induced on $\mathbb{R}$ by the distribution function

$$P \circ X^{-1}(-\infty, x] = P[X \leq x].$$

Thus when X is a random element of $\mathcal{B}'$, we can make probability statements about X, since $X^{-1}(B') \in \mathcal{B}$ and the probability measure P knows how to assign probabilities to elements of $\mathcal{B}$. The concept of measurability is logically necessary in order to be able to assign probabilities to sets determined by random elements.

Example. Consider the experiment of tossing two dice and let

$$\Omega = \{(i, j) : 1 \leq i, j \leq 6\}.$$

Define

$$X : \Omega \mapsto \{2, 3, \ldots, 12\} =: \Omega'$$

by

$$X((i, j)) = i + j.$$

Then

$$X^{-1}(\{4\}) = [X = 4] = \{(1, 3), (3, 1), (2, 2)\} \subset \Omega$$

and

$$X^{-1}(\{2, 3\}) = [X \in \{2, 3\}] = \{(1, 1), (1, 2), (2, 1)\}.$$

The distribution of X is the probability measure on Ω' specified by

$$P \circ X^{-1}(\{i\}) = P[X = i], \quad i \in \Omega'.$$

For example,

$$P[X = 2] = \frac{1}{36}$$
$$P[X = 3] = \frac{2}{36}$$
$$P[X = 4] = \frac{3}{36},$$

and so on. □

The definition of measurability makes it seem like we have to check $X^{-1}(A') \in \mathcal{B}$ for every $A' \in \mathcal{B}'$; that is

$$X^{-1}(\mathcal{B}') \subset \mathcal{B}.$$

In fact, it usually suffices to check that X^{-1} is well behaved on a smaller class than $\mathcal{B}'$.

Proposition 3.2.1 (Test for measurability) *Suppose*

$$X : \Omega \mapsto \Omega'$$

where $(\Omega, \mathcal{B})$, *and* $(\Omega', \mathcal{B}')$ *are two measurable spaces. Suppose* $\mathcal{C}'$ *generates* $\mathcal{B}'$; *that is*

$$\mathcal{B}' = \sigma(\mathcal{C}').$$

Then X *is measurable iff*

$$X^{-1}(\mathcal{C}') \subset \mathcal{B}.$$

Remark. We do not have to check that

$$X^{-1}(\sigma(\mathcal{C}')) \subset \mathcal{B},$$

which is what using the definition would require.

Corollary 3.2.1 (Special case of random variables) *The real valued function*

$$X : \Omega \mapsto \mathbb{R}$$

is a random variable iff

$$X^{-1}((-\infty, \lambda]) = [X \le \lambda] \in \mathcal{B}, \quad \forall \lambda \in \mathbb{R}.$$

Proof of Proposition 3.2.1. If

$$X^{-1}(\mathcal{C}') \subset \mathcal{B},$$

then by minimality

$$\sigma(X^{-1}(\mathcal{C}')) \subset \mathcal{B}.$$

However, we get

$$X^{-1}(\sigma(\mathcal{C}')) = X^{-1}(\mathcal{B}') = \sigma(X^{-1}(\mathcal{C}')) \subset \mathcal{B},$$

which is the definition of measurability. □

Proof of Corollary 3.2.1. This follows directly from

$$\sigma((-\infty, \lambda], \lambda \in \mathbb{R}) = \mathcal{B}(\mathbb{R}).$$

□

3.2.1 Composition

Verification that a map is measurable is sometimes made easy by decomposing the map into the composition of two (or more) maps. If each map in the composition is measurable, then the composition is measurable.

Proposition 3.2.2 (Composition) *Let X_1, X_2 be two measurable maps*

$$\begin{aligned} X_1 : (\Omega_1, \mathcal{B}_1) &\mapsto (\Omega_2, \mathcal{B}_2), \\ X_2 : (\Omega_2, \mathcal{B}_2) &\mapsto (\Omega_3, \mathcal{B}_3) \end{aligned}$$

where $(\Omega_i, \mathcal{B}_i), i = 1, 2, 3$ are measurable spaces. Define

$$X_2 \circ X_1 : \Omega_1 \mapsto \Omega_3$$

by

$$X_2 \circ X_1(\omega_1) = X_2(X_1(\omega_1)), \quad \omega_1 \in \Omega_1.$$

Then

$$X_2 \circ X_1 \in \mathcal{B}_1/\mathcal{B}_3.$$

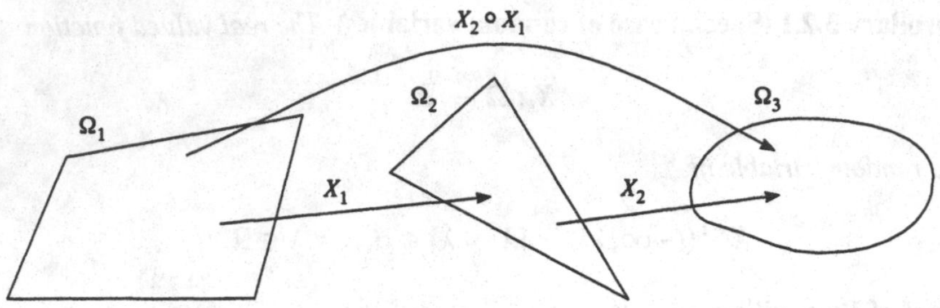

FIGURE 3.2

Proof. It is elementary to check that

$$(X_2 \circ X_1)^{-1} = X_1^{-1}(X_2^{-1}(\cdot))$$

as maps from $\mathcal{P}(\Omega_3) \mapsto \mathcal{P}(\Omega_1)$. The reason is that for any $B_3 \subset \Omega_3$

$$\begin{aligned}(X_2 \circ X_1)^{-1}(B_3) &= \{\omega_1 : X_2 \circ X_1(\omega_1) \in B_3\}\\ &= \{\omega_1 : X_1(\omega_1) \in X_2^{-1}(B_3)\}\\ &= \{\omega_1 : \omega_1 \in X_1^{-1}(X_2^{-1}(B_3))\}.\end{aligned}$$

If $B_3 \in \mathcal{B}_3$, then

$$(X_2 \circ X_1)^{-1}(B_3) = X_1^{-1}(X_2^{-1}(B_3)) \in \mathcal{B}_1,$$

since $X_2^{-1}(B_3) \in \mathcal{B}_2$. Thus

$$X_2 \circ X_1^{-1}(\mathcal{B}_3) \subset \mathcal{B}_1,$$

as required. □

3.2.2 *Random Elements of Metric Spaces*

The most common use of the name random elements is when the range is a metric space.

Let (S, d) be a metric space with metric d so that $d : S \times S \mapsto \mathbb{R}_+$ satisfies

(i) $d(x, y) \geq 0, \quad$ for $x, y \in S$.

(ii) $d(x, y) = 0$ iff $x = y, \quad$ for any $x, y \in S$.

(iii) $d(x, y) = d(y, x), \quad$ for any $x, y \in S$.

(iv) $d(x, z) \leq d(x, y) + d(y, z), \quad$ for any $x, y, z \in S$.

Let $\mathcal{O}$ be the class of open subsets of S. Define the Borel σ-field $\mathcal{S}$ to be the smallest σ-field generated by the open sets

$$\mathcal{S} = \sigma(\mathcal{O}).$$

If

$$X : (\Omega, \mathcal{B}) \mapsto (S, \mathcal{S});$$

that is, $X \in \mathcal{B}/\mathcal{S}$, then call X a *random element* of S. Here are some noteworthy examples of random elements.

1. Suppose $S = \mathbb{R}$ and $d(x, y) = |x - y|$. Then a random element X of S is called a *random variable.*

2. Suppose $S = \mathbb{R}^k$ and

$$d(\mathbf{x}, \mathbf{y}) = \sqrt{\sum_1^k |x_i - y_i|^2}.$$

Then a random element X of S is called a *random vector*. We write $\mathbf{X} = (X_1, \dots, X_k)$.

3. Suppose $S = \mathbb{R}^\infty$, and the metric d is defined by

$$d(\mathbf{x}, \mathbf{y}) = \sum_{k=1}^{\infty} 2^{-k} \left(\frac{\sum_1^k |x_i - y_i|}{1 + \sum_1^k |x_i - y_i|} \right).$$

Then a random element $\mathbf{X}$ of S is called a *random sequence*. We write $\mathbf{X} = (X_1, X_2, \dots)$.

4. Let $S = C[0, \infty)$ be the set of all real valued continuous functions with domain $[0, \infty)$. Define

$$||x(\cdot) - y(\cdot)||_m = \sup_{0 \le t \le m} |x(t) - y(t)|$$

and

$$d(x, y) = \sum_{m=1}^{\infty} 2^{-m} \left(\frac{||x - y||_m}{1 + ||x - y||_m} \right).$$

A random element X of S is called a *random (continuous) function.*

5. Let $(\mathbb{E}, \mathcal{E})$ be a measurable space where $\mathbb{E}$ is a nice metric space and $\mathcal{E}$ are the Borel sets, that is the sets in the σ-field generated by the open sets. Let $S = M_+(\mathbb{E})$ be the set of all measures on $(\mathbb{E}, \mathcal{E})$ which satisfy the property that if $\mu \in M_+(\mathbb{E})$, then $\mu(K) < \infty$ if K is a compact subset of $\mathbb{E}$. Such measures which are finite on compacta are called *Radon* measures. There is a standard metric for S called the *vague* metric. A random element X of S is called a *random measure.*

A special case is where $M_+(\mathbb{E})$ is cut down to the space $M_p(\mathbb{E})$ of non-negative integer valued measures. In this case the random element X is called a *stochastic point process.*

3.2.3 Measurability and Continuity

The idea of distance (measured by a metric) leads naturally to the notion of continuity. A method, which is frequently easy for showing a function from one metric space to another is measurable, is to show the function is continuous.

Proposition 3.2.3 *Suppose (S_i, d_i), $i = 1, 2$ are two metric spaces. Let the Borel σ-fields (generated by open sets) be $\mathcal{S}_i$, $i = 1, 2$. If*

$$X : S_1 \to S_2$$

is continuous, then X is measurable:

$$X \in \mathcal{S}_1/\mathcal{S}_2.$$

Proof. Let $\mathcal{O}_i$ be the class of open subsets of S_i, $i = 1, 2$. If X is continuous, then inverse images of open sets are open, which means that

$$X^{-1}(\mathcal{O}_2) \subset \mathcal{O}_1 \subset \sigma(\mathcal{O}_1) = \mathcal{S}_1.$$

So $X \in \mathcal{S}_1/\mathcal{S}_2$ by Proposition 3.2.1. □

Corollary 3.2.2 *If $\mathbf{X} = (X_1, \dots, X_k)$ is a random vector, and*

$$g : \mathbb{R}^k \mapsto \mathbb{R}, \quad g \in \mathcal{B}(\mathbb{R}^k)/\mathcal{B}(\mathbb{R}),$$

then from Proposition 3.2.2, $g(\mathbf{X})$ is a random variable. In particular, if g is continuous, then g is measurable and the result holds.

Some examples of the sort of g's to which this result could apply:

$$\begin{aligned} g(x_1, \dots, x_k) &= \sum_{i=1}^k x_i, \quad \text{(component sum)} \\ &= \sum_1^k x_i/k, \quad \text{(component average)} \\ &= \bigvee_{i=1}^k x_i, \quad \text{(component extreme)} \\ &= \prod_{i=1}^k x_i, \quad \text{(component product)} \\ &= \sum_{i=1}^k x_i^2, \quad \text{(component sum of squares)}\,. \end{aligned}$$

Another interesting example of g is the projection map

$$g = \pi_i : \mathbb{R}^k \mapsto \mathbb{R}$$

defined by

$$\pi_i(x_1, \ldots, x_k) = x_i.$$

Then π_i is continuous and if $\mathbf{X} = (X_1, \ldots, X_k)$ is a random vector, $\pi_i(\mathbf{X}) = X_i$ is a random variable for $i = 1, \ldots, k$.

This observation leads in a simple way to an important fact about random vectors: A random vector is nothing more than a vector of random variables.

Proposition 3.2.4 $\mathbf{X} = (X_1, \ldots, X_k)$ *is a random vector, that is a measurable map from* $(\Omega, \mathcal{B}) \mapsto (\mathbb{R}^k, \mathcal{B}(\mathbb{R}^k))$, *iff* X_i *is a random variable for each* $i = 1, \ldots, k$.

Proof. If $\mathbf{X}$ is a random vector, then $X_i = \pi_i \circ \mathbf{X}$ is measurable since it is the composition of two measurable functions $\mathbf{X}$ and the continuous function π_i.

The converse is easily proved if we know that

$$\mathcal{B}(\mathbb{R}^k) = \sigma(\mathcal{O}) = \sigma(\text{RECTS})$$

where RECTS is the class of open rectangles. We assume this fact is at our disposal. Suppose $X_1, \ldots, X_k$ are random variables. Write

$$B = I_1 \times \ldots \times I_k$$

for a rectangle whose sides are the intervals $I_1, \ldots, I_k$. Then

$$\mathbf{X}^{-1}(B) = \bigcap_{i=1}^{k} X_i^{-1}(I_i).$$

Since X_i is a random variable, $X_i^{-1}(I_i) \in \mathcal{B}$, so $\mathbf{X}^{-1}(B) \in \mathcal{B}$ and

$$\mathbf{X}^{-1}(\text{RECTS}) \subset \mathcal{B}$$

so $\mathbf{X}^{-1}$ is measurable. □

The corresponding basic fact about random sequences, stated next, is proved in an analogous manner to the proof of the basic fact about random vectors. It says that $\mathbf{X}$ is a random sequence iff each component is a random variable.

Proposition 3.2.5 $\mathbf{X} = (X_1, X_2, \ldots)$ *is a random sequence iff for each* $i = 1, 2, \ldots$ *the* i*th component* X_i *is a random variable. Furthermore,* $\mathbf{X}$ *is a random sequence iff* $(X_1, \ldots, X_k)$ *is a random vector, for any* k.

3.2.4 Measurability and Limits

Limits applied to sequences of measurable functions are measurable.

Proposition 3.2.6 *Let* $X_1, X_2, \ldots$ *be random variables defined on* $(\Omega, \mathcal{B})$. *Then*

(i) $\vee_n X_n$ *and* $\wedge_n X_n$ *are random variables.*

(ii) $\liminf_{n\to\infty} X_n$ *and* $\limsup_{n\to\infty} X_n$ *are random variables.*

(iii) If $\lim_{n\to\infty} X_n(\omega)$ *exists for all* ω, *then* $\lim_{n\to\infty} X_n$ *is a random variable.*

(iv) The set on which $\{X_n\}$ *has a limit is measurable; that is*

$$\{w : \lim X_n(w) \text{ exists }\} \in \mathcal{B}.$$

Proof. (i) We have

$$[\bigvee_n X_n \le x] = \bigcap_n [X_n \le x] \in \mathcal{B},$$

since for each n,

$$[X_n \le x] \in \mathcal{B}.$$

Similarly

$$[\bigwedge_n X_n > x] = \bigcap_n [X_n > x] \in \mathcal{B}.$$

This suffices by Corollary 3.2.1.

(ii) We have that

$$\liminf_{n\to\infty} X_n = \sup_{n\ge 1} \inf_{k\ge n} X_k.$$

By (i) $\inf_{k\ge n} X_k$ is a random variable and hence so is $\sup_{n\ge 1}(\inf_{k\ge n} X_k)$.

(iii) If $\lim_{n\to\infty} X_n(\omega)$ exists for all ω, then

$$\lim_{n\to\infty} X_n(\omega) = \limsup_{n\to\infty} X_n(\omega)$$

is a random variable by (ii).

(iv) Let $\mathbb{Q}$ be the set of all rational real numbers so that $\mathbb{Q}$ is countable. We have

$$\begin{aligned}\{\omega : \lim X_n(\omega) \text{ exists }\}^c &= \{\omega : \liminf X_n(\omega) < \limsup X_n(\omega)\} \\ &= \bigcup_{r\in\mathbb{Q}} \left[\liminf_{n\to\infty} X_n \le r < \limsup_{n\to\infty} X_n\right] \\ &= \bigcup_{r\in\mathbb{Q}} \left[\liminf_{n\to\infty} X_n \le r\right] \bigcap \left[\limsup_{n\to\infty} X_n \le r\right]^c \in \mathcal{B}\end{aligned}$$

since

$$[\liminf_{n\to\infty} X_n \le r] \in \mathcal{B}$$

and

$$[\limsup_{n\to\infty} X_n \le r] \in \mathcal{B}.$$

3.3 σ-Fields Generated by Maps

Let $X : (\Omega, \mathcal{B}) \mapsto (\mathbb{R}, \mathcal{B}(\mathbb{R}))$ be a random variable. The *σ-algebra generated by X*, denoted $\sigma(X)$, is defined as

$$\sigma(X) = X^{-1}(\mathcal{B}(\mathbb{R})). \tag{3.3}$$

Another equivalent description of $\sigma(X)$ is

$$\sigma(X) = \{[X \in A], A \in \mathcal{B}(\mathbb{R})\}.$$

This is the σ-algebra generated by information about X, which is a way of isolating that information in the probability space that pertains to X. More generally, suppose

$$X : (\Omega, \mathcal{B}) \mapsto (\Omega', \mathcal{B}').$$

Then we define

$$\sigma(X) = X^{-1}(\mathcal{B}'). \tag{3.4}$$

If $\mathcal{F} \subset \mathcal{B}$ is a sub-σ-field of $\mathcal{B}$, we say X is *measurable with respect to $\mathcal{F}$*, written $X \in \mathcal{F}$, if $\sigma(X) \subset \mathcal{F}$.

If for each t in some index set T

$$X_t : (\Omega, \mathcal{B}) \mapsto (\Omega', \mathcal{B}'),$$

then we denote by

$$\sigma(X_t, t \in T) = \bigvee_{t \in T} \sigma(X_t)$$

the smallest σ-algebra containing all $\sigma(X_t)$.

Extreme example: Let $X(\omega) \equiv 17$ for all ω. Then

$$\begin{aligned}\sigma(X) &= \{[X \in B], B \in \mathcal{B}(\mathbb{R})\} \\ &= \sigma(\emptyset, \Omega) = \{\emptyset, \Omega\}.\end{aligned}$$

Less extreme example: Suppose $X = 1_A$ for some $A \in \mathcal{B}$. Note X has range $\{0, 1\}$. Then

$$X^{-1}(\{0\}) = A^c, \quad X^{-1}(\{1\}) = A$$

and therefore

$$\sigma(X) = \{\emptyset, \Omega, A, A^c\}.$$

To verify this last assertion, call

$$\{\emptyset, \Omega, A, A^c\} = \text{RHS}$$

and

$$\sigma(X) = \text{LHS}.$$

Then RHS is a σ-field and is certainly contained in LHS. We need to show that for any $B \in \mathcal{B}(\mathbb{R})$,

$$[1_A \in B] \in \text{RHS}.$$

There are four cases to consider: (i) $1 \in B, 0 \notin B$; (ii) $1 \in B, 0 \in B$; (iii) $0 \in B$, $1 \notin B$; (iv) $0 \notin B, 1 \notin B$. For example, in case (i) we find

$$[1_A \in B] = A$$

and the other cases are handled similarly.

Useful example: Simple function. A random variable is *simple* if it has a finite range. Suppose the range of X is $\{a_1, \ldots, a_k\}$, where the a's are distinct. Then define

$$A_i := X^{-1}(\{a_i\}) = [X = a_i].$$

Then $\{A_i,\ i = 1, \ldots, k\}$ partitions Ω, meaning

$$A_i \cap A_j = \emptyset,\ i \neq j,\quad \sum_{i=1}^{k} A_i = \Omega.$$

We may represent X as

$$X = \sum_{i=1}^{k} a_i 1_{A_i},$$

and

$$\sigma(X) = \sigma(A_1, \ldots, A_k) = \left\{ \sum_{i \in I} A_i : I \subset \{1, \ldots, k\} \right\}.$$

In stochastic process theory, we frequently keep track of potential information that can be revealed to us by observing the evolution of a stochastic process by an increasing family of σ-fields. If $\{X_n, n \geq 1\}$ is a (discrete time) stochastic process, we may define

$$\mathcal{B}_n := \sigma(X_1, \ldots, X_n), \quad n \geq 1.$$

Thus, $\mathcal{B}_n \subset \mathcal{B}_{n+1}$ and we think of $\mathcal{B}_n$ as the information potentially available at time n. This is a way of cataloguing what information is contained in the probability model. Properties of the stochastic process are sometimes expressed in terms of $\{\mathcal{B}_n, n \geq 1\}$. For instance, one formulation of the Markov property is that the conditional distribution of X_{n+1} given $\mathcal{B}_n$ is the same as the conditional distribution of X_{n+1} given X_n. (See Chapter 10.)

We end this chapter with the following comment on the σ-field generated by a random variable.

Proposition 3.3.1 *Suppose X is a random variable and $\mathcal{C}$ is a class of subsets of $\mathbb{R}$ such that*

$$\sigma(\mathcal{C}) = \mathcal{B}(\mathbb{R}).$$

Then

$$\sigma(X) = \sigma([X \in B], B \in \mathcal{C}).$$

Proof. We have

$$\begin{aligned}\sigma([X \in B], B \in \mathcal{C}) &= \sigma(X^{-1}(B), B \in \mathcal{C})\\ &= \sigma(X^{-1}(\mathcal{C})) = X^{-1}(\sigma(\mathcal{C}))\\ &= X^{-1}(\mathcal{B}(\mathbb{R})) = \sigma(X).\end{aligned}$$

□

A special case of this result is

$$\sigma(X) = \sigma([X \le \lambda], \lambda \in \mathbb{R}).$$

3.4 Exercises

1. In the measurable space $(\Omega, \mathcal{B})$, show $A \in \mathcal{B}$ iff $1_A \in \mathcal{B}$.

2. Let $(\Omega, \mathcal{B}, P) = \big((0, 1], \mathcal{B}((0, 1]), \lambda\big)$ where λ is Lebesgue measure. Define

$$\begin{aligned}X_1(\omega) &= 0, \quad \forall \omega \in \Omega,\\ X_2(\omega) &= 1_{\{1/2\}}(\omega),\\ X_3(\omega) &= 1_{\mathbb{Q}}(\omega)\end{aligned}$$

   where $\mathbb{Q} \subset (0, 1]$ are the rational numbers in $(0, 1]$. Note

$$P[X_1 = X_2 = X_3 = 0] = 1$$

   and give

$$\sigma(X_i), \quad i = 1, 2, 3.$$

3. Suppose

$$f : \mathbb{R}^k \mapsto \mathbb{R}, \text{ and } f \in \mathcal{B}(\mathbb{R}^k)/\mathcal{B}(\mathbb{R}).$$

   Let $X_1, \dots, X_k$ be random variables on $(\Omega, \mathcal{B})$. Then

$$f(X_1, \dots, X_k) \in \sigma(X_1, \dots, X_k).$$

4. Suppose $X : \Omega \mapsto \mathbb{R}$ has a countable range $\mathcal{R}$. Show $X \in \mathcal{B}/\mathcal{B}(\mathbb{R})$ iff

$$X^{-1}(\{x\}) \in \mathcal{B}, \quad \forall x \in \mathcal{R}.$$

5. If
$$F(x) = P[X \le x]$$
is continuous in x, show that $Y = F(X)$ is measurable and that Y has a uniform distribution
$$P[Y \le y] = y, \quad 0 \le y \le 1.$$

6. If X is a random variable satisfying $P[|X| < \infty] = 1$, then show that for any $\epsilon > 0$, there exists a bounded random variable Y such that
$$P[X \neq Y] < \epsilon.$$
(A random variable Y is bounded if for all ω
$$|Y(\omega)| \le K$$
for some constant K independent of ω.)

7. If X is a random variable, so is $|X|$. The converse may be false.

8. Let X and Y be random variables and let $A \in \mathcal{B}$. Prove that the function
$$Z(\omega) = \begin{cases} X(\omega), & \text{if } \omega \in A, \\ Y(\omega), & \text{if } \omega \in A^c \end{cases}$$
is a random variable.

9. Suppose that $\{B_n, n \ge 1\}$ is a countable partition of Ω and define $\mathcal{B} = \sigma(B_n, n \ge 1)$. Show a function $X : \Omega \mapsto (-\infty, \infty]$ is $\mathcal{B}$-measurable iff X is of the form
$$X = \sum_{i=1}^{\infty} c_i 1_{B_i},$$
for constants $\{c_i\}$. (What is $\mathcal{B}$?)

10. **Circular Lebesgue measure.** Define $C := \{e^{2\pi i\theta} : \theta \in (0, 1]\}$ to be the unit circle in the complex plane. Define
$$T : (0, 1] \mapsto C, \quad T(\theta) = e^{2\pi i\theta}.$$
Specify a σ-field $\mathcal{B}(C))$ of subsets of C by
$$\mathcal{B}(C) := \{A \subset C : T^{-1}(A) \in \mathcal{B}((0, 1]).$$
(Why is this a σ-field?) Define a probability measure μ on $\mathcal{B}(C)$ by $\mu = \lambda \circ T^{-1}$ and call this measure μ *circular Lebesgue measure.*

    (a) Identify the complex plane with $\mathbb{R}^2$. Show
$$\mathcal{B}(C) = \mathcal{B}(\mathbb{R}^2) \cap C.$$

(b) Show that $\mathcal{B}(C)$ is generated by arcs of C.

(c) Show μ is invariant with respect to rotations. This means, if S_{θ_0} : $C \mapsto C$ via

$$S_{\theta_0}(e^{2\pi i\theta}) = e^{2\pi i(\theta+\theta_0)},$$

then $\mu = \mu \circ S_{\theta_0}^{-1}$.

(d) If you did not define μ as the induced image of Lebesgue measure on the unit interval, how could you define it by means of the extension theorems?

11. Let $(\Omega, \mathcal{B}, P)$ be $([0, 1], \mathcal{B}([0, 1]), \lambda)$ where λ is Lebesgue measure on $[0, 1]$. Define the process $\{X_t, 0 \le t \le 1\}$ by

$$X_t(\omega) = \begin{cases} 0, & \text{if } t \neq \omega, \\ 1, & \text{if } t = \omega. \end{cases}$$

Show that each X_t is a random variable. What is the σ-field generated by $\{X_t, 0 \le t \le 1\}$?

12. Show that a monotone real function is measurable.

13. (a) If X is a random variable, then $\sigma(X)$ is a countably generated σ-field.

(b) Conversely, if $\mathcal{B}$ is any countably generated σ-field, show

$$\mathcal{B} = \sigma(X)$$

for some random variable X.

14. A real function f on the line is upper semi-continuous (usc) at x, if, for each ϵ, there is a δ such that $|x - y| < \delta$ implies that

$$f(y) < f(x) + \epsilon.$$

Check that if f is everywhere usc, then it is measurable. (Hint: What kind of set is $\{x : f(x) < t\}$?)

15. Suppose $-\infty < a \le b < \infty$. Show that the indicator function $1_{(a,b]}(x)$ can be approximated by bounded and continuous functions; that is, show that there exist a sequence of continuous functions $0 \le f_n \le 1$ such that $f_n \to 1_{(a,b]}$ pointwise.

Hint: Approximate the rectangle of height 1 and base $(a, b]$ by a trapezoid of height 1 with base $(a, b + n^{-1}]$ whose top line extends from $a + n^{-1}$ to b.

16. Suppose $\mathcal{B}$ is a σ-field of subsets of $\mathbb{R}$. Show $\mathcal{B}(\mathbb{R}) \subset \mathcal{B}$ iff every real valued continuous function is measurable with respect to $\mathcal{B}$ and therefore $\mathcal{B}(\mathbb{R})$ is the smallest σ-field with respect to which all the continuous functions are measurable.

17. Functions are often defined in pieces (for example, let $f(x)$ be x^3 or x^{-1} as $x \geq 0$ or $x < 0$), and the following shows that the function is measurable if the pieces are.

    Consider measurable spaces $(\Omega, \mathcal{B})$ and $(\Omega', \mathcal{B}')$ and a map $T : \Omega \mapsto \Omega'$. Let $A_1, A_2, \ldots$ be a countable covering of Ω by $\mathcal{B}$ sets. Consider the σ-field $\mathcal{B}_n = \{A : A \subset A_n, A \in \mathcal{B}\}$ in A_n and the restriction T_n of T to A_n. Show that T is measurable $\mathcal{B}/\mathcal{B}'$ iff T_n is measurable $\mathcal{B}_n/\mathcal{B}'$ for each n.

18. **Coupling.** If X and Y are random variables on $(\Omega, \mathcal{B})$, show

$$\sup_{A \in \mathcal{B}} |P[X \in A] - P[Y \in A]| \leq P[X \neq Y].$$

19. Suppose $T : (\Omega_1, \mathcal{B}_1) \mapsto (\Omega_2, \mathcal{B}_2)$ is a measurable mapping and X is a random variable on Ω_1. Show $X \in \sigma(T)$ iff there is a random variable Y on $(\Omega_2, \mathcal{B}_2)$ such that

$$X(\omega_1) = Y(T(\omega_1)), \quad \forall \omega_1 \in \Omega_1.$$

20. Suppose $\{X_t, t \geq 0\}$ is a continuous time stochastic process on the probability space $(\Omega, \mathcal{B}, P)$ whose paths are continuous. We can understand this to mean that for each $t \geq 0$, $X_t : \Omega \mapsto \mathbb{R}$ is a random variable and, for each $\omega \in \Omega$, the function $t \mapsto X_t(\omega)$ is continuous; that is a member of $C[0, \infty)$. Let $\tau : \Omega \mapsto [0, \infty)$ be a random variable and define the process stopped at τ as the function $X_\tau : \Omega \mapsto [0, \infty)$ defined by

$$X_\tau(\omega) := X_{\tau(\omega)}(\omega), \quad \omega \in \Omega.$$

    Prove X_τ is a random variable.

21. **Dyadic expansions and Lebesgue measure.** Let $\mathbb{S} = \{0, 1\}$ and

$$\mathbb{S}^\infty = \{(x_1, x_2, \ldots) : x_i \in \mathbb{S}, i = 1, 2, \ldots\}$$

    be sequences consisting of 0's and 1's. Define $\mathcal{B}(\mathbb{S}) = \mathcal{P}(\mathbb{S})$ and define $\mathcal{B}(\mathbb{S}^\infty)$ to be the smallest σ-field of subsets of $\mathbb{S}^\infty$ containing all sets of the form

$$\{i_1\} \times \{i_2\} \times \cdots \times \{i_k\} \times \mathbb{S}^\infty$$

    for $k = 1, 2, \ldots$ and $i_1, i_2, \ldots, i_k$ some string of 0's and 1's.

    For $x \in [0, 1]$, let

$$x = (d_k(x), k \geq 1)$$

    be the non-terminating dyadic expansion ($d_k(0) = 0$ and $d_k(x) = 0$ or 1.) Define $U : [0, 1] \mapsto \mathbb{S}^\infty$ by

$$U(x) = (d_1(x), d_2(x), \ldots).$$

Define $V : \mathbb{S}^\infty \mapsto [0,1]$ by $(\mathbf{x} = (i_1, i_2, \dots))$

$$V(x) = \sum_{n=1}^{\infty} \frac{i_n}{2^n}.$$

Show $U \in \mathcal{B}([0,1])/\mathcal{B}(\mathbb{S}^\infty)$ and $V \in \mathcal{B}(\mathbb{S}^\infty)/\mathcal{B}([0,1])$.

22. Suppose $\{X_n, n \geq 1\}$ are random variables on the probability space $(\Omega, \mathcal{B}, P)$ and define the induced random walk by

$$S_0 = 0, \quad S_n = \sum_{i=1}^{n} X_i, \ n \geq 1.$$

Let

$$\tau := \inf\{n > 0 : S_n > 0\}$$

be the *first upgoing ladder time*. Prove τ is a random variable. Assume we know $\tau(\omega) < \infty$ for all $\omega \in \Omega$. Prove S_τ is a random variable.

23. Suppose $\{X_1, \dots, X_n\}$ are random variables on the probability space $(\Omega, \mathcal{B}, P)$ such that

$$P[\text{ Ties }] := P\{\bigcup_{\substack{i \neq j \\ 1 \leq i,j \leq n}} [X_i = X_j]\} = 0.$$

Define the *relative rank* R_n of X_n among $\{X_1, \dots, X_n\}$ to be

$$R_n = \begin{cases} \sum_{i=1}^{n} 1_{[X_i \geq X_n]} & \text{on [Ties]}^c, \\ 17, & \text{on [Ties]}. \end{cases}$$

Prove R_n is a random variable.

24. Suppose $(S_1, \mathcal{S}_1)$ is a measurable space and suppose $T : S_1 \mapsto S_2$ is a mapping into another space S_2. For an index set Γ, suppose

$$h_\gamma : S_2 \mapsto \mathbb{R}, \quad \gamma \in \Gamma$$

and define

$$\mathcal{G} := \sigma(h_\gamma, \gamma \in \Gamma)$$

to be the σ-field of subsets of S_2 generated by the real valued family $\{h_\gamma, \gamma \in \Gamma\}$, that is, generated by $\{h_\gamma^{-1}(B), \gamma \in \Gamma, B \in \mathcal{B}(\mathbb{R})\}$. Show $T \in \mathcal{S}_1/\mathcal{G}$ iff $h_\gamma \circ T$ is a random variable on $(S_1, \mathcal{S}_1)$.

25. **Egorov's theorem:** Suppose X_n, X are real valued random variables defined on the probability space $(\Omega, \mathcal{B}, P)$. Suppose for all $\omega \in \Lambda \in \mathcal{B}$, we have $X_n(\omega) \to X(\omega)$. Show for every $\epsilon > 0$, there exists a set Λ_ϵ such that $P(\Lambda_\epsilon) < \epsilon$ and

$$\sup_{\omega \in \Lambda \setminus \Lambda_\epsilon} |X(\omega) - X_n(\omega)| \to 0 \quad (n \to \infty).$$

Thus, convergence is uniform off a small set.

Hints:

(a) Define

$$B_n^{(k)} = \left[\bigvee_{i \geq n} |X(\omega) - X_i(\omega)|\right] \cap \Lambda.$$

(b) Show $B_n^{(k)} \downarrow \emptyset$ as $n \to \infty$.

(c) There exists $\{n_k\}$ such that $P(B_{n_k}^{(k)}) < \epsilon/2^k$.

(d) Set $B = \cup_k B_{n_k}^{(k)}$ so that $P(B) < \epsilon$.

26. Review Exercise 12 of Chapter 2. Suppose $\mathcal{C}$ is a class of subsets of Ω such that, for a real function X defined on Ω, we have $X \in \mathcal{B}(\mathcal{C})$. Show there exists a countable subclass $\mathcal{C}^* \subset \mathcal{C}$ such that X is measurable $\mathcal{C}^*$.

4

Independence

Independence is a basic property of events and random variables in a probability model. Its intuitive appeal stems from the easily envisioned property that the occurrence or non-occurrence of an event has no effect on our estimate of the probability that an independent event will or will not occur. Despite the intuitive appeal, it is important to recognize that *independence* is a technical concept with a technical definition which must be checked with respect to a specific probability model. There are examples of dependent events which intuition insists must be independent, and examples of events which intuition insists cannot be independent but still satisfy the definition. One really must check the technical definition to be sure.

4.1 Basic Definitions

We give a series of definitions of independence in increasingly sophisticated circumstances.

Definition 4.1.1 (Independence for two events) Suppose $(\Omega, \mathcal{B}, P)$ is a fixed probability space. Events $A, B \in \mathcal{B}$ are *independent* if

$$P(AB) = P(A)P(B).$$

Definition 4.1.2 (Independence of a finite number of events) The events $A_1, \dots, A_n$ $(n \geq 2)$ are *independent* if

$$P(\bigcap_{i \in I} A_i) = \prod_{i \in I} P(A_i), \quad \text{for all finite } I \subset \{1, \dots, n\}. \tag{4.1}$$

(Note that (4.1) represents

$$\sum_{k=2}^{n} \binom{n}{k} = 2^n - n - 1$$

equations.)

Equation (4.1) can be rephrased as follows: The events $A_1, \ldots, A_n$ are independent if

$$P(B_1 \cap B_2 \cdots \cap B_n) = \prod_{i=1}^{n} P(B_i) \tag{4.2}$$

where for each $i = 1, \ldots, n$,

$$B_i \text{ equals } A_i \text{ or } \Omega.$$

Definition 4.1.3 (Independent classes) Let $\mathcal{C}_i \subset \mathcal{B}, i = 1, \ldots, n$. The classes $\mathcal{C}_i$ are *independent*, if for any choice $A_1, \ldots, A_n$, with $A_i \in \mathcal{C}_i, i = 1, \ldots, n$, we have the events $A_1, \ldots, A_n$ independent events (according to Definition 4.1.2).

Here is a basic criterion for proving independence of σ-fields.

Theorem 4.1.1 (Basic Criterion) *If for each $i = 1, \ldots, n$, $\mathcal{C}_i$ is a non-empty class of events satisfying*

1. *$\mathcal{C}_i$ is a π-system,*
2. *$\mathcal{C}_i$, $i = 1, \ldots, n$ are independent,*

then

$$\sigma(\mathcal{C}_1), \ldots, \sigma(\mathcal{C}_n)$$

are independent.

Proof. We begin by proving the result for $n = 2$. Fix $A_2 \in \mathcal{C}_2$. Let

$$\mathcal{L} = \{A \in \mathcal{B} : P(AA_2) = P(A)P(A_2)\}.$$

Then we claim that $\mathcal{L}$ is a λ-system. We verify the postulates.

(a) We have $\Omega \in \mathcal{L}$ since

$$P(\Omega A_2) = P(A_2) = P(\Omega)P(A_2).$$

(b) If $A \in \mathcal{L}$, then $A^c \in \mathcal{L}$ since

$$\begin{aligned} P(A^c A_2) &= P((\Omega \setminus A)A_2) = P(A_2 \setminus AA_2) \\ &= P(A_2) - P(AA_2) = P(A_2) - P(A)P(A_2) \\ &= P(A_2)(1 - P(A)) = P(A^c)P(A_2). \end{aligned}$$

(c) If $B_n \in \mathcal{L}$ are disjoint $(n \geq 1)$, then $\sum_{n=1}^{\infty} B_n \in \mathcal{L}$ since

$$P((\bigcup_{n=1}^{\infty} B_n)A_2) = P(\bigcup_{n=1}^{\infty} B_n A_2) = \sum_{n=1}^{\infty} P(B_n A_2)$$
$$= \sum_{n=1}^{\infty} P(B_n)P(A_2) = P(\bigcup_{n=1}^{\infty} B_n)P(A_2).$$

Also $\mathcal{L} \supset \mathcal{C}_1$, so $\mathcal{L} \supset \sigma(\mathcal{C}_1)$ by Dynkin's theorem 2.2.2 in Chapter 2. Thus $\sigma(\mathcal{C}_1), \mathcal{C}_2$ are independent.

Now extend this argument to show $\sigma(\mathcal{C}_1), \sigma(\mathcal{C}_2)$ are independent. Also, we may use induction to extend the argument for $n = 2$ to general n. □

We next define independence of an arbitrary collection of classes of events.

Definition 4.1.4 (Arbitrary number of independent classes) Let T be an arbitrary index set. The classes $\mathcal{C}_t, t \in T$ are independent families if for each finite $I, I \subset T, \mathcal{C}_t, t \in I$ is independent.

Corollary 4.1.1 *If $\{\mathcal{C}_t, t \in T\}$ are non-empty π-systems that are independent, then $\{\sigma(\mathcal{C}_t), t \in T\}$ are independent.*

The proof follows from the Basic Criterion Theorem 4.1.1.

4.2 Independent Random Variables

We now turn to the definition of independent random variables and some criteria for independence of random variables.

Definition 4.2.1 (Independent random variables) $\{X_t, t \in T\}$ is an independent family of random variables if $\{\sigma(X_t), t \in T\}$ are independent σ-fields.

The random variables are independent if their induced σ-fields are independent. The information provided by any individual random variable should not affect behavior of other random variables in the family. Since

$$\sigma(1_A) = \{\phi, \Omega, A, A^c\},$$

we have $1_{A_1}, \ldots, 1_{A_n}$ independent iff $A_1, \ldots, A_n$ are independent.

We now give a criterion for independence of random variables in terms of distribution functions. For a family of random variables $\{X_t, t \in T\}$ indexed by a set T, the *finite dimensional distribution functions* are the family of multivariate distribution functions

$$F_J(x_t, t \in J) = P[X_t \leq x_t, t \in J] \tag{4.3}$$

for all finite subsets $J \subset T$.

Theorem 4.2.1 (Factorization Criterion) *A family of random variables $\{X_t, t \in T\}$ indexed by a set T, is independent iff for all finite $J \subset T$*

$$F_J(x_t, t \in J) = \prod_{t \in J} P[X_t \le x_t], \quad \forall x_t \in \mathbb{R}. \tag{4.4}$$

Proof. Because of Definition 4.1.4, it suffices to show for a finite index set J that $\{X_t, t \in J\}$ is independent iff (4.4) holds. Define

$$\mathcal{C}_t = \{[X_t \le x], x \in \mathbb{R}\}.$$

Then

(i) $\mathcal{C}_t$ is a π-system since

$$[X_t \le x] \bigcap [X_t \le y] = [X_t \le x \wedge y]$$

and

(ii) $\sigma(\mathcal{C}_t) = \sigma(X_t)$.

Now (4.4) says $\{\mathcal{C}_t, t \in J\}$ is an independent family and therefore by the Basic Criterion 4.1.1, $\{\sigma(\mathcal{C}_t) = \sigma(X_t), t \in J\}$ are independent. □

Corollary 4.2.1 *The finite collection of random variables $X_1, \dots, X_k$ is independent iff*

$$P[X_1 \le x_1, \dots, X_k \le x_k] = \prod_{i=1}^{k} P[X_i \le x_i],$$

for all $x_i \in \mathbb{R}$, $i = 1, \dots, k$.

For the next result, we define a random variable to be *discrete* if it has a countable range.

Corollary 4.2.2 *The discrete random variables $X_1, \dots, X_k$ with countable range $\mathcal{R}$ are independent iff*

$$P[X_i = x_i, i = 1, \dots, k] = \prod_{i=1}^{k} P[X_i = x_i], \tag{4.5}$$

for all $x_i \in \mathcal{R}$, $i = 1, \dots, k$.

Proof. If $X_1, \dots, X_n$ is an independent family, then $\sigma(X_i), i = 1, \dots, k$ is independent. Since

$$[X_i = x_i] \in \sigma(X_i)$$

we have $[X_i = x_i], i = 1, \dots, k$ are independent events and (4.5) follows.

Conversely, suppose (4.5) holds. Define $\mathbf{z} \le \mathbf{x}$ to mean $z_i \le x_i, i = 1, \dots, k$. Then

$$
\begin{aligned}
P[X_i \le x_i, i = 1, \dots, k] &= \sum_{\substack{\mathbf{z}\le\mathbf{x}\\ z_i\in\mathcal{R}\\ i=1,\dots,k}} P[X_i = z_i, i = 1, \dots, k] \\
&= \sum_{\substack{\mathbf{z}\le\mathbf{x}\\ z_i\in\mathcal{R}\\ i=1,\dots,k}} \prod_{i=1}^{k} P[X_i = z_i] \\
&= \sum_{\substack{z_2\le x_2,\dots,z_k\le x_k\\ z_i\in\mathcal{R}, i=2,\dots,k}} \sum_{\substack{z_1\le x_1\\ z_1\in\mathcal{R}}} P[X_1 = z_1] \prod_{i=2}^{k} P[X_i = z_i] \\
&= \sum_{\substack{z_2\le x_2,\dots,z_k\le x_k\\ z_i\in\mathcal{R}, i=2,\dots,k}} P[X_1 \le x_1] \prod_{i=2}^{k} P[X_i = z_i] \\
&= \cdots = \prod_{i=1}^{n} P[X_i \le x_i].
\end{aligned}
$$

□

4.3 Two Examples of Independence

This section provides two interesting examples of unexpected independence:

- Ranks and records.
- Dyadic expansions of uniform random numbers.

4.3.1 *Records, Ranks, Renyi Theorem*

Let $\{X_n, n \ge 1\}$ be iid with common continuous distribution function $F(x)$. The continuity of F implies

$$P[X_i = X_j] = 0, \tag{4.6}$$

so that if we define

$$[\text{ Ties }] = \bigcup_{i\ne j} [X_i = X_j],$$

then

$$P[\text{ Ties }] = 0.$$

Call X_n a *record* of the sequence if

$$X_n > \bigvee_{i=1}^{n-1} X_i,$$

and define

$$A_n = [X_n \text{ is a record }].$$

A result due to Renyi says that the events $\{A_j, j \geq 1\}$ are independent and

$$P(A_j) = 1/j, \quad j \geq 2.$$

This is a special case of a result about *relative ranks*.

Let R_n be the relative rank of X_n among $X_1, \dots, X_n$ where

$$R_n = \sum_{j=1}^{n} 1_{[X_j \geq X_n]}.$$

So

$$\begin{aligned} R_n &= 1 \text{ iff } X_n \text{ is a record,} \\ &= 2 \text{ iff } X_n \text{ is the second largest of } X_1, \dots, X_n, \end{aligned}$$

and so on.

Theorem 4.3.1 (Renyi Theorem) *Assume $\{X_n, n \geq 1\}$ are iid with common, continuous distribution function $F(x)$.*

(a) The sequence of random variables $\{R_n, n \geq 1\}$ is independent and

$$P[R_n = k] = \frac{1}{n},$$

for $k = 1, \dots, n$.

(b) The sequence of events $\{A_n, n \geq 1\}$ is independent and

$$P(A_n) = \frac{1}{n}.$$

Proof. (b) comes from (a) since $A_n = [R_n = 1]$.

(a) These are $n!$ orderings of $X_1, \dots, X_n$. (For a given ω, one such ordering is $X_1(\omega) < \dots < X_n(\omega)$. Another possible ordering is $X_2(\omega) < \dots < X_n(\omega) < X_1(\omega)$, and so on.) By symmetry, since $X_1, \dots, X_n$ are identically distributed and independent, all possible orderings have the same probability $\frac{1}{n!}$, so for example,

$$P[X_2 < X_3 < \dots < X_n < X_1] = \frac{1}{n!}.$$

Each realization of $R_1, \dots, R_n$ uniquely determines an ordering: For example, if $n = 3$, suppose $R_1(\omega) = 1$, $R_2(\omega) = 1$, and $R_3(\omega) = 1$. This tells us that

$$X_1(\omega) < X_2(\omega) < X_3(\omega),$$

and if $R_1(\omega) = 1$, $R_2(\omega) = 2$, and $R_3(\omega) = 3$, then this tells us

$$X_3(\omega) < X_2(\omega) < X_1(\omega).$$

Each realization of $R_1, \dots, R_n$ has the same probability as a particular ordering of $X_1, \dots, X_n$. Hence

$$P[R_1 = r_1, \dots, R_n = r_n] = \frac{1}{n!},$$

for $r_i \in \{1, \dots, i\}$, $i = 1, \dots, n$.

Note that

$$\begin{aligned} P[R_n = r_n] &= \sum_{r_1, \dots, r_{n-1}} P[R_1 = r_1, \dots, R_{n-1} = r_{n-1}, R_n = r_n] \\ &= \sum_{r_1, \dots, r_{n-1}} \frac{1}{n!}. \end{aligned}$$

Since r_i ranges over i values, the number of terms in the sum is

$$1 \cdot 2 \cdot 3 \cdot \dots \cdot n - 1 = (n-1)!.$$

Thus

$$P[R_n = r_n] = \frac{(n-1)!}{n!} = \frac{1}{n}, \quad n = 1, 2, \dots .$$

Therefore

$$\begin{aligned} P[R_1 &= r_1, \dots, R_n = r_n] = \frac{1}{n!} \\ &= P[R_1 = r_1] \cdots P[R_n = r_n]. \end{aligned}$$

□

Postscript: If $\{X_n, n \geq 1\}$ is iid with common continuous distribution $F(x)$, why is the probability of ties zero? We have

$$P[\text{ Ties }] = P(\bigcup_{i \neq j} [X_i = X_j])$$

and by subadditivity, this probability is bounded above by

$$\sum_{i \neq j} P[X_i = X_j].$$

Thus it suffices to show that

$$P[X_1 = X_2] = 0.$$

Note the set containment: For every n,

$$[X_1 = X_2] \subset \bigcup_{k=-\infty}^{\infty} [\frac{k-1}{2^n} < X_1, \ X_2 \leq \frac{k}{2^n}].$$

By monotonicity and subadditivity

$$\begin{aligned} P[X_1 = X_2] &\le \sum_{k=-\infty}^{\infty} P[\frac{k-1}{2^n} < X_1 \le \frac{k}{2^n}, \frac{k-1}{2^n} < X_2 \le \frac{k}{2^n}] \\ &= \sum_{k=-\infty}^{\infty} \left(P[\frac{k-1}{2^n} < X_1 \le \frac{k}{2^n}] \right)^2 . \end{aligned} \tag{4.7}$$

Write

$$F[a, b] = F(b) - F(a)$$

and the above (4.7) is equal to

$$\begin{aligned} &\sum_{k=-\infty}^{\infty} F(\frac{k-1}{2^n}, \frac{k}{2^n}] \; F(\frac{k-1}{2^n}, \frac{k}{2^n}] \\ &\qquad \le \max_{-\infty<k<\infty} F(\frac{k-1}{2^n}, \frac{k}{2^n}] \sum_{k=-\infty}^{\infty} F(\frac{k-1}{2^n}, \frac{k}{2^n}] \\ &\qquad \le \max_{-\infty<k<\infty} F(\frac{k-1}{2^n}, \frac{k}{2^n}] \cdot 1 \\ &\qquad = \max_{-\infty<k<\infty} F(\frac{k-1}{2^n}, \frac{k}{2^n}]. \end{aligned}$$

Since F is continuous on $\mathbb{R}$, because F is a also a probability distribution, F is *uniformly* continuous on $\mathbb{R}$. (See Exercise 22 in Chapter 2.) Thus given any $\varepsilon > 0$, for $n \ge n_0(\varepsilon)$ and all k, we have

$$F(\frac{k-1}{2^n}, \frac{k}{2^n}] = F(\frac{k}{2^n}) - F(\frac{k-1}{2^n}) \le \varepsilon.$$

Thus for any $\varepsilon > 0$,

$$P[X_1 = X_2] \le \varepsilon, \tag{4.8}$$

and since ε is arbitrary, the probability in (4.8) must be 0.

4.3.2 Dyadic Expansions of Uniform Random Numbers

Here we consider

$$(\Omega, \mathcal{B}, P) = \big((0, 1], \mathcal{B}((0, 1]), \lambda\big),$$

where λ is Lebesgue measure. We write $\omega \in (0, 1]$ using its dyadic expansion

$$\omega = \sum_{n=1}^{\infty} \frac{d_n(\omega)}{2^n} = .d_1(\omega)d_2(\omega)d_3(\omega)\cdots ,$$

where each $d_n(\omega)$ is either 0 or 1.

We write 1 as

$$0.11111\cdots = \sum_{n=1}^{\infty} 2^{-n} = \frac{1/2}{1-1/2} = 1,$$

and if a number such as $\frac{1}{2}$ has two possible expansions, we agree to use the non-terminating one. Thus, even though $1/2$ has two expansions

$$\frac{1}{2} = \sum_{n=2}^{\infty} \frac{1}{2^n} = .01111\cdots,$$

and

$$\frac{1}{2} = \frac{1}{2} + 0 + 0 + \cdots = .1000\cdots,$$

by our convention, we use the first expansion.

Fact 1. Each d_n is a random variable. Since d_n is discrete with possible values 0, 1, it suffices to check

$$[d_n = 0] \in \mathcal{B}((0,1]), \quad [d_n = 1] \in \mathcal{B}((0,1]),$$

for any $n \geq 1$. In fact, since $[d_n = 0] = [d_n = 1]^c$, it suffices to check $[d_n = 1] \in \mathcal{B}((0,1])$.

To verify this, we start gently by considering a relatively easy case as a warm-up. For $n = 1$,

$$[d_1 = 1] = (.1000\cdots, .1111\cdots] = (\frac{1}{2}, 1] \in \mathcal{B}((0,1]).$$

The left endpoint is open because of the convention that we take the non-terminating expansion. Note $P[d_1 = 1] = P[d_1 = 0] = 1/2$.

After understanding this warmup, we proceed to the general case. For any $n \geq 2$

$$\begin{aligned}
&[d_n = 1] \\
&= \bigcup_{(u_1,u_2,\dots,u_{n-1})\in\{0,1\}^{n-1}} (.u_1u_2\dots u_{n-1}1000\dots, .u_1u_2\dots u_{n-1}1111\cdots] \qquad (4.9) \\
&= \text{disjoint union of } 2^{n-1} \text{ intervals } \in \mathcal{B}((0,1]).
\end{aligned}$$

For example

$$[d_2 = 1] = (\frac{1}{4}, \frac{1}{2}] \cup (\frac{3}{4}, 1].$$

Fact 2. We may also use (4.9) to compute the mass function of d_n. We have

$$\begin{aligned}
&P[d_n = 1] \\
&= \sum_{(u_1,u_2,\dots,u_{n-1})\in\{0,1\}^{n-1}} P\big((.u_1u_2\dots u_{n-1}1000\cdots, .u_1u_2\dots u_{n-1}1111\cdots]\big) \\
&=2^{n-1}\Big\{\sum_{i=1}^{n-1}\frac{u_i}{2^i} + \sum_{i=n}^{\infty}2^{-i} - \sum_{i=1}^{n-1}\frac{u_i}{2^i} + \frac{1}{2^n}\Big\} \\
&=2^{n-1}\Big\{\sum_{i=n+1}^{\infty}\frac{1}{2^i}\Big\} = \frac{1}{2}.
\end{aligned}$$

The factor 2^{n-1} results from the number of intervals whose length we must sum. We thus conclude that

$$P[d_n = 0] = P[d_n = 1] = \frac{1}{2}. \tag{4.10}$$

Fact 3. The sequence $\{d_n, n \geq 1\}$ is iid. The previous fact proved in (4.10) that $\{d_n\}$ is identically distributed and thus we only have to prove $\{d_n\}$ is independent. For this, it suffices to pick $n \geq 1$ and prove $\{d_1, \dots, d_n\}$ is independent.

For $(u_1, \dots, u_n) \in \{0, 1\}^n$, we have

$$\bigcap_{i=1}^{n}[d_i = u_i] = (.u_1u_2\dots u_n000\dots, .u_1u_2\dots u_n111\dots].$$

Again, the left end of the interval is open due to our convention decreeing that we take non-terminating expansions when a number has two expansions. Since the probability of an interval is its length, we get

$$\begin{aligned}
P\Big(\bigcap_{i=1}^{n}[d_i = u_i]\Big) &= \sum_{i=1}^{n}\frac{u_i}{2^i} + \sum_{i=n+1}^{\infty}\frac{1}{2^i} - \sum_{i=1}^{n}\frac{u_i}{2^i} \\
&= \frac{2^{-(n+1)}}{1-\frac{1}{2}} = \frac{1}{2^n} \\
&= \prod_{i=1}^{n} P[d_i = u_i]
\end{aligned}$$

where the last step used (4.10). So the joint mass function of $d_1, \dots, d_n$ factors into a product of individual mass functions and we have proved independence of the finite collection, and hence of $\{d_n, n \geq 1\}$.

4.4 More on Independence: Groupings

It is possible to group independent events or random variables according to disjoint subsets of the index set to achieve independent groupings. This is a useful property of independence.

Lemma 4.4.1 (Grouping Lemma) *Let $\{\mathcal{B}_t, t \in T\}$ be an independent family of σ-fields. Let S be an index set and suppose for $s \in S$ that $T_s \subset T$ and $\{T_s, s \in S\}$ is pairwise disjoint. Now define*

$$\mathcal{B}_{T_s} = \bigvee_{t \in T_s} \mathcal{B}_t.$$

Then

$$\{\mathcal{B}_{T_s}, s \in S\}$$

is an independent family of σ-fields.

Remember that $\bigvee_{t \in T_s} \mathcal{B}_t$ is the smallest σ-field containing all the $\mathcal{B}_t$'s.

Before discussing the proof, we consider two examples. For these and other purposes, it is convenient to write

$$X \perp\!\!\!\perp Y$$

when X and Y are independent random variables. Similarly, we write $\mathcal{B}_1 \perp\!\!\!\perp \mathcal{B}_2$ when the two σ-fields $\mathcal{B}_1$ and $\mathcal{B}_2$ are independent.

(a) Let $\{X_n, n \geq 1\}$ be independent random variables. Then

$$\sigma(X_j, j \leq n) \perp\!\!\!\perp \sigma(X_j, j > n),$$

$$\sum_{i=1}^{n} X_i \perp\!\!\!\perp \sum_{i=n+1}^{n+k} X_i,$$

$$\bigvee_{i=1}^{n} X_i \perp\!\!\!\perp \bigvee_{j=n+1}^{n+k} X_j.$$

(b) Let $\{A_n\}$ be independent events. Then $\bigcup_{n=1}^{N} A_j$ and $\bigcup_{j=N+1}^{\infty} A_j$ are independent.

Proof. Without loss of generality we may suppose S is finite. Define

$$\mathcal{C}_{T_s} := \{\bigcap_{\alpha \in K} B_\alpha : B_\alpha \in \mathcal{B}_\alpha, K \subset T_s, \ K \text{ is finite.}\}$$

Then $\mathcal{C}_{T_s}$ is a π-system for each s, and $\{\mathcal{C}_{T_s}, s \in S\}$ are independent classes. So by the Basic Criterion 4.1.1 we are done, provided you believe

$$\sigma(\mathcal{C}_{T_s}) = \mathcal{B}_{T_s}.$$

Certainly it is the case that

$$\mathcal{C}_{T_s} \subset \mathcal{B}_{T_s}$$

and hence

$$\sigma(\mathcal{C}_{T_s}) \subset \mathcal{B}_{T_s}.$$

Also,

$$\mathcal{C}_{T_s} \supset \mathcal{B}_\alpha, \quad \forall \alpha \in T_s$$

(we can take $K = \{\alpha\}$) and hence

$$\sigma(\mathcal{C}_{T_s}) \supset \mathcal{B}_\alpha, \quad \forall \alpha \in T_s.$$

It follows that

$$\sigma(\mathcal{C}_{T_s}) \supset \bigcup_{\alpha \in T_s} \mathcal{B}_\alpha,$$

and hence

$$\sigma(\mathcal{C}_{T_s}) \supset \sigma\left(\bigcup_{\alpha \in T_s} \mathcal{B}_\alpha\right) =: \bigvee_{\alpha \in T_s} \mathcal{B}_\alpha. \qquad \square$$

4.5 Independence, Zero-One Laws, Borel-Cantelli Lemma

There are several common zero-one laws which identify the possible range of a random variable to be trivial. There are also several zero-one laws which provide the basis for all proofs of almost sure convergence. We take these up in turn.

4.5.1 Borel-Cantelli Lemma

The Borel-Cantelli Lemma is very simple but still is the basic tool for proving almost sure convergence.

Proposition 4.5.1 (Borel-Cantelli Lemma.) *Let $\{A_n\}$ be any events. If*

$$\sum_n P(A_n) < \infty,$$

then

$$P([A_n \text{ i.o. }]) = P(\limsup_{n\to\infty} A_n) = 0.$$

Proof. We have

$$\begin{aligned} P([A_n \text{ i.o. }]) &= P(\lim_{n\to\infty} \bigcup_{j\geq n} A_j) \\ &= \lim_{n\to\infty} P(\bigcup_{j\geq n} A_j) \quad \text{(continuity of } P) \\ &\leq \limsup_{n\to\infty} \sum_{j=n}^{\infty} P(A_j) \quad \text{(subadditivity)} \\ &= 0, \end{aligned}$$

since $\sum_n P(A_n) < \infty$ implies $\sum_{j=n}^{\infty} P(A_j) \to 0$, as $n \to \infty$. $\square$

Example 4.5.1 Suppose $\{X_n, n \geq 1\}$ are Bernoulli random variables with

$$P[X_n = 1] = p_n = 1 - P[X_n = 0].$$

Note we have not supposed $\{X_n\}$ independent and certainly not identically distributed. We assert that

$$P[\lim_{n\to\infty} X_n = 0] = 1, \tag{4.11}$$

if

$$\sum_n p_n < \infty. \tag{4.12}$$

To verify that (4.12) is sufficient for (4.11), observe that if

$$\sum_n p_n = \sum_n P[X_n = 1] < \infty,$$

then by the Borel-Cantelli Lemma

$$P([X_n = 1] \text{ i.o. }) = 0.$$

Taking complements, we find

$$1 = P(\limsup_{n\to\infty}[X_n = 1]^c) = P(\liminf_{n\to\infty}[X_n = 0]) = 1.$$

Since with probability 1, the two valued functions $\{X_n\}$ are zero from some point on, with probability 1 the variables must converge to zero. □

4.5.2 *Borel Zero-One Law*

The Borel-Cantelli Lemma does not require independence. The next result does.

Proposition 4.5.2 (Borel Zero-One Law) *If $\{A_n\}$ is a sequence of independent events, then*

$$P([A_n \text{ i.o.}]) = \begin{cases} 0, & \text{iff } \sum_n P(A_n) < \infty, \\ 1, & \text{iff } \sum_n P(A_n) = \infty. \end{cases}$$

Proof. From the Borel-Cantelli Lemma, if

$$\sum_n P(A_n) < \infty,$$

then

$$P([A_n \text{ i.o. }]) = 0.$$

Conversely, suppose $\sum_n P(A_n) = \infty$. Then

$$\begin{aligned}
P([A_n \text{ i.o. }]) &= P(\limsup_{n\to\infty} A_n) \\
&= 1 - P(\liminf_{n\to\infty} A_n^c) \\
&= 1 - P(\lim_{n\to\infty} \bigcap_{k\geq n} A_k^c) \\
&= 1 - \lim_{n\to\infty} P(\bigcap_{k\geq n} A_k^c) \\
&= 1 - \lim_{n\to\infty} P(\lim_{m\to\infty} \downarrow \bigcap_{k=n}^{m} A_k^c) \\
&= 1 - \lim_{n\to\infty} \lim_{m\to\infty} P(\bigcap_{k=n}^{m} A_k^c) \\
&= 1 - \lim_{n\to\infty} \lim_{m\to\infty} \prod_{k=n}^{m} (1 - P(A_k)),
\end{aligned}$$

where the last equality resulted from independence. It suffices to show

$$\lim_{n\to\infty} \lim_{m\to\infty} \prod_{k=n}^{m} (1 - P(A_k)) = 0. \tag{4.13}$$

To prove (4.13), we use the inequality

$$1 - x \leq e^{-x}, \quad 0 < x < 1. \tag{4.14}$$

To verify (4.14), note for $0 < x < 1$ that

$$-\log(1-x) = \sum_{n=1}^{\infty} \frac{x^n}{n} \geq x$$

so exponentiating both sides yields

$$\frac{1}{1-x} \geq e^x$$

or

$$e^{-x} \geq 1 - x.$$

Now for (4.13). We have

$$\begin{aligned}
\lim_{m\to\infty} \prod_{k=n}^{m} (1 - P(A_k)) &\leq \lim_{m\to\infty} \prod_{k=n}^{m} e^{-P(A_k)} \\
&= \lim_{m\to\infty} e^{-\sum_{k=n}^{m} P(A_k)} \\
&= e^{-\sum_{k=n}^{\infty} P(A_k)} = e^{-\infty} = 0,
\end{aligned}$$

since $\sum_n P(A_n) = \infty$. This is true for all n, and so

$$\lim_{n\to\infty} \lim_{m\to\infty} \prod_{k=n}^{m} (1 - P(A_k)) = 0.$$

□

Example 4.5.1 (continued) Suppose $\{X_n, n \geq 1\}$ are independent in addition to being Bernoulli, with

$$P[X_k = 1] = p_k = 1 - P[X_k = 0].$$

Then we assert that

$$P[X_n \to 0] = 1 \text{ iff } \sum_n p_k < \infty.$$

To verify this assertion, we merely need to observe that

$$P\{[X_n = 1] \text{ i.o. }\} = 0$$

iff

$$\sum_n P[X_n = 1] = \sum_n p_n < \infty.$$

Example 4.5.2 (Behavior of exponential random variables) We assume that $\{E_n, n \geq 1\}$ are iid unit exponential random variables; that is,

$$P[E_n > x] = e^{-x}, \quad x > 0.$$

Then

$$P[\limsup_{n\to\infty} E_n / \log n = 1] = 1. \tag{4.15}$$

This result is sometimes considered surprising. There is a (mistaken) tendency to think of iid sequences as somehow roughly constant, and therefore the division by $\log n$ should send the ratio to 0. However, every so often, the sequence $\{E_n\}$ spits out a large value and the growth of these large values approximately matches that of $\{\log n, n \geq 1\}$.

To prove (4.15), we need the following simple fact: If $\{B_k\}$ are any events satisfying $P(B_k) = 1$, then $P(\bigcap_k B_k) = 1$. See Exercise 11 of Chapter 2.

Proof of (4.15). For any $\omega \in \Omega$,

$$\limsup_{n\to\infty} \frac{E_n(\omega)}{\log n} = 1$$

means

(a) $\forall \varepsilon > 0$, $\frac{E_n(\omega)}{\log n} \leq 1 + \varepsilon$, for all large n,
and

(b) $\forall \varepsilon > 0$, $\frac{E_n(\omega)}{\log n} > 1-\varepsilon$, for infinitely many n.

Note (a) says that for any ε, there is no subsequential limit bigger than $1+\varepsilon$ and (b) says that for any ε, there is always some subsequential limit bounded below by $1-\varepsilon$.

We have the following set equality: Let $\varepsilon_k \downarrow 0$ and observe

$$[\limsup_{n\to\infty} \frac{E_n}{\log n} = 1]$$
$$= \bigcap_k \left\{ \liminf_{n\to\infty} [\frac{E_n}{\log n} \le 1+\varepsilon_k] \right\} \bigcap \bigcap_k \left\{ [\frac{E_n}{\log n} > 1-\varepsilon_k] \text{i.o.} \right\} \tag{4.16}$$

To prove that the event on the left side of (4.16) has probability 1, it suffices to prove every braced event on the right side of (4.16) has probability 1. For fixed k

$$\begin{aligned}\sum_n P[\frac{E_n}{\log n} > 1-\varepsilon_k] &= \sum_n P[E_n > (1-\varepsilon_k)\log n] \\ &= \sum_n \exp\{-(1-\varepsilon_k)\log n\} \\ &= \sum_n \frac{1}{n^{1-\varepsilon_k}} = \infty.\end{aligned}$$

So the Borel Zero-One Law 4.5.2 implies

$$P\left\{[\frac{E_n}{\log n} > 1-\varepsilon_k] \text{ i.o.}\right\} = 1.$$

Likewise

$$\begin{aligned}\sum_n P[\frac{E_n}{\log n} > 1+\varepsilon_k] &= \sum_n \exp\{-(1+\varepsilon_k)\log n\} \\ &= \sum_n \frac{1}{n^{1+\varepsilon_k}} < \infty,\end{aligned}$$

so

$$P\left(\limsup_{n\to\infty}\left[\frac{E_n}{\log n} > 1+\varepsilon_k\right]\right) = 0$$

implies

$$P\left\{\liminf_{n\to\infty}\left[\frac{E_n}{\log n} \le 1+\varepsilon_k\right]\right\} = 1 - P\left\{\limsup_{n\to\infty}\left[\frac{E_n}{\log n} \le 1+\varepsilon_k\right]^c\right\} = 1.$$

□

4.5.3 Kolmogorov Zero-One Law

Let $\{X_n\}$ be a sequence of random variables and define

$$\mathcal{F}'_n = \sigma(X_{n+1}, X_{n+2}, \dots), \quad n = 1, 2, \dots .$$

The *tail σ-field* $\mathcal{T}$ is defined as

$$\mathcal{T} = \bigcap_n \mathcal{F}'_n = \lim_{n\to\infty} \downarrow \sigma(X_n, X_{n+1}, \dots).$$

These are events which depend on the tail of the $\{X_n\}$ sequence. If $A \in \mathcal{T}$, we will call A a *tail event* and similarly a random variable measurable with respect to $\mathcal{T}$ is called a *tail random variable.*

We now give some examples of tail events and random variables.

1. Observe that
$$\{\omega : \sum_{n=1}^{\infty} X_n(\omega) \text{ converges }\} \in \mathcal{T}.$$
To see this note that, for any m, the sum $\sum_{n=1}^{\infty} X_n(\omega)$ converges if and only if $\sum_{n=m}^{\infty} X_n(\omega)$ converges. So
$$[\sum_n X_n \text{ converges }] = [\sum_{n=m+1}^{\infty} X_n \text{ converges }] \in \mathcal{F}'_m.$$
This holds for all m and after intersecting over m.

2. We have
$$\limsup_{n\to\infty} X_n \in \mathcal{T},$$
$$\liminf_{n\to\infty} X_n \in \mathcal{T},$$
$$\{\omega : \lim_{n\to\infty} X_n(\omega) \text{ exists }\} \in \mathcal{T}.$$
This is true since the lim sup of the sequence $\{X_1, X_2, \dots\}$ is the same as the lim sup of the sequence $\{X_m, X_{m+1}, \dots\}$ for all m.

3. Let $S_n = X_1 + \dots + X_n$. Then
$$\left\{\omega : \lim_{n\to\infty} \frac{S_n(\omega)}{n} = 0\right\} \in \mathcal{T}$$
since for any m,
$$\lim_{n\to\infty} \frac{S_n(\omega)}{n} = \lim_{n\to\infty} \frac{\sum_{i=1}^{n} X_i(\omega)}{n} = \lim_{n\to\infty} \frac{\sum_{i=m+1}^{n} X_i(\omega)}{n},$$
and so for any m,
$$\lim_{n\to\infty} \frac{S_n(\omega)}{n} \in \mathcal{F}'_m.$$

Call a σ-field, all of whose events have probability 0 or 1 *almost trivial*. One example of an almost trivial σ-field is the σ-field $\{\emptyset, \Omega\}$. Kolmogorov's Zero-One Law characterizes tail events and random variables of independent sequences as *almost trivial.*

Theorem 4.5.3 (Kolmogorov Zero-One Law) *If $\{X_n\}$ are independent random variables with tail σ-field $\mathcal{T}$, then $\Lambda \in \mathcal{T}$ implies $P(\Lambda) = 0$ or 1 so that the tail σ-field $\mathcal{T}$ is almost trivial.*

Before proving Theorem 4.5.3, we consider some implications. To help us do this, we need the following lemma which provides further information on almost trivial σ-fields.

Lemma 4.5.1 (Almost trivial σ-fields) *Let $\mathcal{G}$ be an almost trivial σ-field and let X be a random variable measurable with respect to $\mathcal{G}$. Then there exists c such that $P[X = c] = 1$.*

Proof of Lemma 4.5.1. Let

$$F(x) = P[X \leq x].$$

Then F is non-decreasing and since $[X \leq x] \in \sigma(X) \subset \mathcal{G}$,

$$F(x) = 0 \text{ or } 1$$

for each $x \in \mathbb{R}$. Let

$$c = \sup\{x : F(x) = 0\}.$$

The distribution function must have a jump of size 1 at c and thus

$$P[X = c] = 1.$$

□

With this in mind, we can consider some consequences of the Kolmogorov Zero-One Law.

Corollary 4.5.1 (Corollaries of the Kolmogorov Zero-One Law) *Let $\{X_n\}$ be independent random variables. Then the following are true.*

(a) *The event*

$$[\sum_n X_n \text{ converges}]$$

has probability 0 *or* 1.

(b) *The random variables* $\limsup_{n\to\infty} X_n$ *and* $\liminf_{n\to\infty} X_n$ *are constant with probability* 1.

(c) *The event*

$$\{\omega : S_n(\omega)/n \to 0\}$$

has probability 0 *or* 1.

We now commence the proof of Theorem 4.5.3.

Proof of the Kolmogorov Zero-One Law. Suppose $\Lambda \in \mathcal{T}$. We show Λ is independent of itself so that

$$P(\Lambda) = P(\Lambda \cap \Lambda) = P(\Lambda)P(\Lambda)$$

and thus $P(\Lambda) = (P(\Lambda))^2$. Therefore $P(\Lambda) = 0$ or 1.

To show Λ is independent of itself, we define

$$\mathcal{F}_n = \sigma(X_1, \dots, X_n) = \bigvee_{j=1}^{n} \sigma(X_j),$$

so that $\mathcal{F}_n \uparrow$ and

$$\mathcal{F}_\infty = \sigma(X_1, X_2, \dots) = \bigvee_{j=1}^{\infty} \sigma(X_j) = \bigvee_{n=1}^{\infty} \mathcal{F}_n.$$

Note that

$$\Lambda \in \mathcal{T} \subset \mathcal{F}'_n = \sigma(X_{n+1}, X_{n+2}, \dots) \subset \sigma(X_1, X_2, \dots) = \mathcal{F}_\infty. \tag{4.17}$$

Now for all n, we have

$$\Lambda \in \mathcal{F}'_n,$$

so since $\mathcal{F}_n \perp\!\!\!\perp \mathcal{F}'_n$, we have

$$\Lambda \perp\!\!\!\perp \mathcal{F}_n$$

for all n, and therefore

$$\Lambda \perp\!\!\!\perp \bigcup_n \mathcal{F}_n.$$

Let $\mathcal{C}_1 = \{\Lambda\}$, and $\mathcal{C}_2 = \bigcup_n \mathcal{F}_n$. Then $\mathcal{C}_i$ is a π-system, $i = 1, 2$, $\mathcal{C}_1 \perp\!\!\!\perp \mathcal{C}_2$ and therefore the Basic Criterion 4.1.1 implies

$$\sigma(\mathcal{C}_1) = \{\phi, \Omega, \Lambda, \Lambda^c\} \text{ and } \sigma(\mathcal{C}_2) = \bigvee_n \mathcal{F}_n = \mathcal{F}_\infty$$

are independent. Now

$$\Lambda \in \sigma(\mathcal{C}_1)$$

and

$$\Lambda \in \bigvee_n \mathcal{F}_n = \mathcal{F}_\infty$$

by (4.17). Thus Λ is independent of Λ. □

4.6 Exercises

1. Let $B_1, \dots, B_n$ be independent events. Show

$$P\Big(\bigcup_{i=1}^{n} B_i\Big) = 1 - \prod_{i=1}^{n}(1 - P(B_i)).$$

2. What is the minimum number of points a sample space must contain in order that there exist n independent events $B_1, \dots, B_n$, none of which has probability zero or one?

3. If $\{A_n, n \geq 1\}$ is an independent sequence of events, show

$$P\Big(\bigcap_{n=1}^{\infty} A_n\Big) = \prod_{n=1}^{\infty} P(A_n).$$

4. Suppose $(\Omega, \mathcal{B}, P)$ is the uniform probability space; that is, $([0,1], \mathcal{B}, \lambda)$ where λ is the uniform probability distribution. Define

$$X(\omega) = \omega.$$

(a) Does there exist a bounded random variable that is both independent of X and not constant almost surely?

(b) Define $Y = X(1 - X)$. Construct a random variable Z which is not almost surely constant and such that Z and Y are independent.

5. Suppose X is a random variable.

(a) X is independent of itself if and only if there is some constant c such that $P[X = c] = 1$.

(b) If there exists a measurable

$$g : (\mathbb{R}, \mathcal{B}(\mathbb{R})) \mapsto (\mathbb{R}, \mathcal{B}(\mathbb{R})),$$

such that X and $g(X)$ are independent, then prove there exists $c \in \mathbb{R}$ such that

$$P[g(X) = c] = 1.$$

6. Let $\{X_k, k \geq 1\}$ be iid random variables with common continuous distribution F. Let π be a permutation of $1, \dots, n$. Show

$$(X_1, \dots, X_n) \stackrel{d}{=} (X_{\pi(1)}, \dots, X_{\pi(n)})$$

where $\stackrel{d}{=}$ means the two vectors have the same joint distribution.

7. If A, B, C are independent events, show directly that both $A \cup B$ and $A \setminus B$ are independent of C.

8. If X and Y are independent random variables and f, g are measurable and real valued, why are $f(X)$ and $g(Y)$ independent? (No calculation is necessary.)

9. Suppose $\{A_n\}$ are independent events satisfying $P(A_n) < 1$, for all n. Show

$$P(\bigcup_{n=1}^{\infty} A_n) = 1 \text{ iff } P(A_n \text{ i.o. }) = 1.$$

Give an example to show that the condition $P(A_n) < 1$ cannot be dropped.

10. Suppose $\{X_n, n \geq 1\}$ are independent random variables. Show

$$P[\sup_n X_n < \infty] = 1$$

iff

$$\sum_n P[X_n > M] < \infty, \text{ for some } M.$$

11. Use the Borel-Cantelli Lemma to prove that given any sequence of random variables $\{X_n, n \geq 1\}$ whose range is the real line, there exist constants $c_n \to \infty$ such that

$$P[\lim_{n\to\infty} \frac{X_n}{c_n} = 0] = 1.$$

Give a careful description of how you choose c_n.

12. For use with the Borel Zero-One Law, the following is useful: Suppose we have two non-negative sequences $\{a_n\}$ and $\{b_n\}$ satisfying $a_n \sim b_n$ as $n \to \infty$; that is,

$$\lim_{n\to\infty} \frac{a_n}{b_n} = 1.$$

Show

$$\sum_n a_n < \infty \text{ iff } \sum_n b_n < \infty.$$

13. Let $\{X_n, n \geq 1\}$ be iid with $P[X_1 = 1] = p = 1 - P[X_1 = 0]$. What is the probability that the pattern 1,0,1 appears infinitely often?

Hint: Let

$$A_k = [X_k = 1, X_{k+1} = 0, X_{k+2} = 1]$$

and consider $A_1, A_4, A_7, \ldots$.

14. In a sequence of independent Bernoulli random variables $\{X_n, n \geq 1\}$ with

$$P[X_n = 1] = p = 1 - P[X_n = 0],$$

let A_n be the event that a run of n consecutive 1's occurs between the 2^n and 2^{n+1}st trial. If $p \geq 1/2$, then there is probability 1 that infinitely many A_n occur.

Hint: Prove something like

$$P(A_n) \geq 1 - (1 - p^n)^{2^n/2n} > 1 - e^{-(2p)^n/2n}.$$

15. (a) A finite family $\mathcal{B}_i, i \in I$ of σ-algebras is independent iff for every choice of non-negative $\mathcal{B}_i$–measurable random variable $Y_i, i \in I$, we have

$$E\Big(\prod_{i \in I} Y_i\Big) = \prod_{i \in I} E(Y_i).$$

(One direction is immediate. For the opposite direction, prove the result first for positive simple functions and then extend.)

(b) If $\{\mathcal{B}_t, t \in T\}$ is an arbitrary independent family of σ-algebras in $(\Omega, \mathcal{B}, P)$, the family $\{\mathcal{B}'_t, t \in T\}$ is again independent if $\mathcal{B}_t \supset \mathcal{B}'_t$, $(t \in T)$. Deduce from this that $\{f_t(X_t), t \in T\}$ is a family of independent random variables if the family $\{X_t, t \in T\}$ is independent and the f_t are measurable. In order for the family $\{X_t, t \in T\}$ of random variables to be independent, it is necessary and sufficient that

$$E\left(\prod_J f_j(X_j)\right) = \prod_J E\left(f_j(X_j)\right)$$

for every finite family $\{f_j, j \in J\}$ of bounded measurable functions.

16. The probability of convergence of a sequence of independent random variables is equal to 0 or 1. If the sequence $\{X_n\}$ is iid, and not constant with probability 1, then

$$P[X_n \text{ converges }] = 0.$$

17. Review Example 4.5.2

(a) Suppose $\{X_n, n \geq 1\}$ are iid random variables and suppose $\{a_n\}$ is a sequence of constants. Show

$$P\{[X_n > a_n] \text{ i.o. }\} = \begin{cases} 0, & \text{iff } \sum_n P[X_1 > a_n] < \infty, \\ 1, & \text{iff } \sum_n P[X_1 > a_n] = \infty. \end{cases}$$

(b) Suppose $\{X_n, n \geq 1\}$ are iid N(0,1) random variables. Show

$$P[\limsup_{n \to \infty} \frac{|X_n|}{\sqrt{\log n}} = \sqrt{2}] = 1.$$

Hint: Review, or look up Mill's Ratio which says

$$\lim_{x \to \infty} \frac{P[X_n > x]}{n(x)/x} = 1,$$

where $n(x)$ is the standard normal density.

(c) Suppose $\{X_n, n \geq 1\}$ are iid and Poisson distributed with parameter λ. Prove

$$\frac{\lambda^n}{n!}e^{-\lambda} \leq P[X_1 \geq n] \leq \frac{\lambda^n}{n!},$$

and therefore

$$P[\limsup_{n\to\infty} \frac{X_n}{\log n/\log(\log n)} = 1] = 1.$$

18. If the event A is independent of the π-system $\mathcal{P}$ and $A \in \sigma(\mathcal{P})$, then $P(A)$ is either 0 or 1.

19. Give a simple example to show that 2 random variables may be independent according to one probability measure but dependent with respect to another.

20. Counterexamples and examples:

a) Let $\Omega = \{1, 2, 3, 4\}$ with each point carrying probability 1/4. Let $A_1 = \{1, 2\}$, $A_2 = \{1, 3\}$, $A_3 = \{1, 4\}$. Then any two of A_1, A_2, A_3 are independent, but A_1, A_2, A_3 are not independent.

b) Let $\{A_i, 1 \leq i \leq 5\}$ be a measurable partition of Ω such that $P(A_1) = P(A_2) = P(A_3) = 15/64$, $P(A_4) = 1/64$, $P(A_5) = 18/64$. Define $B = A_1 \cup A_4$, $C = A_2 \cup A_4$, $D = A_3 \cup A_4$. Check that

$$P(BCD) = P(B)P(C)P(D)$$

but that B, C, D are not independent.

c) Let X_1, X_2 be independent random variables each assuming only the values $+1$ and -1 with probability 1/2. Are X_1, X_2, X_1X_2 pairwise independent? Are X_1, X_2, X_1X_2 an independent collection?

21. Suppose $\{A_n\}$ is a sequence of events.

(a) If $P(A_n) \to 1$ as $n \to \infty$, prove there exists a subsequence $\{n_k\}$ tending to infinity such that $P(\cap_k A_{n_k}) > 0$. (Hint: Use Borel-Cantelli.)

(b) Show the following is false: Given $\epsilon > 0$ such that $P(A_n) \geq \epsilon$, it follows that there exists a subsequence $\{n_k\}$ tending to infinity such that $P(\cap_k A_{n_k}) > 0$.

22. Suppose $\{A_n\}$ are independent events such that

$$\sum_{n=1}^{\infty} \Big(P(A_n) \bigwedge (1 - P(A_n))\Big) = \infty.$$

Show P is non-atomic.

23. Suppose $\{A_n\}$ are events.

(a) If for each k

$$\sum_{n=k}^{\infty} P(A_n | \bigcap_{i=k}^{n-1} A_i^c) = \infty,$$

show

$$P(\limsup_{n\to\infty} A_n) = 1.$$

(b) What is the relevance to the Borel Zero-One Law?

(c) Is it enough to assume

$$\sum_{n=1}^{\infty} P(A_n | \bigcap_{i=1}^{n-1} A_i^c) = \infty?$$

(d) Show

$$P(\limsup_{n\to\infty} A_n) = 1$$

iff

$$\sum_{n=1}^{\infty} P(AA_n) = \infty$$

for all events A such that $P(A) > 0$.

24. If $P(A_n) \geq \epsilon > 0$, for all large n, then $P(A_n \text{ i.o.}) \geq \epsilon$.

25. Use Renyi's theorem to prove that if $\{X_n, n \geq 1\}$ is iid with common continuous distribution

$$P\{[X_n = \bigvee_{i=1}^{n} X_i] \text{ i.o.}\} = 1.$$

26. (Barndorff-Nielsen) Suppose $\{E_n\}$ is a sequence of events such that

$$\lim_{n\to\infty} P(E_n) = 0, \quad \sum_n P(E_n E_{n+1}^c) < \infty.$$

Prove

$$P(E_n \text{ i.o.}) = 0.$$

Hint: Decompose $\cup_{j=n}^{m} E_j$ for $m > n$.

27. If $\{X_n, n \geq 1\}$ are independent random variables, show that the radius of convergence of the power series $\sum_{n=1}^{\infty} X_n z^n$ is a constant (possibly infinite) with probability one.

Hint: The radius of convergence of $\sum_{n=1}^{\infty} c_n z^n$ is given by

$$R^{-1} = \limsup_{n\to\infty} |c_n|^{1/n}.$$

28. Show $\{X_n, n \geq 1\}$ are independent if

$$\sigma(X_1, \dots, X_{n-1}) \perp\!\!\!\perp \sigma(X_n)$$

are independent for each $n \geq 2$.

29. Let

$$\Omega = \{1, \dots, r\}^n = \{(x_1, \dots, x_n) : x_i \in \{1, \dots, r\}, i = 1, \dots, n\}$$

and assume an assignment of probabilities such that each point of Ω is equally likely. Define the coordinate random variables

$$X_i((x_1, \dots, x_n)) = x_i, \quad 1 = 1, \dots, n.$$

Prove that the random variables $X_1, \dots, X_n$ are independent.

30. Refer to Subsection 4.3.2.

(a) Define

$$A = \{[d_{2n} = 0] \text{ i.o. }\}, \quad B = \{[d_{2n+1} = 1] \text{ i.o. }\}.$$

Show $A \perp\!\!\!\perp B$.

(b) Define

$$\begin{aligned} l_n(\omega) &:= \text{length of the run of 0's starting at } d_n(\omega), \\ &= \begin{cases} k \geq 1, & \text{if } d_n(\omega) = 0, \dots, d_{n+k-1}(\omega) = 0, d_{n+k}(\omega) = 1, \\ 0, & \text{if } d_n(\omega) = 1. \end{cases} \end{aligned}$$

Show

$$P[l_n = k] = \left(\frac{1}{2}\right)^{k+1}, \quad P[l_n \geq r] = \left(\frac{1}{2}\right)^r. \tag{4.18}$$

(c) Show $\{[l_n = 0], n \geq 1\}$ are independent events.

(d) Show $P\{[l_n = 0] \text{ i.o. }\} = 1$. (Use the Borel Zero-One Law.)

(e) The events $\{[l_n = 1], n \geq 1\}$ are not independent but the events $\{[l_{2n} = 1], n \geq 1\}$ are, and therefore prove

$$P\{[l_{2n} = 1] \text{ i.o. }\} = 1$$

so that

$$P\{[l_n = 1] \text{ i.o. }\} = 1.$$

(f) Let $\log_2 n$ be the logarithm to the base 2 of n. Show

$$P[\limsup_{n\to\infty} \frac{l_n}{\log_2 n} \leq 1] = 1. \tag{4.19}$$

Hint: Show

$$\sum_n P[l_n > (1+\epsilon)\log_2 n] < \infty$$

and use Borel-Cantelli. Then replace ϵ by $\epsilon_k \downarrow 0$.

(g) Show

$$P[\limsup_{n\to\infty} \frac{l_n}{\log_2 n} \geq 1] = 1.$$

Combine this with (4.19).

Hint: Set $r_n = [\log_2 n]$ and define n_k by $n_1 = 1, n_2 = 1+r_1, \dots, n_{k+1} = n_k + r_{n_k}$ so that $n_{k+1} - n_k = r_{n_k}$. Then

$$[l_{n_k} \geq r_{n_k}] \in \mathcal{B}(d_i, n_k \leq i < n_{k+1})$$

and hence $\{[l_{n_k} \geq r_{n_k}], k \geq 1\}$ are independent events. Use the Borel Zero-One Law to show

$$P\{[l_{n_k} \geq r_{n_k}] \text{ i.o. }\} = 1$$

and hence

$$P\{[l_n \geq r_n] \text{ i.o. }\} = 1.$$

31. Suppose $\{B_n, n \geq 1\}$ is a sequence of events such that for some $\delta > 0$

$$P(B_n) \geq \delta > 0,$$

for all $n \geq 1$. Show $\limsup_{n\to\infty} B_n \neq \emptyset$. Use this to help show with minimum calculation that in an infinite sequence of independent Bernoulli trials, there is an infinite number of successes with probability one.

32. **The Renyi representation.** Suppose $E_1, \dots, E_n$ are iid exponentially distributed random variables with parameter $\lambda > 0$ so that

$$P[E_1 \leq x] = 1 - e^{-\lambda x}, \quad x > 0.$$

Let

$$E_{1,n} \leq E_{2,n} \leq \cdots \leq E_{n,n}$$

be the order statistics. Prove the n spacings

$$E_{1,n}, E_{2,n} - E_{1,n}, \dots, E_{n,n} - E_{n-1,n}$$

are independent exponentially distributed random variables where $E_{k+1,n} - E_{k,n}$ has parameter $(n-k)\lambda$. Intuitively, this results from the forgetfulness property of the exponential distribution.

5

Integration and Expectation

One of the more fundamental concepts of probability theory and mathematical statistics is the *expectation* of a random variable. The expectation represents a central value of the random variable and has a measure theory counterpart in the theory of integration.

5.1 Preparation for Integration

5.1.1 *Simple Functions*

Many integration results are proved by first showing they hold true for *simple* functions and then extending the result to more general functions. Recall that a function on the probability space $(\Omega, \mathcal{B}, P)$

$$X : \Omega \mapsto \mathbb{R}$$

is *simple* if it has a finite range. Henceforth, assume that a simple function is $\mathcal{B}/\mathcal{B}(\mathbb{R})$ measurable. Such a function can always be written in the form

$$X(\omega) = \sum_{i=1}^{k} a_i 1_{A_i}(\omega),$$

where $a_i \in \mathbb{R}$ and $A_i \in \mathcal{B}$ and $A_1, \ldots, A_k$ are disjoint and $\sum_{i=1}^{k} A_i = \Omega$.

Recall

$$\sigma(X) = \sigma(A_i, i = 1, \ldots, k) = \left\{ \bigcup_{i \in I} A_i : I \subset \{1, \ldots, k\} \right\}.$$

Let $\mathcal{E}$ be the set of all simple functions on Ω. We have the following important properties of $\mathcal{E}$.

1. $\mathcal{E}$ is a vector space. This means the following two properties hold.
   (a) If $X = \sum_{i=1}^k a_i 1_{A_i} \in \mathcal{E}$, then $\alpha X = \sum_{i=1}^k \alpha a_i 1_{A_i} \in \mathcal{E}$.
   (b) If $X = \sum_{i=1}^k a_i 1_{A_i}$ and $Y = \sum_{i=1}^m b_j 1_{B_j}$ and $X, Y \in \mathcal{E}$, then

$$X + Y = \sum_{i,j} (a_i + b_j) 1_{A_i \cap B_j}$$

   and $\{A_i B_j, 1 \le i \le k,\ 1 \le j \le m\}$ is a partition of Ω. So $X+Y \in \mathcal{E}$.

2. If $X, Y \in \mathcal{E}$, then $XY \in \mathcal{E}$ since

$$XY = \sum_{i,j} a_i b_j 1_{A_i \cap B_j}.$$

3. If $X, Y \in \mathcal{E}$, then $X \vee Y$, $X \wedge Y \in \mathcal{E}$, since, for instance,

$$X \bigvee Y = \sum_{i,j} a_i \vee b_j 1_{A_i B_j}.$$

5.1.2 *Measurability and Simple Functions*

The following result shows that any measurable function can be approximated by a simple function. It is the reason why it is often the case that an integration result about random varables is proven first for simple functions.

Theorem 5.1.1 (Measurability Theorem) *Suppose $X(\omega) \ge 0$, for all ω. Then $X \in \mathcal{B}/\mathcal{B}(\mathbb{R})$ iff there exist simple functions $X_n \in \mathcal{E}$ and*

$$0 \le X_n \uparrow X.$$

Proof. If $X_n \in \mathcal{E}$, then $X_n \in \mathcal{B}/\mathcal{B}(\mathbb{R})$, and if $X = \lim_{n\to\infty} \uparrow X_n$, then $X \in \mathcal{B}/\mathcal{B}(\mathbb{R})$ since taking limits preserves measurability.

Conversely, suppose $0 \le X \in \mathcal{B}/\mathcal{B}(\mathbb{R})$. Define

$$X_n := \sum_{k=1}^{n2^n} \left(\frac{k-1}{2^n}\right) 1_{[\frac{k-1}{2^n} \le X < \frac{k}{2^n}]} + n 1_{[X \ge n]}.$$

Because $X \in \mathcal{B}/\mathcal{B}(\mathbb{R})$, it follows that $X_n \in \mathcal{E}$. Also $X_n \le X_{n+1}$ and if $X(\omega) < \infty$, then for all large enough n

$$|X(\omega) - X_n(\omega)| \le \frac{1}{2^n} \to 0.$$

If $X(\omega) = \infty$, then $X_n(\omega) = n \to \infty$. □

Note if

$$M := \sup_{\omega\in\Omega} |X(\omega)| < \infty,$$

then

$$\sup_{\omega\in\Omega} |X(\omega) - X_n(\omega)| \to 0.$$

5.2 Expectation and Integration

This section takes up the definition of expectation, or in measure theory terms, the *Lebesgue-Stieltjes integral.* Suppose $(\Omega, \mathcal{B}, P)$ is a probability space and

$$X : (\Omega, \mathcal{B}) \mapsto (\bar{\mathbb{R}}, \mathcal{B}(\bar{\mathbb{R}}))$$

where $\bar{\mathbb{R}} = [-\infty, \infty]$ so X might have $\pm\infty$ in its range. We will define the expectation of X, written $E(X)$ or

$$\int_\Omega X dP$$

or

$$\int_\Omega X(\omega)P(d\omega),$$

as the Lebesgue-Stieltjes integral of X with respect to P. We will do this in stages, the first stage being to define the integral for simple functions.

5.2.1 Expectation of Simple Functions

Suppose X is a simple random variable of the form

$$X = \sum_{i=1}^{n} a_i 1_{A_i},$$

where $|a_i| < \infty$, and $\sum_{i=1}^{k} A_i = \Omega$. Define for $X \in \mathcal{E}$ the expectation as

$$E(X) \equiv \int X dP =: \sum_{i=1}^{k} a_i P(A_i). \tag{5.1}$$

Note this definition coincides with your knowledge of discrete probability from more elementary courses. For a simple function (which includes all random variables on finite probabilty spaces) the expectation is computed by taking a possible value, multiplying by the probability of the possible value and then summing over all possible values.

We now discuss the properties arising from this definition (5.1).

1. We have that
$$E(1) = 1, \text{ and } E(1_A) = P(A).$$
This follows since $1 = 1_\Omega$ so $E(1) = P(\Omega) = 1$ and
$$1_A = 1 \cdot 1_A + 0 \cdot 1_{A^c},$$
so
$$E(1_A) = 1P(A) + 0P(A^c).$$

2. If $X \geq 0$ and $X \in \mathcal{E}$ then $E(X) \geq 0$.
To verify this, note that if $X \geq 0$, then
$$X = \sum_{i=1}^k a_i 1_{A_i}, \text{ and } a_i \geq 0,$$
and therefore $E(X) = \sum_{i=1}^k a_i P(A_i) \geq 0$.

3. The expectation operator E is linear in the sense that if $X, Y \in \mathcal{E}$, then
$$E(\alpha X + \beta Y) = \alpha E(X) + \beta E(Y)$$
for $\alpha, \beta \in \mathbb{R}$.
To check this, suppose
$$X = \sum_{i=1}^k a_i 1_{A_i}, \quad Y = \sum_{j=1}^m b_j 1_{B_j},$$
and then
$$\alpha X + \beta Y = \sum_{i,j} (\alpha a_i + \beta b_j) 1_{A_i B_j},$$
so that
$$\begin{aligned} E(\alpha X + \beta Y) &= \sum_{i,j} (\alpha a_i + \beta b_j) P(A_i B_j) \\ &= \sum_{i,j} \alpha a_i P(A_i B_j) + \sum_{i,j} \beta b_j P(A_i B_j) \\ &= \alpha \sum_{i=1}^k a_i \sum_{j=1}^m P(A_i B_j) + \beta \sum_{j=1}^m b_j \sum_{i=1}^k P(A_i B_j) \\ &= \alpha \sum_{i=1}^k a_i P(A_i) + \beta \sum_{j=1}^m b_j P(B_j) \\ &= \alpha E(X) + \beta E(Y). \end{aligned}$$

4. The expectation operator E is monotone on $\mathcal{E}$ in the sense that if $X \leq Y$ and $X, Y \in \mathcal{E}$, then $E(X) \leq E(Y)$.
   To prove this, we observe that we have $Y - X \geq 0$ and $Y - X \in \mathcal{E}$. So $E(Y - X) \geq 0$ from property 2, and thus
$$E(Y) = E(Y - X + X) = E(Y - X) + E(X) \geq EX$$
   since $E(Y - X) \geq 0$.

5. If $X_n, X \in \mathcal{E}$ and either $X_n \uparrow X$ or $X_n \downarrow X$, then
$$E(X_n) \uparrow E(X) \text{ or } E(X_n) \downarrow E(X).$$

Suppose $X_n \in \mathcal{E}$, and $X_n \downarrow 0$. We prove $E(X_n) \downarrow 0$. As a consequence of being simple, X_1 has a finite range. We may suppose without loss of generality that
$$\sup_{\omega \in \Omega} X_1(\omega) = K < \infty.$$
Since $\{X_n\}$ is non-increasing, we get that
$$0 \leq X_n \leq K$$
for all n. Thus for any $\epsilon > 0$,
$$\begin{aligned} 0 \leq X_n &= X_n 1_{[X_n > \epsilon]} + X_n 1_{[X_n \leq \epsilon]} \\ &\leq K 1_{[X_n > \epsilon]} + \epsilon 1_{[X_n \leq \epsilon]}, \end{aligned}$$
and therefore by the monotonicity property 4,
$$\begin{aligned} 0 \leq E(X_n) &\leq K P[X_n > \epsilon] + \epsilon P[X_n \leq \epsilon] \\ &\leq K P[X_n > \epsilon] + \epsilon. \end{aligned}$$
Since $X_n \downarrow 0$, we have, as $n \to \infty$,
$$[X_n > \epsilon] \downarrow \emptyset,$$
and by continuity of P
$$P[X_n > \epsilon] \downarrow 0.$$
So $E(X_n) \geq E(X_{n+1})$ and
$$\limsup_{n \to \infty} E(X_n) \leq \epsilon.$$
Since ϵ is arbitrary, $E(X_n) \downarrow 0$.
If $X_n \downarrow X$, then $X_n - X \downarrow 0$, so
$$E(X_n) - E(X) = E(X_n - X) \downarrow 0$$
from the previous step.
If $X_n \uparrow X$, then $X - X_n \downarrow 0$ and
$$E(X) - E(X_n) = E(X - X_n) \downarrow 0.$$
□

5.2.2 *Extension of the Definition*

We now extend the definition of the integral beyond simple functions. The program is to define expectation for all positive random variables and then for all *integrable* random variables. The term *integrable* will be explained later.

It is convenient and useful to assume our random variables take values in the extended real line $\bar{\mathbb{R}}$ (cf. Exercise 33). In stochastic modeling, for instance, we often deal with waiting times for an event to happen or return times to a state or set. If the event never occurs, it is natural to say the waiting time is infinite. If the process never returns to a state or set, it is natural to say the return time is infinite.

Let $\mathcal{E}_+$ be the non-negative valued simple functions, and define

$$\bar{\mathcal{E}_+} := \{X \geq 0 : X : (\Omega, \mathcal{B}) \mapsto (\bar{\mathbb{R}}, \mathcal{B}(\bar{\mathbb{R}}))\}$$

to be non-negative, measurable functions with domain Ω. If $X \in \bar{\mathcal{E}_+}$ and $P[X = \infty] > 0$, define $E(X) = \infty$.

Otherwise by the measurability theorem (Theorem 5.1.1, page 118), we may find $X_n \in \mathcal{E}_+$, such that

$$0 \leq X_n \uparrow X.$$

We call $\{X_n\}$ the *approximating sequence* to X. The sequence $\{E(X_n)\}$ is non-decreasing by monotonicity of expectations applied to $\mathcal{E}_+$. Since limits of monotone sequences always exist, we conclude that $\lim_{n\to\infty} E(X_n)$ exists. We define

$$E(X) := \lim_{n\to\infty} E(X_n). \tag{5.2}$$

This extends expectation from $\mathcal{E}$ to $\bar{\mathcal{E}_+}$.

The next result discusses this definition further.

Proposition 5.2.1 (Well definition) *E is well defined on $\bar{\mathcal{E}_+}$, since if $X_n \in \mathcal{E}_+$ and $Y_m \in \mathcal{E}_+$ and $X_n \uparrow X$, $\quad Y_m \uparrow X$, then*

$$\lim_{n\to\infty} E(X_n) = \lim_{n\to\infty} E(Y_m).$$

Proof. Because of the symmetric roles of X_n and Y_m, it suffices to prove the following assertion. If $X_n, Y_m \in \mathcal{E}_+$ and both sequences $\{X_n\}$ and $\{Y_m\}$ are non-decreasing, then the assumption

$$\lim_{n\to\infty} \uparrow X_n \leq \lim_{m\to\infty} \uparrow Y_m \tag{5.3}$$

implies

$$\lim_{n\to\infty} \uparrow E(X_n) \leq \lim_{m\to\infty} \uparrow E(Y_m). \tag{5.4}$$

To prove (5.4), note that as $m \to \infty$

$$\mathcal{E}_+ \ni X_n \wedge Y_m \uparrow X_n \in \mathcal{E}_+,$$

since

$$\lim_{m\to\infty} Y_m \geq \lim_{m\to\infty} X_m \geq X_n.$$

So from monotonicity of expectations on $\mathcal{E}_+$,

$$E(X_n) = \lim_{m\to\infty} \uparrow E(X_n \wedge Y_m) \leq \lim_{m\to\infty} E(Y_m).$$

This is true for all n, so let $n \to \infty$ to obtain

$$\lim_{n\to\infty} \uparrow E(X_n) \leq \lim_{n\to\infty} \uparrow E(Y_n).$$

This proves (5.4). □

5.2.3 *Basic Properties of Expectation*

We now list some properties of the expectation operator applied to random variables in $\bar{\mathcal{E}}_+$.

1. We have
$$0 \leq E(X) \leq \infty,$$
and if $X, Y \in \bar{\mathcal{E}}_+$ and $X \leq Y$, then $E(X) \leq E(Y)$.
The proof is contained in (5.4).

2. E is linear: For $\alpha > 0$ and $\beta > 0$,
$$E(\alpha X + \beta Y) = \alpha E(X) + \beta E(Y).$$
To check this, suppose $X_n \uparrow X$, $Y_n \uparrow Y$ and $X_n, Y_n \in \mathcal{E}_+$. For $c > 0$
$$\begin{aligned} E(cX) &= \lim_{n\to\infty} E(cX_n) \\ &= \lim_{n\to\infty} cE(X_n) \quad \text{(linearity on } \mathcal{E}_+) \\ &= cE(X). \end{aligned}$$
We also have
$$\begin{aligned} E(X+Y) &= \lim_{n\to\infty} E(X_n + Y_n) \\ &= \lim_{n\to\infty} (E(X_n) + E(Y_n)) \quad \text{(linearity on } \mathcal{E}_+) \\ &= E(X) + E(Y). \end{aligned}$$

3. **Monotone Convergence Theorem (MCT).** If
$$0 \leq X_n \uparrow X, \tag{5.5}$$

then

$$E(X_n) \uparrow E(X),$$

or equivalently,

$$E\left(\lim_{n\to\infty} \uparrow X_n\right) = \lim_{n\to\infty} \uparrow E(X_n).$$

We now focus on proving this version of the Monotone Convergence Theorem, which allows the interchange of limits and expectations.

Proof of MCT. Suppose we are given $X_n, X \in \bar{\mathcal{E}}_+$ satisfying (5.5). We may find simple functions $Y_m^{(n)} \in \mathcal{E}_+$, to act as approximations to X_n such that

$$Y_m^{(n)} \uparrow X_n, \quad m \to \infty.$$

We need to find a sequence of simple functions $\{Z_m\}$ approximating X

$$Z_m \uparrow X,$$

which can be expressed in terms of the approximations to $\{X_n\}$. So define

$$Z_m = \bigvee_{n \le m} Y_m^{(n)}.$$

Note that $\{Z_m\}$ is non-decreasing since

$$\begin{aligned} Z_m &\le \bigvee_{n\le m} Y_{m+1}^{(n)} \quad (\text{since } Y_m^{(n)} \le Y_{m+1}^{(n)}) \\ &\le \bigvee_{n\le m+1} Y_{m+1}^{(n)} = Z_{m+1}. \end{aligned}$$

Next observe that for $n \le m$,

(A) $Y_m^{(n)} \le \bigvee_{j\le m} Y_m^{(j)} = Z_m$;

(B) $Z_m \le \bigvee_{j\le m} X_j = X_m$,
since $Y_m^{(j)} \le X_j$, which is monotone in j and so

(C) $Y_m^{(n)} \le Z_m \le X_m$.
By taking limits on m in (C), we conclude that for all n

$$X_n = \lim_{m\to\infty} Y_m^{(n)} \le \lim_{m\to\infty} Z_m \le \lim_{m\to\infty} X_m.$$

So

$$X = \lim_{n\to\infty} X_n \le \lim_{m\to\infty} Z_m \le \lim_{m\to\infty} X_m = X.$$

Therefore

(D) $X = \lim_{n\to\infty} X_n = \lim_{m\to\infty} Z_m$
and it follows that $\{Z_m\}$ is a simple function approximation to X.

(E) Now because expectation is monotone on $\mathcal{E}$,

$$\begin{aligned} E(X_n) &= \lim_{m\to\infty} \uparrow E(Y_m^{(n)}) && \text{(expectation definition)} \\ &\le \lim_{m\to\infty} \uparrow E(Z_m) && \text{(from (C))} \\ &\le \lim_{m\to\infty} \uparrow E(X_m) && \text{(from (C))}. \end{aligned}$$

However $Z_m \in \mathcal{E}_+$ and $\{Z_m\}$ is a simple function approximation to X. Therefore, we get from the definition of expectation on $\bar{\mathcal{E}}_+$ and (D)

$$E(X) = E(\lim_{m\to\infty} \uparrow Z_m) = \lim_{m\to\infty} \uparrow E(Z_m).$$

So (E) implies for all n that

$$E(X_n) \le E(X) \le \lim_{m\to\infty} \uparrow E(X_m),$$

and taking the limit on n

$$\lim_{n\to\infty} E(X_n) \le E(X) \le \lim_{m\to\infty} E(X_m),$$

hence the desired equality follows. □

We now further extend the definition of $E(X)$ beyond $\bar{\mathcal{E}}_+$. For a random variable X, define

$$X^+ = X \vee 0, \quad X^- = (-X) \vee 0. \tag{5.6}$$

Thus

$$\begin{aligned} X^+ &= X, \quad \text{if } X \ge 0 \text{ (and then } X^- = 0), \\ X^- &= -X, \ \text{if } X \le 0 \text{ (and then } X^+ = 0). \end{aligned}$$

Therefore

$$X^\pm \ge 0,$$

and

$$|X| = X^+ + X^-$$

and

$$X \in \mathcal{B}/\mathcal{B}(\mathbb{R}) \text{ iff both } X^\pm \in \mathcal{B}/\mathcal{B}(\mathbb{R}).$$

Call X *quasi-integrable* if at least one of $E(X^+), E(X^-)$ is finite. In this case, define

$$E(X) := E(X^+) - E(X^-).$$

If $E(X^+)$ and $E(X^-)$ are both finite, call X *integrable*. This is the case iff $E|X| < \infty$. The set of integrable random variables is denoted by L_1 or $L_1(P)$ if the probability measure needs to be emphasized. So

$$L_1(P) = \{\text{random variables } X : E|X| < \infty\}.$$

If $E(X^+) < \infty$ but $E(X^-) = \infty$ then $E(X) = -\infty$. If $E(X^+) = \infty$ but $E(X^-) < \infty$, then $E(X) = \infty$. If $E(X^+) = \infty$ and $EX^- = \infty$, then $E(X)$ **does not exist.**

Example 5.2.1 (Heavy Tails) We will see that when the distribution function of a random variable X has a density $f(x)$, the expectation, provided it exists, can be computed by the familiar formula

$$E(X) = \int x f(x)dx.$$

If

$$f(x) = \begin{cases} x^{-2}, & \text{if } x > 1, \\ 0, & \text{otherwise,} \end{cases}$$

then $E(X)$ exists and $E(X) = \infty$.

On the other hand, if

$$f(x) = \begin{cases} \frac{1}{2}|x|^{-2}, & \text{if } |x| > 1, \\ 0, & \text{otherwise ,} \end{cases}$$

then

$$E(X^+) = E(X^-) = \infty,$$

and $E(X)$ does not exist. The same conclusion would hold if f were the Cauchy density

$$f(x) = \frac{1}{\pi(1+x^2)}, \quad x \in \mathbb{R}.$$

□

We now list some properties of the expectation operator E.

1. If X is integrable, then $P[X = \pm\infty] = 0$.

   For example, if $P[X = \infty] > 0$, then $E(X^+) = \infty$ and X is not integrable.

2. If $E(X)$ exists,

   $$E(cX) = cE(X).$$

   If either

   $$E(X^+) < \infty \text{ and } E(Y^+) < \infty,$$

or

$$E(X^-) < \infty \text{ and } E(Y^-) < \infty,$$

then $X + Y$ is quasi-integrable and

$$E(X+Y) = E(X) + E(Y).$$

We verify additivity when $X, Y \in L_1$. Observe that

$$|X+Y| \in \bar{\mathcal{E}}_+,$$

and since

$$|X+Y| \leq |X| + |Y|,$$

we have from monotonicity of expectation on $\bar{\mathcal{E}}_+$ that

$$E|X+Y| \leq E(|X|+|Y|) = E|X| + E|Y| < \infty,$$

the last equality following from linearity on $\bar{\mathcal{E}}_+$. Hence $X + Y \in L_1$.

Next, we have

$$(X+Y)^+ - (X+Y)^- = X + Y = X^+ - X^- + Y^+ - Y^-, \tag{5.7}$$

so

$$\text{LHS} := (X+Y)^+ + X^- + Y^- = (X+Y)^- + X^+ + Y^+ =: \text{RHS}.$$

The advantage of this over (5.7) is that both LHS and RHS are sums of positive random variables. Since expectation is linear on $\bar{\mathcal{E}}_+$, we have

$$\begin{aligned} E(\text{ LHS }) &= E(X+Y)^+ + E(X^-) + E(Y^-) \\ &= E(\text{ RHS }) = E(X+Y)^- + E(X^+) + E(Y^+). \end{aligned}$$

Rearranging we get

$$E(X+Y)^+ - E(X+Y)^- = E(X^+) - E(X^-) + E(Y^+) - E(Y^-),$$

or equivalently,

$$E(X+Y) = E(X) + E(Y).$$

□

3. If $X \geq 0$, then $E(X) \geq 0$ since $X = X^+$. If $X, Y \in L_1$, and $X \leq Y$, then

$$E(X) \leq E(Y).$$

This is readily seen. We have $E(Y - X) \geq 0$ since $Y - X \geq 0$, and thus by property (2) from this list,

$$E(Y-X) = E(Y) - E(X) \geq 0.$$

4. Suppose $\{X_n\}$ is a sequence of random variables such that $X_n \in L_1$ for some n. If either
$$X_n \uparrow X$$
or
$$X_n \downarrow X,$$
then according to the type of monotonicity
$$E(X_n) \uparrow E(X)$$
or
$$E(X_n) \downarrow E(X).$$

To see this in the case $X_n \uparrow X$, note $X_n^- \downarrow X^-$ so $E(X^-) < \infty$. Then
$$0 \le X_n^+ = X_n + X_n^- \le X_n + X_1^- \uparrow X + X_1^-.$$
From the MCT given in equation (5.5)
$$0 \le E(X_n + X_1^-) \uparrow E(X + X_1^-).$$
From property 2 we have
$$E(X_n + X_1^-) = E(X_n) + E(X_1^-).$$
Since $E(X^-) < \infty$ and $E(X_1^-) < \infty$, we also have
$$E(X + X_1^-) = E(X) + E(X_1^-),$$
and thus
$$\lim_{n\to\infty} E(X_n) = E(X).$$

If $X_n \downarrow X$, proceed similarly by considering $-X_n + X_1^+$.

5. **Modulus Inequality.** If $X \in L_1$,
$$|E(X)| \le E(|X|).$$

This has an easy proof. We have
$$|E(X)| = |E(X^+) - E(X^-)| \le E(X^+) + E(X^-) = E(|X|).$$

6. **Variance and Covariance.** Suppose $X^2 \in L_1$, which we write as $X \in L_2$. Recall that we define
$$\mathrm{Var}(X) := E\big(X - E(X)\big)^2,$$

and because of linearity of expectation, this is equal to

$$= E(X^2) - \left(E(X)\right)^2.$$

For random variables $X, Y \in L_2$, write

$$\text{Cov}(X, Y) = E\left((X - E(X))(Y - E(Y))\right),$$

and again because of linearity of expectation, this is equal to

$$= E(XY) - E(X)E(Y).$$

Note that if $X = Y$, then $\text{Cov}(X, Y) = \text{Var}(X)$. We will prove in Example 5.9.2 that if $X \perp\!\!\!\perp Y$ and $X, Y \in L_2$, then $E(XY) = E(X)E(Y)$ and consequently $\text{Cov}(X, Y) = 0$.

The covariance is a bilinear function in the sense that if $X_1, \ldots, X_k$ and $Y_1, \ldots, Y_l$ are L_2 random variables, then for constants $a_1, \ldots, a_k$, and $b_1, \ldots, b_l$

$$\text{Cov}\Big(\sum_{i=1}^{k} a_i X_i, \sum_{j=1}^{l} b_j Y_j\Big) = \sum_{i=1}^{k}\sum_{j=1}^{l} a_i b_j \text{Cov}(X_i, Y_j). \tag{5.8}$$

This also follows from linearity of expectations as follows. Without loss of generality, we may suppose that $E(X_i) = E(Y_j) = 0$ for $i = 1, \ldots, k$, $j = 1, \ldots, l$. Then

$$\begin{aligned}\text{Cov}\Big(\sum_{i=1}^{k} a_i X_i, \sum_{j=1}^{l} b_j Y_j\Big) &= E\left(\sum_{i=1}^{k} a_i X_i \sum_{j=1}^{l} b_j Y_j\right)\\ &= \sum_{i=1}^{k}\sum_{j=1}^{l} a_i b_j E(X_i Y_j)\\ &= \sum_{i=1}^{k}\sum_{j=1}^{l} a_i b_j \text{Cov}(X_i, Y_j).\end{aligned}$$

A special case of this formula is used for computing the variance of a sum of L_2 random variables $X_1, \ldots, X_n$. We have

$$\text{Var}\Big(\sum_{i=1}^{n} X_i\Big) = \text{Cov}\Big(\sum_{i=1}^{n} X_i, \sum_{j=1}^{n} X_j\Big) = \sum_{i=1}^{n}\sum_{j=1}^{n} \text{Cov}(X_i, X_j).$$

Split the index set

$$\{(i, j) : 1 \le i, j \le n\} = \{(i, i) : 1 \le i \le n\} \cup \{(i, j) : i \ne j\},$$

and we get

$$\begin{aligned}\mathrm{Var}(\sum_{i=1}^{n} X_i) &= \sum_{i=1}^{n} \mathrm{Cov}(X_i, X_i) + 2 \sum_{1 \le i < j \le n} \mathrm{Cov}(X_i, X_j) \\ &= \sum_{i=1}^{n} \mathrm{Var}(X_i) + 2 \sum_{1 \le i < j \le n} \mathrm{Cov}(X_i, X_j). \qquad (5.9)\end{aligned}$$

If $\mathrm{Cov}(X_i, X_j) = 0$ for $i \neq j$, that is, if $X_1, \dots, X_n$ are *uncorrelated*, then

$$\mathrm{Var}\Big(\sum_{i=1}^{n} X_i\Big) = \sum_{i=1}^{n} \mathrm{Var}(X_i). \qquad (5.10)$$

In particular, this is true when $X_1, \dots, X_n$ are independent.

7. **Markov Inequality.** Suppose $X \in L_1$. For any $\lambda > 0$

$$P[|X| \ge \lambda] \le \lambda^{-1} E(|X|).$$

This proof is also simple. Note

$$1 \cdot 1_{[\frac{|X|}{\lambda} \ge 1]} \le \frac{|X|}{\lambda} \cdot 1_{[|\frac{X}{\lambda}| \ge 1]} \le \frac{|X|}{\lambda}.$$

Take expectations through the inequalities.

8. **Chebychev Inequality.** We have

$$P[|X - E(X)| \ge \lambda] \le \mathrm{Var}(X)/\lambda^2,$$

assuming $E|X| < \infty$ and $\mathrm{Var}(X) < \infty$.
This follows from the Markov Inequality. We have

$$\begin{aligned} P[|X - E(X)| \ge \lambda] &= P[|X - E(X)|^2 \ge \lambda^2] \\ &\le \lambda^{-2} E(X - E(X))^2, \end{aligned}$$

where the last inequality is an application of the Markov Inequality.

9. **Weak Law Large Numbers (WLLN).** This is one way to express the fact that the sample average of an iid sequence approximates the mean. Let $\{X_n, n \ge 1\}$ be iid with finite mean and variance and suppose $E(X_n) = \mu$ and $\mathrm{Var}(X_n) = \sigma^2 < \infty$. Then for any $\epsilon > 0$,

$$\lim_{n \to \infty} P[|n^{-1} \sum_{i=1}^{n} X_i - \mu| > \epsilon] = 0.$$

To see this, just use Chebychev and (5.10):

$$\begin{aligned} P[\left|n^{-1}\sum_{i=1}^{n} X_i - \mu\right| > \epsilon] &\le \epsilon^{-2}\mathrm{Var}\left(\frac{\sum_{i=1}^{n} X_i}{n}\right) \\ &= \frac{\sum_{i=1}^{n}\mathrm{Var}(X_i)}{n^2\epsilon^2} = \frac{n\mathrm{Var}(X_i)}{n^2\epsilon^2} \\ &= \frac{\sigma^2}{\epsilon^2}\left(\frac{1}{n}\right) \to 0. \end{aligned}$$

□

5.3 Limits and Integrals

This section presents a sequence of results which describe how expectation and limits interact. Under certain circumstances we are allowed to interchange expectation and limits. In this section we will learn when this is safe.

Theorem 5.3.1 (Monotone Convergence Theorem (MCT)) *If*

$$0 \le X_n \uparrow X$$

then

$$0 \le E(X_n) \uparrow E(X).$$

This was proved in the previous subsection 5.2.3. See 3 page 123.

Corollary 5.3.1 (Series Version of MCT) *If $\xi_j \ge 0$ are non-negative random variables for $n \ge 1$, then*

$$E(\sum_{j=1}^{\infty} \xi_j) = \sum_{j=1}^{\infty} E(\xi_j),$$

so that the expectation and infinite sum can be interchanged.

To see this, just write

$$\begin{aligned} E(\sum_{j=1}^{\infty} \xi_j) &= E(\lim_{n\to\infty} \sum_{j=1}^{n} \xi_j) \\ &= \lim_{n\to\infty} \uparrow E(\sum_{j=1}^{n} \xi_j) \qquad \text{(MCT)} \\ &= \lim_{n\to\infty} \uparrow \sum_{j=1}^{n} E(\xi_j) \\ &= \sum_{j=1}^{\infty} E(\xi_j). \end{aligned}$$

□

Theorem 5.3.2 (Fatou Lemma) *If $X_n \geq 0$, then*

$$E(\liminf_{n\to\infty} X_n) \leq \liminf_{n\to\infty} E(X_n).$$

More generally, if there exists $Z \in L_1$ and $X_n \geq Z$, then

$$E(\liminf_{n\to\infty} X_n) \leq \liminf_{n\to\infty} E(X_n).$$

Proof of Fatou. If $X_n \geq 0$, then

$$\begin{aligned} E(\liminf_{n\to\infty} X_n) &= E\left(\lim_{n\to\infty} \uparrow \left(\bigwedge_{k=n}^{\infty} X_k\right)\right) \\ &= \lim_{n\to\infty} \uparrow E\left(\bigwedge_{k=n}^{\infty} X_k\right) \qquad \text{(from MCT 5.3.1)} \\ &\leq \liminf_{n\to\infty} E(X_n). \end{aligned}$$

For the case where we assume $X_n \geq Z$, we have $X_n - Z \geq 0$ and

$$E\left(\liminf_{n\to\infty}(X_n - Z)\right) \leq \liminf_{n\to\infty} E(X_n - Z)$$

so

$$E(\liminf_{n\to\infty} X_n) - E(Z) \leq \liminf_{n\to\infty} E(X_n) - E(Z).$$

The result follows by cancelling $E(Z)$ from both sides of the last relation. □

Corollary 5.3.2 (More Fatou) *If $X_n \leq Z$ where $Z \in L_1$, then*

$$E(\limsup_{n\to\infty} X_n) \geq \limsup_{n\to\infty} E(X_n).$$

Proof. This follows quickly from the previous Fatou Lemma 5.3.2. If $X_n \leq Z$, then $-X_n \geq -Z \in L_1$, and the Fatou Lemma 5.3.2 gives

$$E(\liminf_{n\to\infty}(-X_n)) \leq \liminf_{n\to\infty} E(-X_n),$$

so that

$$E(-\liminf_{n\to\infty}(-X_n)) \geq -\liminf_{n\to\infty}(-EX_n).$$

The proof is completed by using the relation

$$-\liminf - = \limsup.$$

□

Canonical Example. This example is typical of what can go wrong when limits and integrals are interchanged without any dominating condition. Usually something very nasty happens on a small set and the degree of nastiness overpowers the degree of smallness.

Let

$$(\Omega, \mathcal{B}, P) = ([0,1], \mathcal{B}([0,1]), \lambda)$$

where, as usual, λ is Lebesgue measure. Define

$$X_n = n^2 1_{(0,1/n)}.$$

For any $\omega \in [0,1]$,

$$1_{(0,1/n)}(\omega) \to 0,$$

so

$$X_n \to 0.$$

However

$$E(X_n) = n^2 \cdot \frac{1}{n} = n \to \infty,$$

so

$$E(\liminf_{n\to\infty} X_n) = 0 < \liminf_{n\to\infty}(EX_n) = \infty$$

and

$$E(\limsup_{n\to\infty} X_n) = 0, \quad \limsup_{n\to\infty} E(X_n) = \infty.$$

So the second part of the Fatou Lemma given in Corollary 5.3.2 fails. So obviously we cannot hope for Corollary 5.3.2 to hold without any restriction. □

Theorem 5.3.3 (Dominated Convergence Theorem (DCT)) *If*

$$X_n \to X,$$

and there exists a dominating random variable $Z \in L_1$ *such that*

$$|X_n| \le Z,$$

then

$$E(X_n) \to E(X) \text{ and } E|X_n - X| \to 0.$$

Proof of DCT. This is an easy consequence of the Fatou Lemma. We have

$$-Z \le X_n \le Z$$

and $-Z \in L_1$ as well as $Z \in L_1$. So both parts of Fatou's lemma apply:

$$\begin{aligned} E(X) &= E(\liminf_{n\to\infty} X_n) \\ &\le \liminf_{n\to\infty} E(X_n) && \text{(Fatou Lemma 5.3.2)} \\ &\le \limsup_{n\to\infty} E(X_n) && \text{(since inf < sup)} \\ &\le E(\limsup_{n\to\infty} X_n) && \text{(Corollary 5.3.2)} \\ &= E(X). \end{aligned}$$

Thus all inequalities are equality. The rest follows from $|X_n - X| \le 2Z$. □

5.4 Indefinite Integrals

Indefinite integrals allow for integration over only part of the Ω-space. They are simply defined in terms of indicator functions.

Definition 5.4.1 If $X \in L_1$, we define

$$\int_A XdP := E(X1_A)$$

and call $\int_A XdP$ the integral of X over A. Call X the *integrand*.

Suppose $X \geq 0$. For positive integrands, the integral has the following properties:

(1) We have

$$0 \leq \int_A XdP \leq E(X).$$

This is a direct consequence of the monotonicity property of expectations.

(2) We have

$$\int_A XdP = 0$$

iff

$$P(A \cap [X > 0]) = 0.$$

This proof of this important result is assigned in Exercise 6 at the end of the chapter.

(3) If $\{A_n, n \geq 1\}$ is a sequence of disjoint events

$$\int_{\cup_n A_n} XdP = \sum_{n=1}^{\infty} \int_{A_n} XdP. \tag{5.11}$$

To prove (5.11), observe

$$\begin{aligned}\int_{\cup_n A_n} XdP &= E(X1_{\cup_n A_n}) \\ &= E(\sum_{n=1}^{\infty} X1_{A_n}) \\ &= (\sum_{n=1}^{\infty} E(X1_{A_n}) \qquad \text{(from Corollary 5.3.1)} \\ &= \sum_{n=1}^{\infty} \int_{A_n} XdP.\end{aligned}$$

□

(4) If

$$A_1 \subset A_2,$$

then

$$\int_{A_1} X dP \le \int_{A_2} X dP.$$

(5) Suppose $X \in L_1$ and $\{A_n\}$ is a monotone sequence of events. If

$$A_n \nearrow A,$$

then

$$\int_{A_n} X dP \nearrow \int_A X dP$$

while if

$$A_n \searrow A,$$

then

$$\int_{A_n} X dP \searrow \int_A X dP.$$

Property (4) is proved using the monotonicity property of expectations and Property (5) is a direct consequence of the MCT 5.3.1. □

5.5 The Transformation Theorem and Densities

Suppose we are given two measurable spaces $(\Omega, \mathcal{B})$ and $(\Omega', \mathcal{B}')$, and

$$T : (\Omega, \mathcal{B}) \mapsto (\Omega', \mathcal{B}')$$

is a measurable map. P is a probability measure on $\mathcal{B}$. Define $P' := P \circ T^{-1}$ to be the probability measure on $\mathcal{B}'$ given by

$$P'(A') = P(T^{-1}(A')), \quad A' \in \mathcal{B}'.$$

Theorem 5.5.1 (Transformation Theorem) *Suppose*

$$X' : (\Omega', \mathcal{B}') \mapsto (\mathbb{R}, \mathcal{B}(\mathbb{R}))$$

is a random variable with domain Ω'. (Then $X' \circ T : \Omega \to \mathbb{R}$ is also a random variable by composition.)

(i) If $X' \ge 0$, then

$$\int_\Omega X'(T(\omega))P(d\omega) = \int_{\Omega'} X'(\omega')P'(d\omega'), \tag{5.12}$$

where $P' = P \circ T^{-1}$. Equation (5.12) can also be expressed as

$$E(X' \circ T) = E'(X'), \tag{5.13}$$

where E' is the expectation operator computed with respect to P'.

(ii) We have

$$X' \in L_1(P') \text{ iff } X' \circ T \in L_1(P)$$

in which case

$$\int_{T^{-1}(A')} X'(T(\omega))P(d\omega) = \int_{A'} X'(\omega')P'(d\omega'). \tag{5.14}$$

Proof. (i) Typical of many integration proofs, we proceed in a series of steps, starting with X as an indicator function, proceeding to X as a simple function and concluding with X being general.

(a) Suppose

$$X' = 1_{A'}, \quad A' \in \mathcal{B}'.$$

Note

$$X'(T(\omega)) = 1_{A'}(T(w)) = 1_{T^{-1}A'}(\omega),$$

so

$$\begin{aligned}
\int_\Omega X'(T(\omega))P(d\omega) &= \text{Left side of (5.12)} \\
&= \int_\Omega 1_{A'}(T(\omega))P(d\omega) \\
&= \int_\Omega 1_{T^{-1}(A')}(\omega)P(d\omega) \\
&= P(T^{-1}(A')) = P'(A') \\
&= \int_{\Omega'} 1_{A'}(\omega)P'(d\omega') \\
&= \text{Right side (5.12)}.
\end{aligned}$$

(b) Let X' be simple:

$$X' = \sum_{i=1}^{k} a_i' 1_{A_i'}$$

so that

$$\begin{aligned}\int_{\Omega} X'(T\omega)P(d\omega) &= \int_{\Omega}\sum_{i=1}^{k} a_i' 1_{A_i'}(T(\omega))P(d\omega)\\ &= \sum_{i=1}^{k} a_i' \int_{\Omega} 1_{T^{-1}(A_i')}(\omega)P(d\omega)\\ &= \sum_{i=1}^{k} a_i' P(T^{-1}(A_i'))\\ &= \sum_{i=1}^{k} a_i' P'(A_i')\\ &= \int_{\Omega}\sum_{i=1}^{k} a_i' 1_{A_i'}(\omega')P'(d\omega').\end{aligned}$$

(c) Let $X' \geq 0$ be measurable. There exists a sequence of simple functions $\{X_n'\}$, such that

$$X_n' \uparrow X'.$$

Then it is also true that

$$X_n' \circ T \uparrow X' \circ T$$

and

$$\begin{aligned}\text{Left side (5.12)} &= \int_{\Omega} X'(T\omega)P(d\omega)\\ &= \lim_{n\to\infty} \uparrow \int_{\Omega} X_n'(T\omega)P(d\omega) && \text{(MCT)}\\ &= \lim_{n\to\infty} \uparrow \int_{\Omega'} X_n'(\omega')P'(d\omega') && \text{(from Step (b))}\\ &= \int_{\Omega'} X'(\omega')P'(d\omega') && \text{(from MCT).}\end{aligned}$$

The proof of (ii) is similar. To get (5.14) we replace X' in (5.12) by $X'1_{A'}$. □

5.5.1 *Expectation is Always an Integral on $\mathbb{R}$*

Let X be a random variable on the probability space $(\Omega, \mathcal{B}, P)$. Recall that the *distribution* of X is the measure

$$F := P \circ X^{-1}$$

on $(\mathbb{R}, \mathcal{B}(\mathbb{R}))$ defined by ($A \in \mathcal{B}(\mathbb{R})$):

$$F(A) = P \circ X^{-1}(A) = P[X \in A].$$

The *distribution function* of X is

$$F(x) := F((-\infty, x]) = P[X \leq x].$$

Note that the letter "F" is used in two ways. Usually, there will be no confusion.

Here is an excellent example of the use of the Transformation Theorem which allows us to compute the abstract integral

$$E(X) = \int_{\Omega} X dP$$

as

$$E(X) = \int_{\mathbb{R}} xF(dx),$$

which is an integral on $\mathbb{R}$.

Corollary 5.5.1 *(i) If X is an integrable random variable with distribution F, then*

$$E(X) = \int_{\mathbb{R}} xF(dx).$$

(ii) Suppose

$$X : (\Omega, \mathcal{B}) \mapsto (\mathbb{E}, \mathcal{E})$$

is a random element of $\mathbb{E}$ with distribution $F = P \circ X^{-1}$ and suppose

$$g : (\mathbb{E}, \mathcal{E}) \mapsto (\mathbb{R}_+, \mathcal{B}(\mathbb{R}_+))$$

is a non-negative measurable function. The expectation of $g(X)$ is

$$E(g(X)) = \int_{\Omega} g(X(\omega))P(d\omega) = \int_{x \in \mathbb{E}} g(x)F(dx).$$

Proof. (i) For ease of applying the Transformation Theorem, we make the following notational identifications:

$$\begin{aligned} X &: (\Omega, \mathcal{B}) \mapsto (\mathbb{R}, \mathcal{B}(\mathbb{R})), \\ X' &: (\Omega', \mathcal{B}') = (\mathbb{R}, \mathcal{B}(\mathbb{R})) \mapsto (\mathbb{R}, \mathcal{B}(\mathbb{R})), \\ X'(x) &= x, \\ T &= X \\ P' &= P \circ X^{-1} =: F. \end{aligned}$$

According to the conclusion of the Transformation Theorem, we get the equation

$$\int_{\Omega} X'(T(\omega))P(d\omega) = \int_{\Omega'} X'(\omega')P'(d\omega')$$

and with the identifications listed above, the equation becomes

$$\int_{\Omega} X(\omega)P(d\omega) = \int_{\mathbb{R}} xF(dx).$$

(ii) We proceed in stages using the usual progression: start with an indicator function, proceed to simple functions and graduate to general non-negative functions. Here are the stages in more detail.

(a) If $A \in \mathcal{E}$ and $g(x) = 1_A(x)$, then (i) is true. This follows from $F = P \circ X^{-1}$.

(b) Check (i) holds for g simple.

(c) Finish with an application of the MCT. □

The concluding message: Instead of computing expectations on the abstract space Ω, you can always compute them on $\mathbb{R}$ using F, the distribution of X.

5.5.2 *Densities*

Let $\mathbf{X} : (\Omega, \mathcal{B}) \mapsto (\mathbb{R}^k, \mathcal{B}(\mathbb{R}^k))$ be a random vector on $(\Omega, \mathcal{B}, P)$ with distribution F. We say X or F is *absolutely continuous (AC)* if there exists a non-negative function

$$f : (\mathbb{R}^k, \mathcal{B}(\mathbb{R}^k)) \mapsto (\mathbb{R}_+, \mathcal{B}(\mathbb{R}_+))$$

such that

$$F(A) = \int_A f(\mathbf{x}) d\mathbf{x},$$

where $d\mathbf{x}$ stands for Lebesgue measure and the integral is a Lebesgue-Stieltjes integral.

Proposition 5.5.2 *Let* $g : (\mathbb{R}^k, \mathcal{B}(\mathbb{R}^k)) \mapsto (\mathbb{R}_+, \mathcal{B}(\mathbb{R}_+))$ *be a non-negative measurable function. Suppose* $\mathbf{X}$ *is a random vector with distribution* F. *If* F *is AC with density* f, *we have for the expectation of* $g(\mathbf{X})$

$$Eg(\mathbf{X}) = \int_{\mathbb{R}^k} g(\mathbf{x}) f(\mathbf{x}) d\mathbf{x}.$$

Proof. Repeat (a), (b), (c) of the proof of Corollary 5.5.1 (ii) for the case where there is a density. □

5.6 The Riemann vs Lebesgue Integral

Every probability student is first taught to compute expectations using densities and Riemann integrals. How does this the Riemann integral compare with the Lebesgue integral?

Suppose $(-\infty < a < b < \infty)$ and let f be real valued on $(a, b]$. Generally speaking, if f is Riemann integrable on $(a, b]$, then f is Lebesgue integrable on $(a, b]$ and the two integrals are equal. The precise statement is next.

Theorem 5.6.1 (Riemann and Lebesgue) *Suppose* $f : (a, b] \mapsto \mathbb{R}$ *and*

(a) f is $\mathcal{B}((a,b])/\mathcal{B}(\mathbb{R})$ measurable,

(b) f is Riemann-integrable on $(a,b]$.

Let λ be Lebesgue measure on $(a,b]$. Then

(i) $f \in L_1([a,b],\lambda)$. In fact f is bounded.

(ii) The Riemann integral of f equals the Lebesgue integral.

Proof. If f is Riemann integrable, then from analysis we know that f is bounded on $(a,b]$ (and also continuous almost everywhere on $(a,b]$). For an interval I, define

$$f^{\vee}(I) = \sup_{x\in I} f(x), \quad f^{\wedge}(I) = \inf_{x\in I} f(x).$$

Chop up $(a,b]$ into n subintervals $I_1^{(n)},\dots,I_n^{(n)}$ where

$$\begin{aligned} I_1^{(n)} &= (a, a+\frac{b-a}{n}], \\ I_2^{(n)} &= (a+\frac{b-a}{n}, a+\frac{2(b-a)}{n}], \\ &\vdots \\ I_n^{(n)} &= (b-\frac{a-b}{n}, b]. \end{aligned}$$

Define

$$\begin{aligned} \bar{f}_n(x) &= \sum_{j=1}^{n} f^{\vee}(I_j^{(n)}) 1_{I_j^{(n)}}(x), \\ \underline{f}_n(x) &= \sum_{j=1}^{n} f^{\wedge}(I_j^{(n)}) 1_{I_j^{(n)}}(x) \end{aligned}$$

so that $\bar{f}_n, \underline{f}_n$ are simple (and hence measurable) and

$$\underline{f}_n \le f \le \bar{f}_n. \tag{5.15}$$

Define

$$\begin{aligned} \bar{\sigma}_n &= \int_{(a,b]} \bar{f}_n(x)\lambda(dx) = \sum_{j=1}^{n} f^{\vee}(I_j^{(n)})\lambda(I_j^{(n)}) \\ \underline{\sigma}_n &= \int_{(a,b]} \underline{f}_n(x)\lambda(dx) = \sum_{j=1}^{n} f^{\wedge}(I_j^{(n)})\lambda(I_j^{(n)}) \end{aligned}$$

where $\lambda(I_j^{(n)}) = (b-a)/n$. Let

$$I = \int_a^b f(x)dx$$

be the Riemann integral of f. I has upper Riemann approximating sum $\bar{\sigma}_n$ and lower Riemann approximating sum $\underline{\sigma}_n$. Since f is Riemann-integrable, given ϵ, there exists $n_0 = n_0(\epsilon)$ such that $n \geq n_0$ implies

$$|I - \bar{\sigma}_n| \bigvee |I - \underline{\sigma}_n| \leq \epsilon. \tag{5.16}$$

Because of (5.15) and monotonicity of expectations

$$\underline{\sigma}_n = \int_{(a,b]} \underline{f}_n d\lambda \leq \int_{(a,b]} f d\lambda \leq \int_{(a,b]} \bar{f}_n d\lambda = \bar{\sigma}_n,$$

and from (5.16)

$$\bar{\sigma}_n \leq I + \epsilon, \quad \underline{\sigma}_n \geq I - \epsilon$$

so

$$I - \epsilon \leq \int_{(a,b]} f d\lambda \leq I + \epsilon;$$

that is,

$$|\int_{(a,b]} f d\lambda - I| \leq \epsilon.$$

This completes the proof. □

We need the next lemma in order to discuss the subsequent example.

Lemma 5.6.1 (Integral Comparison Lemma) *Suppose X and X' are random variables on the probability space $(\Omega, \mathcal{B}, P)$ and suppose $X \in L_1$.*

(a) If

$$P[X = X'] = 1,$$

then

$$X' \in L_1 \text{ and } E(X) = E(X').$$

(b) We have

$$P[X = X'] = 1$$

iff

$$\int_A X dP = \int_A X' dP, \quad \forall A \in \mathcal{B}.$$

The condition "for all $A \in \mathcal{B}$" can be replaced by "for all $A \in \mathcal{P}$" where $\mathcal{P}$ is a π-system generating $\mathcal{B}$.

Proof. Part (a) basically follows from Exercise 6. Here, we restrict ourselves to commenting on why if $X \in L_1$ and $P[X = X'] = 1$, it follows that $X' \in L_1$. Write $N = [X \neq X']$ so that $P(N) = 0$ and then

$$\begin{aligned} E(|X'|) &= E(|X|1_{[|X|=|X'|]}) + E(|X'|1_N) \\ &\leq E(|X|) + 0 < \infty, \end{aligned}$$

where we have applied Exercise 6. A modification of this argument shows $E(X) = E(X')$.

Now consider (b) which uses the following result also following from Exercise 6:

$$\text{If } X \geq 0, \text{ then } E(X) = 0 \text{ implies } P[X = 0] = 1, \tag{5.17}$$

or equivalently

$$\text{if } X \geq 0, \text{ then } P[X > 0] > 0 \text{ implies } E(X) > 0. \tag{5.18}$$

Suppose for all $A \in \mathcal{B}$ that

$$\int_A X dP = \int_A X' dP.$$

To get a contradiction suppose $P[X \neq X'] > 0$. So either $P[X > X'] > 0$ or $P[X < X'] > 0$. If $P[X > X'] > 0$, then set $A = [X > X']$ and $(X - X')1_A \geq 0$, and $P[(X - X')1_A > 0] \geq P(A) > 0$. So from (5.18) we have

$$E\big((X - X')1_A\big) > 0;$$

that is,

$$\int_A X - \int_A X' > 0,$$

a contradiction. So $P(A) = 0$.

Conversely, if $P[X = X'] = 1$, then set $N = [X \neq X']$ and for any $A \in \mathcal{B}$

$$\begin{aligned} \int_A X dP &= \int_{A \cap N} X dP + \int_{A \cap N^c} X dP \\ &= 0 + \int_{A \cap N^c} X' dP = \int_A X' dP, \end{aligned}$$

with the 0 resulting from Exercise 6. □

Example 5.6.1 For this example we set $\Omega = [0, 1]$, and $P = \lambda =$ Lebesgue measure. Let $X(s) = 1_{\mathbb{Q}}(s)$ where $\mathbb{Q}$ are the rational real numbers. Note that

$$\lambda(\mathbb{Q}) = \lambda(\cup_{r \in \mathbb{Q}}\{r\}) = \sum_{r \in \mathbb{Q}} \lambda(\{r\}) = 0$$

so that

$$\lambda([X=0]) = 1 = \lambda([0,1]\setminus \mathbb{Q}).$$

Therefore from the Integral Comparison Lemma 10.1 $E(X) = E(0) = 0$ since $\lambda[X=0]=1$. Note also that X is not Riemann-integrable, since for every n,

$$\sum_1^n X^\vee(\frac{j-l}{n},\frac{j}{n}]\frac{1}{n} = \sum_1^n 1\cdot\frac{1}{n} = \frac{n}{n} = 1$$

$$\sum_1^n X^\wedge(\frac{j-l}{n},\frac{j}{n}]\frac{1}{n} = \sum_1^n 0\cdot\frac{1}{n} = 0$$

and thus the upper and lower Riemann approximating sums do not converge to each other. We conclude that the Riemann integral does not exist but the Lebesgue integral does and is equal to 0.

For a function to be Riemann-integrable, it is necessary and sufficient that the function be bounded and continuous almost everywhere. However,

$$\{\omega\in[0,1]: 1_{\mathbb{Q}}(\cdot) \text{ is discontinuous at } \omega\} = \{\omega\in[0,1]\} = [0,1]$$

and thus

$$\lambda\{\omega: 1_{\mathbb{Q}}(\cdot) \text{ is continuous at } \omega\} = 0.$$

□

5.7 Product Spaces

This section shows how to build independence into a model and is also important for understanding concepts such as Markov dependence.

Let Ω_1, Ω_2 be two sets. Define the *product space*

$$\Omega_1\times\Omega_2 = \{(\omega_1,\omega_2): \omega_i\in\Omega_i, i=1,2\}$$

and define the *coordinate* or *projection* maps by $(i=1,2)$

$$\pi_i(\omega_1,\omega_2) = \omega_i$$

so that

$$\pi_i: \Omega_1\times\Omega_2 \mapsto \Omega_i.$$

If $A\subset\Omega_1\times\Omega_2$ define

$$\begin{aligned} A_{\omega_1} &= \{\omega_2 : (\omega_1,\omega_2)\in A\}\subset\Omega_2 \\ A_{\omega_2} &= \{\omega_1 : (\omega_1,\omega_2)\in A\}\subset\Omega_1. \end{aligned}$$

A_{ω_i} is called the *section* of A at ω_i.

Here are some basic properties of set sections.

(i) If $A \subset \Omega_1 \times \Omega_2$, then $(A^c)_{\omega_1} = (A_{\omega_1})^c$.

(ii) If, for an index set T, we have $A_\alpha \subset \Omega_1 \times \Omega_2$, for all $\alpha \in T$, then

$$(\bigcup_\alpha A_\alpha)_{\omega_1} = \bigcup_\alpha (A_\alpha)_{\omega_1}, \quad (\bigcap_\alpha A_\alpha)_{\omega_1} = \bigcap_\alpha (A_\alpha)_{\omega_1}.$$

Now suppose we have a function X with domain $\Omega_1 \times \Omega_2$ and range equal to some set S. It does no harm to think of S as a metric space. Define the *section* of the function X as

$$X_{\omega_1}(\omega_2) = X(\omega_1, \omega_2)$$

so

$$X_{\omega_1} : \Omega_2 \mapsto S.$$

We think of ω_1 as fixed and the section is a function of varying ω_2. Call X_{ω_1} the *section* of X at ω_1.

Basic properties of sections of functions are the following:

(i) $(1_A)_{\omega_1} = 1_{A_{\omega_1}}$

(ii) If $S = \mathbb{R}^k$ for some $k \geq 1$ and if for $i = 1, 2$ we have

$$X_i : \Omega_1 \times \Omega_2 \mapsto S,$$

then

$$(X_1 + X_2)_{\omega_1} = (X_1)_{\omega_1} + (X_2)_{\omega_1}.$$

(iii) Suppose S is a metric space, $X_n : \Omega_1 \times \Omega_2 \mapsto S$ and $\lim_{n\to\infty} X_n$ exists. Then

$$\lim_{n\to\infty} (X_n)_{\omega_1} = \lim_{n\to\infty} (X_n)_{\omega_1}.$$

A *rectangle* in $\Omega_1 \times \Omega_2$ is a subset of $\Omega_1 \times \Omega_2$ of the form $A_1 \times A_2$ where $A_i \subset \Omega_i$, for $i = 1, 2$. We call A_1 and A_2 the *sides* of the rectangle. The rectangle is *empty* if at least one of the sides is empty.

Suppose $(\Omega_i, \mathcal{B}_i)$ are two measurable spaces ($i = 1, 2$). A rectangle is called *measurable* if it is of the form $A_1 \times A_2$ where $A_i \in \mathcal{B}_i$, for $i = 1, 2$.

An important **fact:** The class of measurable rectangles is a semi-algebra which we call RECT. To verify this, we need to check the postulates defining a semi-algebra. (See definition 2.4.1, page 44.)

(i) $\emptyset, \Omega \in$ RECT

(ii) RECT is a π-class: If $A_1 \times A_2, A_1' \times A_2' \in$ RECT, then $(A_1 \times A_2) \cap (A_1' \times A_2') = A_1A_1' \times A_2A_2' \in$ RECT.

(iii) RECT is closed under complementation. Suppose $A \times A_2 \in$ RECT. Then

$$\Omega_1 \times \Omega_2 \setminus A_1 \times A_2 = (\Omega_1 \setminus A_1) \times A_2 + A_1 \times (\Omega_2 \setminus A_2) + A_1^c \times A_2^c.$$

We now define a σ-field on $\Omega_1 \times \Omega_2$ to be the smallest σ-field containing RECT. We denote this σ-field $\mathcal{B}_1 \times \mathcal{B}_2$ and call it the *product σ-field*. Thus

$$\mathcal{B}_1 \times \mathcal{B}_2 := \sigma(\text{RECT}). \tag{5.19}$$

Note if $\Omega_1 = \Omega_2 = \mathbb{R}$, this defines

$$\mathcal{B}_1 \times \mathcal{B}_2 = \sigma(A_1 \times A_2 : A_i \in \mathcal{B}(\mathbb{R}),\ i = 1, 2).$$

There are other ways of generating the product σ-field on $\mathbb{R}^2$. If $\mathcal{C}^{(]}$ is the class of semi-open intervals (open on the left, closed on the right), an induction argument gives

$$\mathcal{B}_1 \times \mathcal{B}_2 = \sigma(\{I_1 \times I_2 : I_j \in \mathcal{C}^{(]},\ j = 1, 2\}).$$

Lemma 5.7.1 (Sectioning Sets) *Sections of measurable sets are measurable. If $A \in \mathcal{B}_1 \times \mathcal{B}_2$, then for all $\omega_1 \in \Omega_1$,*

$$A_{\omega_1} \in \mathcal{B}_2.$$

Proof. We proceed by set induction. Define

$$\mathcal{C}_{\omega_1} = \{A \subset \Omega_1 \times \Omega_2 : A_{\omega_1} \in \mathcal{B}_2\}.$$

If $A \in$ RECT and $A = A_1 \times A_2$ where $A_i \in \mathcal{B}_i$, then

$$\begin{aligned} A_{\omega_1} &= \{\omega_2 : (\omega_1 \times \omega_2) \in A_1 \times A_2\} \\ &= \begin{cases} A_2 \in \mathcal{B}_2, & \text{if } \omega_1 \in A_1 \\ \emptyset, & \text{if } \omega_1 \notin A_1. \end{cases} \end{aligned}$$

Thus $A_{\omega_1} \in \mathcal{C}_{\omega_1}$, implying that

$$\text{RECT} \subset \mathcal{C}_{\omega_1}.$$

Also $\mathcal{C}_{\omega_1}$ is a λ-system. In order to verify this, we check the λ-system postulates.

(i) We have

$$\Omega_1 \times \Omega_2 \in \mathcal{C}_{\omega_1}$$

since $\Omega_1 \times \Omega_2 \in$ RECT.

(ii) If $A \in \mathcal{C}_{\omega_1}$ then $A^c \in \mathcal{C}_{\omega_1}$ since $(A^c)_{\omega_1} = (A_{\omega_1})^c$ and $A \in \mathcal{C}_{\omega_1}$ implies $A_{\omega_1} \in \mathcal{B}_2$ and hence $(A_{\omega_1})^c \in \mathcal{B}_2$. This necessitates $(A^c)_{\omega_1} \in \mathcal{C}_{\omega_1}$.

c) If $A_n \in \mathcal{C}_{\omega_1}$, for $n \geq 1$ with $\{A_n\}$ disjoint, then $(A_n)_{\omega_1} \in \mathcal{B}_2$ implies $\sum_n (A_n)_{\omega_1} \in \mathcal{B}_2$. But

$$\sum_{n=1}^{\infty}(A_n)_{\omega_1} = (\sum_{n=1}^{\infty} A_n)_{\omega_1} \in \mathcal{B}_2$$

and hence

$$\sum_{n=1}^{\infty} A_n \in \mathcal{C}_{\omega_1}.$$

Now we know that $\mathcal{C}_{\omega_1}$ is a λ-system. Further we know that

$$\mathcal{C}_{\omega_1} \supset \text{RECT}$$

which implies by Dynkin's theorem 2.2.2 that

$$\mathcal{C}_{\omega_1} \supset \sigma(\text{ RECT }) = \mathcal{B}_1 \times \mathcal{B}_2. \qquad \square$$

There is a companion result to Lemma 5.7.1 about sections of measurable functions.

Corollary 5.7.1 *Sections of measurable functions are measurable. That is, if*

$$X : (\Omega_1 \times \Omega_2, \mathcal{B}_1 \times \mathcal{B}_2) \mapsto (S, \mathcal{S})$$

then

$$X_{\omega_1} \in \mathcal{B}_2.$$

Proof. Since X is $\mathcal{B}_1 \times \mathcal{B}_2/\mathcal{S}$ measurable, we have for $\Lambda \in \mathcal{S}$ that

$$\{(\omega_1, \omega_2) : X(\omega, \omega_2) \in \Lambda\} = X^{-1}(\Lambda) \in \mathcal{B}_1 \times \mathcal{B}_2,$$

and hence by the previous results

$$(X^{-1}(\Lambda))_{\omega_1} \in \mathcal{B}_2.$$

However

$$\begin{aligned}(X^{-1}(\Lambda))_{\omega_1} =& \{\omega_2 : X(\omega_1, \omega_2) \in \Lambda\}\\ =& \{\omega_2 : X_{\omega_1}(\omega_2) \in \Lambda\} = (X_{\omega_1})^{-1}(\Lambda),\end{aligned}$$

which says X_{ω_1} is $\mathcal{B}_2/\mathcal{S}$ measurable. $\square$

5.8 Probability Measures on Product Spaces

We now consider how to construct probability measures on the product space $(\Omega_1 \times \Omega_2, \mathcal{B}_1 \times \mathcal{B}_2)$. In particular, we will see how to construct independent random variables.

Transition Functions. Call a function

$$K(\omega_1, A_2) : \Omega_1 \times \mathcal{B}_2 \mapsto [0, 1]$$

a *transition function* if

(i) for each ω_1, $K(\omega_1, \cdot)$ is a probability measure on $\mathcal{B}_2$, and

(ii) for each $A_2 \in \mathcal{B}_2$, $K(\cdot, A_2)$ is $\mathcal{B}_1/\mathcal{B}([0, 1])$ measurable.

Transition functions are used to define discrete time Markov processes where $K(\omega_1, A_2)$ represents the conditional probability that, starting from ω_1, the next movement of the system results in a state in A_2. Here our interest in transition functions comes from the connection with measures on product spaces and Fubini's theorem.

Theorem 5.8.1 *Let P_1 be a probability measure on $\mathcal{B}_1$, and suppose*

$$K : \Omega_1 \times \mathcal{B}_2 \mapsto [0, 1]$$

is a transition function. Then K and P_1, uniquely determine a probability on $\mathcal{B}_1 \times \mathcal{B}_2$ via the formula

$$P(A_1 \times A_2) = \int_{A_1} K(\omega_1, A_2) P_1(d\omega_1), \tag{5.20}$$

for all $A_1 \times A_2 \in RECT$.

The measure P given in (5.20) is specified on the semialgebra RECT and we need to verify that the conditions of the Combo Extension Theorem 2.4.3 on page 48 are applicable so that P can be extended to $\sigma(RECT) = \mathcal{B}_1 \times \mathcal{B}_2$.

We verify that P is σ-additive on RECT and apply the Combo Extension Theorem 2.4.3. Let

$$\{A_1^{(n)} \times A_2^{(n)}, n \geq 1\}$$

be disjoint elements of RECT whose union is in RECT. We show

$$P(\sum_{n=1}^{\infty} A_1^{(n)} \times A_2^{(n)}) = \sum_{n=1}^{\infty} P(A_1^{(n)} \times A_2^{(n)}).$$

Note if $\sum_n A_1^{(n)} \times A_2^{(n)} = A_1 \times A_2$, then

$$1_{A_1}(\omega_1) 1_{A_2}(\omega_2) = 1_{A_1 \times A_2}(\omega_1, \omega_2) = \sum_n 1_{A_1^{(n)} \times A_2^{(n)}}(\omega_1, \omega_2)$$

$$= \sum_n 1_{A_1^{(n)}}(\omega_1) 1_{A_2^{(n)}}(\omega_2).$$

Now making use of the series form of the monotone convergence theorem we have the following string of equalities:

$$\begin{aligned} P(A_1 \times A_2) &= \int_{\Omega_1} 1_{A_1}(\omega_1) K(\omega_1, A_2) P_1(d\omega_1) \\ &= \int_{\Omega_1} [\int_{\Omega_2} 1_{A_1}(\omega_1) 1_{A_2}(\omega_2) K(\omega_1, d\omega_2)] P_1(d\omega_1) \\ &= \int_{\Omega_1} [\int_{\Omega_2} \sum_n 1_{A_1^{(n)}}(\omega_1) 1_{A_2^{(n)}}(\omega_2) K(\omega_1, d\omega_2)] P_1(d\omega_1) \\ &= \int_{\Omega_1} \sum_n [\int_{\Omega_2} 1_{A_1^{(n)}}(\omega_1) 1_{A_2^{(n)}}(\omega_2) K(\omega_1, d\omega_2)] P_1(d\omega_1) \\ &= \sum_n \int_{\Omega_1} 1_{A_1^{(n)}}(\omega_1) [\int_{\Omega_2} 1_{A_2^{(n)}}(\omega_2) K(\omega_1, d\omega_2)] P_1(d\omega_1) \\ &= \sum_n \int_{\Omega_1} 1_{A_1^{(n)}}(\omega_1) K(\omega_1, A_2^{(n)}) P_1(d\omega_1) \\ &= \sum_n \int_{A_1^{(n)}} K(\omega_1, A_2^{(n)}) P_1(d\omega_1) \\ &= \sum_n P(A_1^{(n)} \times A_2^{(n)}). \end{aligned}$$

□

Special case. Suppose for some probability measure P_2 on $\mathcal{B}_2$ that $K(\omega_1, A_2) = P_2(A_2)$. Then the probability measure P, defined in the previous result on $\mathcal{B}_1 \times \mathcal{B}_2$ is

$$P(A_1 \times A_2) = P_1(A_1) P_2(A_2).$$

We denote this P by $P_1 \times P_2$ and call P *product measure.* Define σ-fields in $\Omega_1 \times \Omega_2$ by

$$\begin{aligned} \mathcal{B}_1^{\#} &= \{A_1 \times \Omega_2 : A_1 \in \mathcal{B}_1\} \\ \mathcal{B}_2^{\#} &= \{\Omega_1 \times A_2 : A_2 \in \mathcal{B}_2\}. \end{aligned}$$

With respect to the product measure $P = P_1 \times P_2$, we have $\mathcal{B}_1^{\#} \perp\!\!\!\perp \mathcal{B}_2^{\#}$ since

$$\begin{aligned} P(A_1 \times \Omega_2 \cap \Omega_1 \times A_2) &= P(A_1 \times A_2) = P_1(A_1) P_2(A_2) \\ &= P(A_1 \times \Omega_2) P(\Omega_1 \times A_2). \end{aligned}$$

Suppose $X_i : (\Omega_i, \mathcal{B}_i) \mapsto (\mathbb{R}, \mathcal{B}(\mathbb{R}))$ is a random variable on Ω_i for $i = 1, 2$. Define on $\Omega_1 \times \Omega_2$ the new functions

$$X_1^{\#}(\omega_1, \omega_2) = X_1(\omega_1), \quad X_2^{\#}(\omega_1, \omega_2) = X_2(\omega_2).$$

With respect to $P = P_1 \times P_2$, the variables $X_1^\#$ and $X_2^\#$ are independent since

$$\begin{aligned}
P[X_1^\# \leq x, X_2^\# \leq y] &= P_1 \times P_2(\{(\omega_1, \omega_2) : X_1(\omega_1) \leq x, X_2(\omega_2) \leq y\}) \\
&= P_1 \times P_2(\{\omega_1 : X_1(\omega_1) \leq x\} \times \{\omega_2 : X_2(\omega_2) \leq y\}) \\
&= P_1(\{X_1(\omega_1) \leq x\}) P_2(\{\omega_2 : X_2(\omega_2) \leq y\}) \\
&= P(\{(\omega_1, \omega_2) : X_1^\#(\omega_1, \omega_2) \leq x\}) \\
&\qquad P(\{(\omega_1, \omega_2) : X_2^\#(\omega_1, \omega_2) \leq y\}) \\
&= P[X_1^\# \leq x] P[X_2^\# \leq y].
\end{aligned}$$

The point of these remarks is that independence is automatically built into the model by construction when using product measure. Quantities depending on disjoint components can be asserted to be independent without proof. We can extend these constructions from two factors to $d \geq 2$ factors and define product measure $P_1 \times \cdots \times P_d$. Having independence built into the model is a big economy since otherwise a large number of conditions (of order 2^d) would have to be checked. See Definition 4.1.2 on page 91 and the following discussion.

5.9 Fubini's theorem

Fubini's theorem is a basic result which allows interchange of the order of integration in multiple integrals. We first present a somewhat more general and basic result that uses transition kernels. We continue to work on the product space $(\Omega_1 \times \Omega_2, \mathcal{B}_1 \times \mathcal{B}_2)$.

Theorem 5.9.1 *Let P_1 be a probability measure on $(\Omega_1, \mathcal{B}_1)$ and suppose $K : \Omega_1 \times \mathcal{B}_2 \mapsto [0, 1]$ is a transition kernel. Define P on $(\Omega_1 \times \Omega_2, \mathcal{B}_1 \times \mathcal{B}_2)$ by*

$$P(A_1 \times A_2) = \int_{A_1} K(\omega_1, A_2) P_1(d\omega_1). \tag{5.21}$$

Assume

$$X : (\Omega_1 \times \Omega_2, \mathcal{B}_1 \times \mathcal{B}_2) \mapsto (\mathbb{R}, \mathcal{B}(\mathbb{R}))$$

and furthermore suppose $X \geq 0$ (X is integrable). Then

$$Y(\omega_1) = \int_{\Omega_2} K(\omega_1, d\omega_2) X_{\omega_1}(\omega_2)$$

has the properties

(a) Y is well defined.

(b) $Y \in \mathcal{B}_1$.

(c) $Y \geq 0$ $(Y \in L_1(P_1))$,

and furthermore

$$\int_{\Omega_1\times\Omega_2} XdP = \int_{\Omega_1} Y(\omega_1)P_1(d\omega_1) = \int_{\Omega_1}[\int_{\Omega_2} K(\omega_1, d\omega_2)X_{\omega_1}(\omega_2)]P_1(d\omega_1). \tag{5.22}$$

Proof. For fixed ω_1, we have $X_{\omega_1}(\omega_2)$ is $\mathcal{B}_2$-measurable so Y is well defined. It is not hard to show that Y is $\mathcal{B}_1$ measurable and we skip this and we proceed to show (5.22) under the assumption $X \geq 0$. Define

$$\text{LHS} := \int_{\Omega_1\times\Omega_2} XdP$$

and

$$\text{RHS} := \int_{\Omega_1} Y(\omega_1)P_1(d\omega_1).$$

We begin by supposing

$$X = 1_{A_1\times A_2}$$

where

$$A_1 \times A_2 \in \text{RECT}.$$

Then

$$\text{LHS} = \int_{A_1\times A_2} dP = P(A_1 \times A_2)$$

and

$$\begin{aligned}\text{RHS} &= \int_{\Omega_1}[\int_{\Omega_2} K(\omega_1, d\omega_2)1_{A_1}(\omega_1)1_{A_2}(\omega_2)]P_1(d\omega_1)\\ &= \int_{A_1} K(\omega_1, A_2)P_1(d\omega_1) = P(A_1 \times A_2).\end{aligned}$$

So (5.22) holds for indicators of measurable rectangles. Let

$$\mathcal{C} = \{A \in \mathcal{B}_1 \times \mathcal{B}_2 : \text{(5.22) holds for } X = 1_A\},$$

and we know RECT $\subset \mathcal{C}$.

We claim that $\mathcal{C}$ is a λ-system. We check the postulates.

(i) $\Omega_1 \times \Omega_2 \in \mathcal{C}$ since $\Omega_1 \times \Omega_2 \in$ RECT.

(ii) If $A \in \mathcal{C}$ then for $X = 1_{A^c}$, we have LHS $= P(A^c) = 1 - P(A)$ so that

$$\begin{aligned}
\text{LHS} &= 1 - \int\int K(\omega_1, d\omega_2) 1_{A_{\omega_1}}(\omega_2) P_1(d\omega_1) \\
&= \int\int K(\omega_1, d\omega_2)(1 - 1_{A_{\omega_1}}(\omega_2)) P_1(d\omega_1) \\
&= \int\int K(\omega_1, d\omega_2) 1_{(A_{\omega_1})^c}(\omega_2) P_1(d\omega_2) \\
&= \int\int K(\omega_1, d\omega_2) 1_{(A^c)_{\omega_1}}(\omega_2) P_1(d\omega_2) \\
&= \text{RHS}.
\end{aligned}$$

So $A^c \in \mathcal{C}$.

(iii) If $A_n \in \mathcal{C}$, and $\{A_n, n \geq 1\}$ are disjoint events, then

$$\begin{aligned}
\int_{\Omega_1 \times \Omega_2} 1_{\sum_{n=1}^{\infty} A_n} dP = P(\sum_n A_n) &= \sum_n P(A_n) \\
&= \sum_n \int\int K(\omega_1, d\omega_2) 1_{(A_n)_{\omega_1}}(\omega_2) P_1(d\omega_1)
\end{aligned}$$

because $A_n \in \mathcal{C}$; applying monotone convergence we get

$$\begin{aligned}
&= \int\int K(\omega_1, d\omega_2) \sum_n 1_{(A_n)_{\omega_1}}(\omega_2) P_1(d\omega_1) \\
&= \int\int K(\omega_1, d\omega_2) 1_{(\cup_n A_n)_{\omega_1}}(\omega_2) P_1(d\omega_1) \\
&= \text{RHS},
\end{aligned}$$

so $\sum_n A_n \in \mathcal{C}$.

We have thus verified that $\mathcal{C}$ is a λ-system. Furthermore,

$$\mathcal{C} \supset \text{RECT},$$

which implies

$$\mathcal{C} \supset \sigma(\text{RECT}) = \mathcal{B}_1 \times \mathcal{B}_2.$$

We may therefore conclude that for any $A \in \mathcal{B}_1 \times \mathcal{B}_2$, if $X = 1_A$ then (5.22) holds for this X.

Now express (5.22) as LHS(X) =RHS(X). Both LHS(X) and RHS(X) are linear in X so (5.22) holds for simple functions of the form

$$X = \sum_{i=1}^{k} a_i 1_{A_i}, \quad A_i \in \mathcal{B}_1 \times \mathcal{B}_2.$$

For arbitrary $X \geq 0$, there exists a sequence of simple X_n, such that $X_n \uparrow X$. We have

$$\text{LHS}\,(X_n) = \text{RHS}\,(X_n),$$

and by monotone convergence

$$\text{LHS}\,(X_n) \uparrow \text{LHS}\,(X).$$

Also, we get for RHS, by applying monotone convergence twice, that

$$\begin{aligned}
\lim_{n\to\infty} \uparrow \text{RHS}\,(X_n) &= \lim_{n\to\infty} \uparrow \int_{\Omega_1} \int_{\Omega_2} K(\omega_1, d\omega_2)(X_n)_{\omega_1}(\omega_2)]P_1(d\omega_1) \\
&= \int_{\Omega_1} [\lim_{n\to\infty} \uparrow \int_{\Omega_2} K(\omega_2, d\omega_2)(X_n)_{\omega_1}(\omega_2)]P_1(d\omega_1) \\
&= \int_{\Omega_1} [\int_{\Omega_2} \lim_{n\to\infty} (X_n)_{\omega_1}(\omega_2) K(\omega_1, d\omega_2)]P_1(d\omega_1) \\
&= \int_{\Omega_1} [\int_{\Omega_2} K(\omega_1, d\omega_2) X_{\omega_1}(\omega_2)]P_1(d\omega_1) \\
&= \text{RHS}\,(X).
\end{aligned}$$

□

We can now give the result, called *Fubini's theorem*, which justifies interchange of the order of integration.

Theorem 5.9.2 (Fubini Theorem) *Let $P = P_1 \times P_2$ be product measure. If X is $\mathcal{B}_1 \times \mathcal{B}_2$ measurable and is either non-negative or integrable with respect to P, then*

$$\begin{aligned}
\int_{\Omega_1 \times \Omega_2} X dP &= \int_{\Omega_1} [\int_{\Omega_2} X_{\omega_1}(\omega_2) P_2(d\omega_2)]P_1(d\omega_1) \\
&= \int_{\Omega_2} [\int_{\Omega_1} X_{\omega_2}(\omega_1) P_1(d\omega_1)]P_2(d\omega_2).
\end{aligned}$$

Proof. Let $K(\omega_1, A_2) = P_2(A_2)$. Then P_1 and K determine $P = P_1 \times P_2$ on $\mathcal{B}_1 \times \mathcal{B}_2$ and

$$\begin{aligned}
\int_{\Omega_1 \times \Omega_2} X dP &= \int_{\Omega_1} [\int_{\Omega_2} K(\omega_1, d\omega_2) X_{\omega_1}(\omega_2)]P_1(d\omega_1) \\
&= \int_{\Omega_1} [\int_{\Omega_2} P_2(d\omega_2) X_{\omega_1}(\omega_2)]P_1(d\omega_1).
\end{aligned}$$

Also let

$$\tilde{K}(\omega_2, A_1) = P_1(A_1)$$

be a transition function with

$$\tilde{K} : \Omega_2 \times \mathcal{B}_1 \mapsto [0, 1].$$

Then $\tilde{K}$ and P_2 also determine $P = P_1 \times P_2$ and

$$\int_{\Omega_1\times\Omega_2} XdP = \int_{\Omega_2} [\int_{\Omega_1} \tilde{K}(\omega_2, d\omega_1)X_{\omega_2}(\omega_1)]P_2(d\omega_2)$$
$$= \int_{\Omega_2} [\int_{\Omega_1} P_1(d\omega_1)X_{\omega_2}(\omega_1)]P_2(d\omega_2).$$

□

We now give some simple examples of the use of Fubini's theorem.

Example 5.9.1 (Occupation times) Let $\{X(t,\omega), t \in [0,1]\}$ be a continuous time stochastic process indexed by $[0,1]$ on the probability space $(\Omega, \mathcal{B}, P)$ satisfying

(a) The process $X(\cdot)$ has state space $\mathbb{R}$.

(b) The process X is two-dimensional measurable; that is,

$$X : ([0,1] \times \Omega, \mathcal{B}([0,1]) \times \mathcal{B}) \mapsto \mathcal{B}(\mathbb{R})$$

so that for $\Lambda \in \mathcal{B}(\mathbb{R})$

$$X^{-1}(\Lambda) = \{(t,\omega) : X(t,\omega) \in \Lambda\} \in \mathcal{B}([0,1]) \times \mathcal{B}.$$

Fix a set $\Lambda \in \mathcal{B}(\mathbb{R})$. We ask for the occupation time of X in Λ during times $t \in A$, for $A \in \mathcal{B}([0,1])$. Since $\Lambda \in \mathcal{B}(\mathbb{R})$,

$$1_\Lambda : (\mathbb{R}, \mathcal{B}(\mathbb{R})) \mapsto (\{0,1\}, \{\emptyset, \{0,1\}, \{0\}, \{1\}\})$$

is measurable and therefore

$$1_\Lambda(X(s,\omega)) : ([0,1] \times \Omega, \mathcal{B}([0,1]) \times \mathcal{B}) \mapsto (\{0,1\}, \mathcal{B}(\{0,1\})).$$

Define the random measure

$$\chi(A,\omega) := \int_A 1_\Lambda(X(s,\omega))ds$$

and call it the *occupation time* in Λ during times $t \in A$.

We have

$$E\chi(A,\omega) = \int_\Omega \left[\int_A 1_\Lambda(X(s,\omega))ds\right] dP,$$

which by Fubini's theorem is the same as

$$= \int_A \left[\int_\Omega 1_\Lambda(X(s,\omega))dP\right] ds$$
$$= \int_A P[X(s) \in \Lambda]ds.$$

Thus expected occupation times can be computed by integrating the probability the process is in the set. □

Example 5.9.2 Let $X_i \geq 0, i = 1, 2$ be two independent random variables. Then

$$E(X_1X_2) = E(X_1)E(X_2).$$

To prove this using Fubini's theorem 5.9.2, let $\mathbf{X} = (X_1, X_2)$, and let $g(x_1, x_2) = x_1x_2$. Note $P \circ \mathbf{X}^{-1} = F_1 \times F_2$ where F_i is the distribution of X_i. This follows since

$$\begin{aligned} P \circ \mathbf{X}^{-1}(A_1 \times A_2) &= P[(X_1, X_2) \in A_1 \times A_2] \\ &= P[X_1 \in A_1, X_2 \in A_2] \\ &= P[X_1 \in A_1]P[X_2 \in A_2] \\ &= F_1(A_1)F_2(A_2) \\ &= F_1 \times F_2(A_1 \times A_2). \end{aligned}$$

So $P \circ \mathbf{X}^{-1}$ and $F_1 \times F_2$ agree on RECT and hence on $\sigma(RECT) = \mathcal{B}_1 \times \mathcal{B}_2$. From Corollary 5.5.1 we have

$$\begin{aligned} EX_1X_2 = Eg(\mathbf{X}) &= \int_{\mathbb{R}_+^2} g(\mathbf{x}) P \circ \mathbf{X}^{-1}(d\mathbf{x}) \\ &= \int_{\mathbb{R}_+^2} gd(F_1 \times F_2) \\ &= \int_{\mathbb{R}_+} x_2[\int x_1 F_1(dx_1)]F_2(dx_2) \qquad \text{(Fubini)} \\ &= E(X_1) \int x_2 F_2(dx_2) = E(X_1)E(X_2). \end{aligned}$$

□

Example 5.9.3 (Convolution) Suppose X_1, X_2 are two independent random variables with distributions F_1, F_2. The distribution function of the random variable $X_1 + X_2$ is given by the *convolution* $F_1 * F_2$ of the distribution functions. For $x \in \mathbb{R}$

$$P[X_1 + X_2 \leq x] =: F_1 * F_2(x) = \int_{\mathbb{R}} F_1(x-u)F_2(du) = \int_{\mathbb{R}} F_2(x-u)F_1(du).$$

To see this, proceed as in the previous example. Let $\mathbf{X} = (X_1, X_2)$ which has distribution $F_1 \times F_2$ and set

$$g(x_1, x_2) = 1_{\{(u,v) \in \mathbb{R}^2 : u+v \leq x\}}(x_1, x_2), \quad (x_1, x_2) \in \mathbb{R}^2.$$

From Corollary 5.5.1

$$P[X_1 + X_2 \leq x] = Eg(\mathbf{X}) = \int_{\mathbb{R}^2} gd(F_1 \times F_2).$$

Iterating the multiple integral à la Fubini, we have

$$= \int_{\mathbb{R}} \left[\int_{\mathbb{R}} 1_{\{(u,v)\in\mathbb{R}^2: u+v\le x\}}(x_1, x_2) F_1(dx_1) \right] F_2(dx_2)$$

$$= \int_{\mathbb{R}} \left[\int_{\mathbb{R}} 1_{\{v\in\mathbb{R}: v\le x-x_2\}}(x_1) F_1(dx_1) \right] F_2(dx_2)$$

$$= \int_{\mathbb{R}} F_1(x - x_2) F_2(dx_2).$$

□

5.10 Exercises

1. Consider the triangle with vertices $(-1, 0)$, $(1, 0)$, $(0, 1)$ and suppose (X_1, X_2) is a random vector uniformly distributed on this triangle. Compute $E(X_1 + X_2)$.

2. Argue without a computation that if $X \in L_2$ and $c \in \mathbb{R}$, then $\mathrm{Var}(c) = 0$ and $\mathrm{Var}(X + c) = \mathrm{Var}(X)$.

3. Refer to Renyi's theorem 4.3.1 in Chapter 4. Let

$$L_1 := \inf\{j \ge 2 : X_j \text{ is a record.}\}$$

Check $E(L_1) = \infty$.

4. Let (X, Y) be uniformly distributed on the discrete points $(-1, 0)$, $(1, 0)$, $(0, 1)$, $(0, -1)$. Verify X, Y are not independent but $E(XY) = E(X)E(Y)$.

5. (a) If F is a continuous distribution function, prove that

$$\int_{\mathbb{R}} F(x) F(dx) = \frac{1}{2}.$$

Thus show that if X_1, X_2 are iid with common distribution F, then

$$P[X_1 \le X_2] = \frac{1}{2}$$

and $E(F(X_1)) = 1/2$. (One solution method relies on Fubini's theorem.)

(b) If F is not continuous

$$E(F(X_1)) = \frac{1}{2} + \frac{1}{2} \sum_a (P[X_1 = a])^2,$$

where the sum is over the atoms of F.

(c) If X, Y are random variables with distribution functions $F(x), G(x)$ which have no common discontinuities, then

$$E(F(Y)) + E(G(X)) = 1.$$

Interpret the sum of expectations on the left as a probability.

(d) Even if F and G have common jumps, if $X \perp\!\!\!\perp Y$ then

$$E(F(Y)) + E(G(X)) = 1 + P[X = Y].$$

6. Suppose $X \in L_1$ and A and A_n are events.

(a) Show

$$\int_{[|X|>n]} X dP \to 0.$$

(b) Show that if $P(A_n) \to 0$, then

$$\int_{A_n} X dP \to 0.$$

Hint: Decompose

$$\int_{A_n} |X| dP = \int_{A_n[|X| \le M]} |X| dP + \int_{A_n[|X|>M]} |X| dP$$

for large M.

(c) Show

$$\int_A |X| dP = 0 \text{ iff } P(A \cap [|X| > 0]) = 0.$$

(d) If $X \in L_2$, show $\mathrm{Var}(X) = 0$ implies $P[X = E(X)] = 1$ so that X is equal to a constant with probability 1.

(e) Suppose that $(\Omega, \mathcal{B}, P)$ is a probability space and $A_i \in \mathcal{B}$, $i = 1, 2$. Define the distance $d : \mathcal{B} \times \mathcal{B} \mapsto \mathbb{R}$ by

$$d(A_1, A_2) = P(A_1 \Delta A_2).$$

Check the following continuity result: If A_n, $A \in \mathcal{B}$ and

$$d(A_n, A) \to 0$$

then

$$\int_{A_n} X dP \to \int_A X dP$$

so that the map

$$A \mapsto \int_A X dP$$

is continuous.

7. Suppose X_n, $n \geq 1$ and X are uniformly bounded random variables; i.e. there exists a constant K such that

$$|X_n| \bigvee |X| \leq K.$$

If $X_n \to X$ as $n \to \infty$, show by means of dominated convergence that

$$E|X_n - X| \to 0.$$

8. Suppose $X, X_n, n \geq 1$ are random variables on the space $(\Omega, \mathcal{B}, P)$ and assume

$$\sup_{\substack{\omega \in \Omega \\ n \geq 1}} |X_n(\omega)| < \infty;$$

that is, the sequence $\{X_n\}$ is uniformly bounded.

(a) Show that if in addition

$$\sup_{\omega \in \Omega} |X(\omega) - X_n(\omega)| \to 0, \quad n \to \infty,$$

then $E(X_n) \to E(X)$.

(b) Use Egorov's theorem (Exercise 25, page 89 of Chapter 3) to prove: If $\{X_n\}$ is uniformly bounded and $X_n \to X$, then $E(X_n) \to E(X)$. (Obviously, this follows from dominated convergence as in Exercise 7 above; the point is to use Egorov's theorem and not dominated convergence.)

9. Use Fubini's theorem to show for a distribution function $F(x)$

$$\int_{\mathbb{R}} (F(x+a) - F(x))dx = a,$$

where "dx" can be interpreted as Lebesgue measure.

10. For $X \geq 0$, let

$$X_n^* = \sum_{k=1}^{\infty} \frac{k}{2^n} 1_{[\frac{k-1}{2^n} \leq X < \frac{k}{2^n}]} + \infty 1_{[X=\infty]}.$$

Show

$$E(X_n^*) \downarrow E(X).$$

11. If X, Y are independent random variables and $E(X)$ exists, then for all $B \in \mathcal{B}(\mathbb{R})$, we have

$$\int_{[Y \in B]} X dP = E(X) P[Y \in B].$$

12. Suppose X is an uncountable set and let $\mathcal{B}$ be the σ-field of countable and co-countable (complements are countable) sets. Show that the diagonal

$$\text{DIAG} := \{(x, x) : x \in X\} \notin \mathcal{B} \times \mathcal{B}$$

is **not** in the product σ-field. However, every section of DIAG is measurable. (Although sections of measurable sets are measurable, the converse is thus not true.)

Hints:

- Show that

$$\mathcal{B} \times \mathcal{B} = \sigma\big(\big\{\{x\} \times X, X \times \{x\}, x \in X\big\}\big),$$

so that the product σ-field is generated by horizontal and vertical lines.

- Review Exercise 12 in Chapter 2.
- Proceed by contradiction. Suppose DIAG $\in \mathcal{B} \times \mathcal{B}$. Then there exists countable $S \subset X$ such that

$$\text{DIAG} \in \sigma\big(\big\{\{x\} \times X, X \times \{x\}, x \in S\big\}\big) =: \mathcal{G}.$$

- Define

$$\mathcal{P} := \{\{s\}, s \in S, S^c\}$$

and observe this is a partition of X and that

$$\{\Lambda_1 \times \Lambda_2 : \Lambda_i \in \mathcal{P};\ i = 1, 2\}$$

is a partition of $X \times X$ and that

$$\mathcal{G} = \sigma(\Lambda_1 \times \Lambda_2 : \Lambda_i \in \mathcal{P},\ i = 1, 2).$$

Show elements of $\mathcal{G}$ can be written as unions of sets $\Lambda_j \times \Lambda_k$.

- Show it is impossible for DIAG $\in \mathcal{G}$.

13. Suppose the probability space is the Lebesgue interval

$$(\Omega = [0, 1], \mathcal{B}([0, 1]), \lambda)$$

and define

$$X_n = \frac{n}{\log n} 1_{(0, \frac{1}{n})}.$$

Show $X_n \to 0$ and $E(X_n) \to 0$ even though the condition of domination in the Dominated Convergence Theorem fails.

14. Suppose $X \perp\!\!\!\perp Y$ and $h : \mathbb{R}^2 \mapsto [0, \infty)$ is measurable. Define

$$g(x) = E(h(x, Y))$$

and show

$$E(g(X)) = E(h(X, Y)).$$

15. Suppose X is a non-negative random variable satisfying

$$P[0 \leq X < \infty] = 1.$$

Show

$$\text{(a)} \quad \lim_{n\to\infty} nE\left(\frac{1}{X}1_{[X>n]}\right) = 0,$$

$$\text{(b)} \quad \lim_{n\to\infty} n^{-1}E\left(\frac{1}{X}1_{[X>n^{-1}]}\right) = 0.$$

16. (a) Suppose $-\infty < a \leq b < \infty$. Show that the indicator function $1_{(a,b]}(x)$ can be approximated by bounded and continuous functions; that is, show that there exist a sequence of continuous functions $0 \leq f_n \leq 1$ such that $f_n \to 1_{(a,b]}$ pointwise.

Hint: Approximate the rectangle of height 1 and base $(a, b]$ by a trapezoid of height 1 with base $(a, b + n^{-1}]$ whose top line extends from $a + n^{-1}$ to b.

(b) Show that two random variables X_1 and X_2 are independent iff for every pair f_1, f_2 of non-negative continuous functions, we have

$$E(f_1(X_1)f_2(X_2)) = Ef_1(X_1)Ef_2(X_2).$$

(c) Suppose for each n, that the pair ξ_n and η_n are independent random variables and that pointwise

$$\xi_n \to \xi_\infty, \quad \eta_n \to \eta_\infty.$$

Show that the pair ξ_∞ and η_∞ are independent so that independence is preserved by taking limits.

17. **Integration by parts.** Suppose F and G are two distribution functions with no common points of discontinuity in an interval $(a, b]$. Show

$$\int_{(a,b]} G(x)F(dx)$$

$$= F(b)G(b) - F(a)G(a) - \int_{(a,b]} F(x)G(dx).$$

The formula can fail if F and G have common discontinuities. If F and G are absolutely continuous with densities f and g, try to prove the formula by differentiating with respect to the upper limit of integration. (Why can you differentiate?)

18. Suppose $(\Omega, \mathcal{B}, P) = ((0,1], \mathcal{B}((0,1]), \lambda)$ where λ is Lebesgue measure on $(0,1]$. Let $\lambda \times \lambda$ be product measure on $(0,1] \times (0,1]$. Suppose that $A \subset (0,1] \times (0,1]$ is a rectangle whose sides are **NOT** parallel to the axes. Show that

$$\lambda \times \lambda(A) = \text{area of } A.$$

19. Define $(\Omega_i, \mathcal{B}_i, \mu_i)$, for $i = 1, 2$ as follows: Let μ_1 be Lebesgue measure and μ_2 counting measure so that $\mu_2(A)$ is the number of elements of A. Let

$$\Omega_1 = (0,1), \quad \mathcal{B}_1 = \text{Borel subsets of } (0,1),$$
$$\Omega_2 = (0,1), \quad \mathcal{B}_2 = \text{All subsets of } (0,1).$$

Define

$$f(x,y) = \begin{cases} 1, & \text{if } x = y, \\ 0, & \text{otherwise.} \end{cases}$$

(a) Compute

$$\int_{\Omega_1} [\int_{\Omega_2} f(x,y)\mu_2(dy)]\mu_1(dx)$$

and

$$\int_{\Omega_2} [\int_{\Omega_1} f(x,y)\mu_1(dx)]\mu_2(dy).$$

(b) Are the two integrals equal? Are the measures σ-finite?

20. For a random variable X with distribution F, define the moment generating function $\phi(\lambda)$ by

$$\phi(\lambda) = E(e^{\lambda X}).$$

(a) Prove that

$$\phi(\lambda) = \int_R e^{\lambda x} F(dx).$$

Let

$$\Lambda = \{\lambda \in R : \phi(\lambda) < \infty\}$$

and set

$$\lambda_\infty = \sup \Lambda.$$

(b) Prove for λ in the interior of Λ that $\phi(\lambda) > 0$ and that $\phi(\lambda)$ is continuous on the interior of Λ. (This requires use of the dominated convergence theorem.)

(c) Give an example where (i) $\lambda_\infty \in \Lambda$ and (ii) $\lambda_\infty \notin \Lambda$. (Something like gamma distributions should suffice to yield the needed examples.)

Define the measure F_λ by

$$F_\lambda(I) = \int_I \frac{e^{\lambda x}}{\phi(\lambda)} F(dx), \quad \lambda \in \Lambda.$$

(d) If F has a density f, verify F_λ has a density f_λ. What is f_λ? (Note that the family $\{f_\lambda, \lambda \in \Lambda\}$ is an *exponential family* of densities.)

(e) If $F(I) = 0$, show $F_\lambda(I) = 0$ as well for I a finite interval and $\lambda \in \Lambda$.

21. Suppose $\{p_k, k \geq 0\}$ is a probability mass function on $\{0, 1, \dots\}$ and define the generating function

$$P(s) = \sum_{k=0}^{\infty} p_k s^k, \quad 0 \leq s \leq 1.$$

Prove using dominated convergence that

$$\frac{d}{dx} P(s) = \sum_{k=1}^{\infty} p_k k s^{k-1}, \quad 0 \leq s \leq 1,$$

that is, prove differentiation and summation can be interchanged.

22. (a) For X, a positive random variable, use Fubini's theorem applied to σ-finite measures to prove

$$E(X) = \int_{[0,\infty)} P[X > t]dt.$$

(b) Check also that for any $\alpha > 0$,

$$E(X^\alpha) = \alpha \int_{[0,\infty)} x^{\alpha-1} P[X > x]dx.$$

(c) If $X \geq 0$ is a random variable such that for some $\delta > 0$ and $0 < \beta < 1$

$$P[X > n\delta] \leq (const)\beta^n,$$

then $E(X^\alpha) < \infty$, for $\alpha > 0$.

(d) If $X \geq 0$ is a random variable such that for some $\delta > 0$, $E(X^\delta) < \infty$, then

$$\lim_{x\to\infty} x^\delta P[X > x] = 0.$$

(e) Suppose $X \geq 0$ has a heavy-tailed distribution given by

$$P[X > x] = \frac{\text{const}}{x \log x}, \quad x \geq 17.$$

Show $E(X) = \infty$ but yet $xP[X > x] \to 0$ as $x \to \infty$.

(f) If $E(X^2) < \infty$, then for any $\eta > 0$

$$\lim_{x\to\infty} xP[|X| > \eta\sqrt{x}] = 0.$$

23. Verify that the product σ-algebra is the smallest σ-algebra making the coordinate mappings π_1, π_2 measurable.

24. Suppose X_1, X_2 are iid random variables with common $N(0,1)$ distribution. Define
$$Y_n = \frac{X_1}{\frac{1}{n} + |X_2|}.$$
Use Fubini's theorem to verify that
$$E(Y_n) = 0.$$
Note that as $n \to \infty$,
$$Y_n \to Y := \frac{X_1}{|X_2|}$$
and that the expectation of Y does not exist, so this is one case where random variables converge but means do not.

25. In cases where expectations are not guaranteed to exist, the following is a proposal for defining a central value. Suppose $F(x)$ is a strictly increasing and continuous distribution function. For example F could be the standard normal distribution function. Define
$$g : \mathbb{R} \mapsto (-1, 1)$$
by
$$g(x) = 2(F(x) - \frac{1}{2}).$$
For a random variable X, define $\phi : \mathbb{R} \mapsto (-1, 1)$ by
$$\phi(\gamma) = E\big(g(X - \gamma)\big). \tag{*}$$
The *central value of X with respect to g*, denoted $\gamma(X)$, is defined as the solution of
$$\phi(\gamma) = 0.$$
(a) Show $\phi(\gamma)$ is a continuous function of γ.

(b) Show
$$\lim_{\gamma \to \infty} \phi(\gamma) = -1,$$
$$\lim_{\gamma \to -\infty} \phi(\gamma) = 1.$$
(c) Show $\phi(\gamma)$ is non-increasing.

(d) Show $\gamma(X)$, the solution of
$$\phi(\gamma) = 0$$

is unique.

Show $\gamma(X)$ has some of the properties of expectation, namely the following.

(e) For any $c \in \mathbb{R}$
$$\gamma(X+c) = \gamma(X) + c.$$

(f) Now suppose g in (*) is $g : \mathbb{R} \mapsto (-\pi/2, \pi/2)$ defined by
$$g(x) := \arctan(x),$$
so that $g(-x) = -g(x)$. Show
$$\gamma(-X) = -\gamma(X).$$

26. Suppose $\{X_n, n \geq 1\}$ is a sequence of (not necessarily independent) Bernoulli random variables with
$$P[X_n = 1] = p_n = 1 - P[X_n = 0].$$
Show that $\sum_{n=1}^{\infty} p_n < \infty$ implies $\sum_{n=1}^{\infty} E(X_n) < \infty$ and therefore that $P[X_n \to 0] = 1$. (Compare with Example 4.5.1 in Chapter 4.)

27. **Rapid variation.** A distribution tail $1 - F(x)$ is called rapidly varying if
$$\lim_{t\to\infty} \frac{1-F(tx)}{1-F(t)} = \begin{cases} \infty, & \text{if } 0 < x < 1, \\ 0, & \text{if } x > 1. \end{cases}$$
Verify that if F is normal or gamma, then the distribution tail is rapidly varying.

If $X \geq 0$ is a random variable with distribution tail which is rapidly varying, then X possesses all positive moments: for any $m > 0$ we have $E(X^m) < \infty$.

28. Let $\{X_n, n \geq 1\}$ be a sequence of random variables. Show
$$E\left(\bigvee_{n=1}^{\infty} |X_n|\right) < \infty$$
iff there exists a random variable $0 \leq Y \in L_1$ such that
$$P[|X_n| \leq Y] = 1, \quad \forall n \geq 1.$$

29. Suppose X_n is a sequence of random variables such that
$$P[X_n = \pm n^3] = \frac{1}{2n^2}, \quad P[X_n = 0] = 1 - \frac{1}{n^2}.$$
Show that using Borel-Cantelli that $P[\lim_{n\to\infty} X_n = 0] = 1$. Compute $\lim_{n\to\infty} E(X_n)$. Is it 0? Is the Lebesgue Dominated Convergence Theorem applicable? Why or why not?

30. **Pratt's lemma.** The following variant of dominated convergence and Fatou is useful: Let X_n, Y_n, X, Y be random variables on the probability space $(\Omega, \mathcal{B}, P)$ such that

$$(i)\ 0 \le X_n \le Y_n,$$
$$(ii)\ X_n \to X, \quad Y_n \to Y,$$
$$(iii)\ E(Y_n) \to E(Y),\ EY < \infty.$$

Prove $E(X_n) \to E(X)$. Show the Dominated Convergence Theorem follows.

31. If X is a random variable, call m a *median* of X if

$$\frac{1}{2} \le P[X \ge m], \quad P[X \le m] \ge \frac{1}{2}.$$

(a) Show the median always exists.
(b) Is the median always unique?
(c) If I is a closed interval such that $P[X \in I] \ge 1/2$, show $m \in I$.
(d) When the variance exists, show

$$|m - E(X)| \le \sqrt{2\mathrm{Var}(X)}.$$

(e) If m is a median of $X \in L_1$ show

$$E\left(|X - m|\right) \le E\left(|X - a|\right), \quad \forall a \in \mathbb{R}.$$

Thus the median minimizes L_1 prediction.

(f) If $X \in L_2$, show that for $\mu = E(X)$

$$E\left(|X - \mu|^2\right) \le E\left(|X - a|^2\right), \quad \forall a \in \mathbb{R}.$$

Thus the mean minimizes L_2 prediction.

(g) Suppose $X_1, X_2, \dots, X_n, X_{n+1}$ are L_2 random variables. Find the *best linear predictor* $\widehat{X_{n+1}}$ based on $X_1, \dots, X_n$ of X_{n+1}; that is, find the linear function $\sum_{i=1}^n \alpha_i X_i$ such that

$$E\left(\left(\sum_{i=1}^n \alpha_i X_i - X_{n+1}\right)^2\right)$$

is minimized.

32. (a) Suppose X has possible values $\pm 1, \pm 2$ and that X assumed each of these 4 values with equal probability 1/4. Define $Y = X^2$. Despite this functional relationship, show X and Y are uncorrelated; that is, show

$$\mathrm{Cov}(X, Y) = 0.$$

(b) Suppose U, V have zero mean and the same variance. Check that $X = U + V$ and $Y = U - V$ are uncorrelated.

(c) Toss two dice. Let X be the sum of the faces and Y be the difference. Are X, Y independent? Are they uncorrelated?

33. Suppose X, Y are two L_2 random variables such that (X, Y) and $(-X, Y)$ have the same joint distributions. Show that X and Y are uncorrelated.

34. Suppose $\{X_n, n \geq 1\}$ are iid with $E(X_n) = 0$, $\text{Var}(X_n) = 1$. Compute

$$\text{Cov}(S_n, S_m), \quad n < m,$$

where $S_n = X_1 + \cdots + X_n$.

35. Suppose $X, Y \in L_1$.

(a) Show

$$E(Y) - E(X) = \int_{\mathbb{R}} (P[X < x \leq Y] - P[Y < x \leq X])\, dx.$$

(b) The expected length of the random interval $(X, Y]$ is the integral with respect to x of $P[x \in (X, Y]]$, the probability the random interval covers x.

36. **Beppo Levi Theorem.** Suppose for $n \geq 1$ that $X_n \in L_1$ are random variables such that

$$\sup_{n \geq 1} E(X_n) < \infty.$$

Show that if $X_n \uparrow X$, then $X \in L_1$ and

$$E(X_n) \to E(X).$$

37. **Mean Approximation Lemma.** Suppose that $X \in L_1(\Omega, \mathcal{B}, P)$. For any $\epsilon > 0$, there exists an integrable simple random variable X_ϵ such that

$$E(|X - X_\epsilon|) < \epsilon.$$

Hint: Consider X^+ and X^- separately.

Furthermore, if $\mathcal{A}$ is a field such that $\sigma(\mathcal{A}) = \mathcal{B}$, then X_ϵ can be taken to be of the form

$$X_\epsilon = \sum_{i=1}^{k} c_i 1_{A_i},$$

where $A_i \in \mathcal{A}$ for $i = 1, \ldots, k$. Hint: Review Exercise 5 from Chapter 2.

38. Use the Fatou Lemma to show the following: If $0 \leq X_n \to X$ and $\sup_n E(X_n) = K < \infty$, then $E(X) \leq K$ and $X \in L_1$.

39. A Poisson process $\{N(A,\omega), A \in \mathcal{B}(R^2)$ on R^2 with mean measure μ is defined to have the following properties:

(a) μ is a measure such that if A is a bounded measurable set, $\mu(A) < \infty$.

(b) For any set $A \in \mathcal{B}(R^2)$, the random variable $N(A)$ is Poisson distributed with parameter $\mu(A)$:

$$P[N(A)=k] = \begin{cases} \frac{e^{-\mu(A)}\mu(A)^k}{k!}, & \text{if } \mu(A) < \infty, \\ 0, & \text{if } \mu(A) = \infty. \end{cases}$$

(c) For $A_1, A_2, \dots, A_k$ disjoint regions, the counting random variables $N(A_1), \dots, N(A_k)$ are independent.

Define for a point process the Laplace functional L as follows: L maps non-negative measurable functions $f : R^2 \mapsto [0,\infty)$ into $[0,\infty)$ via the formula

$$\begin{aligned} L(f) :=& E(\exp\{-\int_{R^2} f(\mathbf{x})N(d\mathbf{x})\}) \\ =& \int_\Omega \exp\{-\int_{R^2} f(\mathbf{x})N(d\mathbf{x},\omega)\}P(d\omega). \end{aligned}$$

Show for the Poisson process above that

$$L(f) = e^{-\int_{R^2}(1-e^{-f(\mathbf{x})})\mu(d\mathbf{x})}.$$

Hint: Start with f an indicator function and remember how to compute the generating function or Laplace transform of a Poisson random variable. Then go from indicator variables to simple variables to general non-negative f via monotone convergence.

40. (a) Suppose η is a $N(\mu,\sigma^2)$ random variable satisfying $E(\exp\{\eta\}) = 1$. Show $\mu = -\sigma^2/2$.

(b) Suppose (ξ,η) are jointly normal. If e^ξ and e^η are uncorrelated, then so are ξ and η.

6 Convergence Concepts

Much of classical probability theory and its applications to statistics concerns *limit theorems*; that is, the asymptotic behavior of a sequence of random variables. The sequence could consist of sample averages, cumulative sums, extremes, sample quantiles, sample correlations, and so on. Whereas probability theory discusses limit theorems, the theory of statistics is concerned with *large sample* properties of statistics, where a statistic is just a function of the sample.

There are several different notions of convergence and these are discussed next in some detail.

6.1 Almost Sure Convergence

Suppose we are given a probability space $(\Omega, \mathcal{B}, P)$. We say that a statement about random elements holds *almost surely* (abbreviated *a.s.*) if there exists an event $N \in \mathcal{B}$ with $P(N) = 0$ such that the statement holds if $\omega \in N^c$. Synonyms for almost surely include *almost everywhere* (abbreviated *a.e.*), *almost certainly* (abbreviated *a.c.*). Alternatively, we may say that the statement holds for a.a. (almost all) ω. The set N appearing in the definition is sometimes called the *exception set*.

Here are several examples of statements that hold *a.s.*:

- Let X, X' be two random variables. Then $X = X'$ a.s. means

$$P[X = X'] = 1;$$

that is, there exists an event $N \in \mathcal{B}$, such that $P(N) = 0$ and if $\omega \in N^c$, then $X(\omega) = X'(\omega)$.

- $X \le X'$ a.s. means there exists an event $N \in \mathcal{B}$, such that $P(N) = 0$ and if $\omega \in N^c$ then
$$X(\omega) \le X'(\omega).$$
- If $\{X_n\}$ is a sequence of random variables, then $\lim_{n\to\infty} X_n$ exists a.s. means there exists an event $N \in \mathcal{B}$, such that $P(N) = 0$ and if $\omega \in N^c$ then
$$\lim_{n\to\infty} X_n(\omega)$$
exists. It also means that for a.a. ω,
$$\limsup_{n\to\infty} X_n(\omega) = \liminf_{n\to\infty} X_n(\omega).$$
We will write $\lim_{n\to\infty} X_n = X$ a.s. or $X_n \to X$ a.s. or $X_n \stackrel{a.s.}{\to} X$.
- If $\{X_n\}$ is a sequence of random variables, then $\sum_n X_n$ converges a.s. means there exists an event $N \in \mathcal{B}$, such that $P(N) = 0$, and $\omega \in N^c$ implies $\sum_n X_n(\omega)$ converges.

Most probabilistic properties of random variables are invariant under the relation almost sure equality. For example, if $X = X'$ a.s. then $X \in L_1$ iff $X' \in L_1$ and in this case $E(X) = E(X')$.

Here is an example of a sequence of random variables that converges a.s. but does not converge everywhere. For this example, the exception set N is nonempty.

Example 6.1.1 We suppose the probability space is the Lebesgue unit interval: $([0,1], \mathcal{B}([0,1]), \lambda)$ where λ is Lebesgue measure. Define

$$X_n(s) = \begin{cases} n, & \text{if } 0 \le s \le \frac{1}{n}, \\ 0, & \text{if } \frac{1}{n} < s \le 1. \end{cases}$$

We claim that for this example

$$X_n \to 0 \text{ a.s.}$$

since if $N = \{0\}$, then $s \in N^c$ implies $X_n(s) \to 0$. It is not true for this example that $X_n(s) \to 0$ for all $s \in [0,1]$, since $X_n(0) = n \to \infty$. □

Here is another elementary example of almost sure convergence. This one is taken from extreme value theory.

Proposition 6.1.1 *Let $\{X_n\}$ be iid random variables with common distribution function $F(x)$. Assume that $F(x) < 1$, for all x. Set*

$$M_n = \bigvee_1^n X_i.$$

Then

$$M_n \uparrow \infty \text{ a.s.}$$

Proof. Recall

$$\begin{aligned} P[M_n \le x] &= P[X_1 \le x, \dots, X_n \le x] \\ &= \prod_{i=1}^{n} P[X_i \le x] = F^n(x). \end{aligned}$$

We must prove that there exists $N \in \mathcal{B}$, such that $P(N) = 0$ and, for $\omega \in N^c$, we have that

$$\lim_{n\to\infty} M_n(\omega) = \infty;$$

that is, for all j, there exists $n_0(\omega, j)$ such that if $n \ge n_0(\omega, j)$, then $M_n(\omega) \ge j$. Note

$$\sum_n P[M_n \le j] = \sum_n F^n(j) < \infty$$

since $F(j) < 1$. So the Borel-Cantelli Lemma implies

$$P([M_n \le j] \text{ i.o.}) = P(\limsup_{n\to\infty}[M_n \le j]) = 0$$

and if

$$N_j = \limsup_{n\to\infty}[M_n \le j]$$

we have $P(N_j) = 0$. Note

$$N_j^c = \liminf_{n\to\infty}[M_n > j],$$

so for $\omega \in N_j^c$, we get $M_n(\omega) > j$ for all large n.

Let $N = \bigcup_j N_j$ so

$$P(N) \le \sum_j P(N_j) = 0.$$

If $\omega \in N^c$, we have the property that for any j, $M_n(\omega) > j$ for all sufficiently large n. □

6.2 Convergence in Probability

Suppose $X_n, n \ge 1$ and X are random variables. Then $\{X_n\}$ *converges in probability* (i.p.) to X, written $X_n \xrightarrow{P} X$, if for any $\epsilon > 0$

$$\lim_{n\to\infty} P[|X_n - X| > \epsilon] = 0.$$

Almost sure convergence of $\{X_n\}$ demands that for a.a. ω, $X_n(\omega) - X(\omega)$ gets small and stays small. Convergence i.p. is weaker and merely requires that the probability of the difference $X_n(\omega) - X(\omega)$ being non-trivial becomes small.

It is possible for a sequence to converge in probability but not almost surely.

Example 6.2.1 Here is an example of a sequence which converges i.p. to 0 but does *not* converge a.s. to 0. Let $(\Omega, \mathcal{B}, P) = ([0,1], \mathcal{B}([0,1]), \lambda)$ where λ is Lebesgue measure and define $\{X_n\}$ as follows:

$$\begin{aligned} &X_1 = 1_{[0,1]}, \\ &X_2 = 1_{[0,\frac{1}{2}]}, \quad X_3 = 1_{[\frac{1}{2},1]} \\ &X_4 = 1_{[0,\frac{1}{3}]}, \quad X_5 = 1_{[\frac{1}{3},\frac{2}{3}]}, \quad X_6 = 1_{[\frac{2}{3},1]} \\ &\quad \vdots \qquad\qquad\qquad\qquad\qquad \vdots \end{aligned}$$

and so on.

For any $\omega \in [0,1]$, $X_n(\omega) \not\to 0$ since $X_n(\omega) = 1$ for infinitely many values of n. However $X_n \xrightarrow{P} 0$. □

We next see that a.s. convergence implies convergence i.p. The previous Example 6.2.1 showed the converse false in general.

Theorem 6.2.1 (Convergence a.s. implies convergence i.p.) *Suppose that* $\{X_n, n \geq 1, X\}$ *are random variables on a probability space* $(\Omega, \mathcal{B}, P)$. *If*

$$X_n \to X, \quad a.s.,$$

then

$$X_n \xrightarrow{P} X.$$

Proof. If $X_n \to X$ a.s. then for any ϵ,

$$\begin{aligned} 0 &= P([|X_n - X| > \epsilon] \text{ i.o. }) \\ &= P(\limsup_{n\to\infty}[|X_n - X| > \epsilon]) \\ &= \lim_{N\to\infty} P(\bigcup_{n\geq N}[|X_n - X| > \epsilon]) \\ &\geq \lim_{n\to\infty} P[|X_n - X| > \epsilon]. \end{aligned}$$

□

Remark. The definition of convergence i.p. and convergence a.s. can be readily extended to random elements of metric spaces. If $\{X_n, n \geq 1, X\}$ are random elements of a metric space S with metric d, then $X_n \to X$ a.s. means that $d(X_n, X) \to 0$ a.s. and $X_n \xrightarrow{P} X$ means $d(X_n, X) \xrightarrow{P} 0$.

6.2.1 *Statistical Terminology*

In statistical estimation theory, almost sure and in probability convergence have analogues as strong or weak consistency.

Given a family of probability models $(\Omega, \mathcal{B}, P_\theta)$, $\theta \in \Theta$). Suppose the statistician gets to observe random variables $X_1, \dots, X_n$ defined on Ω and based on these observations must decide which is the correct model; that is, which is the correct value of θ. Statistical *estimation* means: select the correct model.

For example, suppose $\Omega = \mathbb{R}^\infty$, $\mathcal{B} = \mathcal{B}(\mathbb{R}^\infty)$. Let $\omega = (x_1, x_2, \dots)$ and define $X_n(\omega) = x_n$. For each $\theta \in \mathbb{R}$, let P_θ be product measure on $\mathbb{R}^\infty$ which makes $\{X_n, n \geq 1\}$ iid with common $N(\theta, 1)$ distribution. Based on observing $X_1, \dots, X_n$, one estimates θ with an appropriate function of the observations

$$\hat{\theta}_n = \hat{\theta}_n(X_1, \dots, X_n).$$

$\hat{\theta}_n(X_1, \dots, X_n)$ is called a *statistic* and is also an *estimator*. When one actually does the experiment and observes,

$$X_1 = x_1, \dots, X_n = x_n,$$

then $\hat{\theta}(x_1, \dots, x_n)$ is called the *estimate*. So the *estimator* is a random element while the *estimate* is a number or maybe a vector if θ is multidimensional.

In this example, the usual choice of estimator is $\hat{\theta}_n = \sum_{i=1}^n X_i/n$. The estimator $\hat{\theta}_n$ is *weakly consistent* if for all $\theta \in \Theta$

$$P_\theta[|\hat{\theta}_n - \theta| > \epsilon] \to 0, \quad n \to \infty;$$

that is,

$$\hat{\theta}_n \xrightarrow{P_\theta} \theta.$$

This indicates that no matter what the true parameter is or to put it another way, no matter what the true (but unknown) state of nature is, $\hat{\theta}$ does a good job estimating the true parameter. $\hat{\theta}_n$ is *strongly consistent* if for all $\theta \in \Theta$, $\hat{\theta}_n \to \theta$, P_θ–a.s. This is obviously stronger than weak consistency.

6.3 Connections Between a.s. and i.p. Convergence

Here we discuss the basic relations between convergence in probability and almost sure convergence. These relations have certain ramifications such as extension of the dominated convergence principle to convergence in probability.

Theorem 6.3.1 (Relations between i.p. and a.s. convergence) *Suppose that $\{X_n, X, n \geq 1\}$ are real-valued random variables.*

(a) *Cauchy criterion: $\{X_n\}$ converges in probability iff $\{X_n\}$ is Cauchy in probability. Cauchy in probability means*

$$X_n - X_m \xrightarrow{P} 0, \text{ as } n, m \to \infty.$$

or more precisely, given any $\epsilon > 0$, $\delta > 0$, there exists $n_0 = n_0(\epsilon, \delta)$ such that for all $r, s \geq n_0$ we have

$$P[|X_r - X_s| > \epsilon] < \delta. \tag{6.1}$$

(b) $X_n \xrightarrow{P} X$ *iff each subsequence* $\{X_{n_k}\}$ *contains a further subsequence* $\{X_{n_{k(i)}}\}$ *which converges almost surely to* X.

Proof. (i) We first show that if $X_n \xrightarrow{P} X$ then $\{X_n\}$ is Cauchy i.p. For any $\epsilon > 0$,

$$[|X_r - X_s| > \epsilon] \subset [|X_r - X| > \frac{\epsilon}{2}] \cup [|X_s - X| > \frac{\epsilon}{2}]. \tag{6.2}$$

To see this, take complements of both sides and it is clear that if

$$|X_r - X| \leq \frac{\epsilon}{2} \text{ and } |X_s - X| \leq \frac{\epsilon}{2},$$

then by the triangle inequality

$$|X_r - X_s| \leq \epsilon.$$

Thus, taking probabilities and using subadditivity, we get from (6.2)

$$P[|X_r - X_s| > \epsilon] \leq P[|X_r - X| > \frac{\epsilon}{2}] + P[|X_s - X| > \frac{\epsilon}{2}].$$

If

$$P[|X_n - X| > \epsilon] \leq \frac{\delta}{2}$$

for $n \geq n_0(\epsilon, \delta)$, then

$$P[|X_r - X_s| > \epsilon] \leq \delta$$

for $r, s \geq n_0$.

(ii) Next, we prove the following assertion: If $\{X_n\}$ is Cauchy i.p., then there exists a subsequence $\{X_{n_j}\}$ such that $\{X_{n_j}\}$ converges almost surely. Call the almost sure limit X. Then it is also true that also

$$X_n \xrightarrow{P} X.$$

To prove the assertion, define a sequence n_j by $n_1 = 1$ and

$$n_j = \inf\{N > n_{j-1} : P[|X_r - X_s| > 2^{-j}] < 2^{-j} \text{ for all } r, s \geq N\}.$$

(In the definition (6.1) of what it means for a sequence to be Cauchy i.p., we let $\epsilon = \delta = 2^{-j}$.) Note, by construction $n_j > n_{j-1}$ so that $n_j \to \infty$. Consequently, we have

$$P[|X_{n_{j+1}} - X_{n_j}| > 2^{-j}] < 2^{-j},$$

and thus

$$\sum_{j=1}^{\infty} P[|X_{n_{j+1}} - X_{n_j}| > 2^{-j}] < \infty.$$

The Borel–Cantelli Lemma implies

$$P(N) := P\{\limsup_{j\to\infty}[|X_{n_{j+1}} - X_{n_j}| > 2^{-j}]\} = 0.$$

For $\omega \in N^c$,

$$|X_{n_{j+1}}(\omega) - X_{n_j}(\omega)| \le 2^{-j} \tag{6.3}$$

for all large j and thus $\{X_{n_j}(\omega)\}$ is a Cauchy sequence of real numbers. The Cauchy property follows since (6.3) implies for large l that

$$\sum_{j\ge l} |X_{n_{j+1}}(\omega) - X_{n_j}(\omega)| \le \sum_{j\ge l} 2^{-j} = 2\cdot 2^{-l},$$

and therefore for any $k > l$ large, we get

$$|X_{n_k}(\omega) - X_{n_l}(\omega)| \le \sum_{j\ge l} |X_{n_{j+1}}(\omega) - X_{n_j}(\omega)| \le 2\cdot 2^{-l}.$$

Completeness of the real line implies

$$\lim_{j\to\infty} X_{n_j}(\omega)$$

exists; that is

$$\omega \in N^c \text{ implies } \lim_{j\to\infty} X_{n_j}(\omega) \text{ exists.}$$

This means that $\{X_{n_j}\}$ converges a.s. and we call the limit X.

To show $X_n \stackrel{P}{\to} X$ note

$$P[|X_n - X| > \epsilon] \le P[|X_n - X_{n_j}| > \frac{\epsilon}{2}] + P[|X_{n_j} - X| > \frac{\epsilon}{2}].$$

Given any η, pick n_j and n so large that the Cauchy i.p. property guarantees

$$P[|X_n - X_{n_j}| > \frac{\epsilon}{2}] < \frac{\eta}{2}.$$

Since $X_{n_j} \stackrel{a.s.}{\to} X$ implies $X_{n_j} \stackrel{P}{\to} X$,

$$P[|X_{n_j} - X| > \frac{\epsilon}{2}] < \frac{\eta}{2}$$

for large n_j. This finishes the proof of part (a).

We now focus on the proof of (b): Suppose $X_n \stackrel{P}{\to} X$. Pick any subsequence $\{X_{n_k}\}$. Then it is also true that $X_{n_k} \stackrel{P}{\to} X$. From (ii) above, there exists a further subsequence $\{X_{n_{k(i)}}\}$ converging a.s.

Conversely: Suppose every subsequence has a further subsequence converging alomst surely to X. To show $X_n \stackrel{P}{\to} X$, we suppose this fails and get a contradiction. If $\{X_n\}$ fails to converge in probability, there exists a subsequence $\{X_{n_k}\}$ and a $\delta > 0$ and $\epsilon > 0$ such that

$$P[|X_{n_k} - X| > \epsilon] \geq \delta.$$

But every subsequence, such as $\{X_{n_k}\}$ is assumed to have a further subsequence $\{X_{n_{k(i)}}\}$ which converges a.s. and hence i.p. But

$$P[|X_{n_{k(i)}} - X| > \epsilon] \geq \delta$$

contradicts convergence i.p. □

This result relates convergence in probability to point wise convergence and thus allows easy connections to continuous maps.

Corollary 6.3.1 *(i) If $X_n \stackrel{a.s.}{\to} X$ and*

$$g : \mathbb{R} \mapsto \mathbb{R}$$

is continuous, then

$$g(X_n) \stackrel{a.s.}{\to} g(X).$$

(ii) If $X_n \stackrel{P}{\to} X$ and

$$g : \mathbb{R} \mapsto \mathbb{R}$$

is continuous, then

$$g(X_n) \stackrel{P}{\to} g(X).$$

Thus, taking a continuous function of a sequence of random variables which converges either almost surely or in probability, preserves the convergence.

Proof. (i) There exists a null event $N \in \mathcal{B}$ with $P(N) = 0$, such that if $\omega \in N^c$, then

$$X_n(\omega) \to X(\omega)$$

in $\mathbb{R}$, and hence by continuity, if $\omega \in N^c$, then

$$g(X_n(\omega)) \to g(X(\omega)).$$

This is almost sure convergence of $\{g(X_n)\}$.

(ii) Let $\{g(X_{n_k})\}$ be some subsequence of $\{g(X_n)\}$. It suffices to find an a.s. convergence subsequence $\{g(X_{n_{k(i)}})\}$. But we know $\{X_{n_k}\}$ has some a.s. convergent subsequence $\{X_{n_{k(i)}}\}$ such that $X_{n_{k(i)}} \to X$ almost surely. Thus $g(X_{n_{k(i)}}) \stackrel{a.s.}{\to} g(X)$ which finishes the proof. □

Thus we see that if $X_n \stackrel{P}{\to} X$, it is also true that

$$X_n^2 \stackrel{P}{\to} X^2, \text{ and } \arctan X_n \stackrel{P}{\to} \arctan X$$

and so on.

Now for the promised connection with Dominated Convergence: The statement of the Dominated Convergence Theorem holds without change when almost sure convergence is replaced by convergence i.p.

Corollary 6.3.2 (Lebesgue Dominated Convergence) *If $X_n \xrightarrow{P} X$ and if there exists a dominating random variable $\xi \in L_1$ such that*

$$|X_n| \le \xi,$$

then

$$E(X_n) \to E(X).$$

Proof. It suffices to show every convergent subsequence of $E(X_n)$ converges to $E(X)$.

Suppose $E(X_{n_k})$ converges. Then since convergence in probability is assumed, $\{X_{n_k}\}$ contains an a.s. convergent subsequence $\{X_{n_{k(i)}}\}$ such that $X_{n_{k(i)}} \xrightarrow{a.s.} X$. The Lebesgue Dominated Convergence Theorem implies

$$E(X_{n_{k(i)}}) \to E(X).$$

So $E(X_{n_k}) \to E(X)$. □

We now list several easy results related to convergence in probability.

(1) If $X_n \xrightarrow{P} X$ and $Y_n \xrightarrow{P} Y$ then

$$X_n + Y_n \xrightarrow{P} X + Y.$$

To see this, just note

$$[|(X_n + Y_n) - (X + Y)| > \epsilon] \subset [|X_n - X| > \frac{\epsilon}{2}] \cup [|Y_n - Y| > \frac{\epsilon}{2}].$$

Take probabilities, use subadditivity and let $n \to \infty$.

(2) A multiplicative counterpart to (1): If $X_n \xrightarrow{P} X$ and $Y_n \xrightarrow{P} Y$, then

$$X_n Y_n \xrightarrow{P} XY.$$

To see this, observe that given a subsequence $\{n_k\}$, it suffices to find a further subsequence $\{n_{k(i)}\} \subset \{n_k\}$ such that

$$X_{n_{k(i)}} Y_{n_{k(i)}} \xrightarrow{a.s.} XY.$$

Since $X_{n_k} \xrightarrow{P} X$, there exists a subsequence $\{n'_k\} \subset \{n_k\}$ such that

$$X_{n'_k} \xrightarrow{a.s.} X.$$

Since $Y_n \xrightarrow{P} Y$, given the subsequence $\{n'_k\}$, there exists a further subsequence $\{n'_{k(i)}\} \subset \{n'_k\}$ such that

$$X_{n'_{k(i)}} \xrightarrow{a.s.} X, \quad Y_{n'_{k(i)}} \xrightarrow{a.s.} Y$$

and hence, since the product of two convergent sequences is convergent, we have

$$X_{n'_{k(i)}} Y_{n'_{k(i)}} \xrightarrow{a.s.} XY.$$

Thus every subsequence of $\{X_n Y_n\}$ has an a.s. convergent subsequence.

(3) This item is a reminder that Chebychev's inequality implies the Weak Law of Large Numbers (WLLN): If $\{X_n, n \geq 1\}$ are iid with $EX_n = \mu$ and $\text{Var}(X_n) = \sigma^2$, then

$$\sum_{i=1}^{n} X_i/n \xrightarrow{P} \mu.$$

(4) *Bernstein's version of the Weierstrass Approximation Theorem.* Let $f : [0, 1] \mapsto \mathbb{R}$ be continuous and define the Bernstein polynomial of degree n by

$$B_n(x) = \sum_{k=0}^{n} f(\frac{k}{n}) \binom{n}{k} x^k (1-x)^{n-k}, \quad 0 \leq x \leq 1.$$

Then

$$B_n(x) \to f(x)$$

uniformly for $x \in [0, 1]$. The proof of pointwise convergence is easy using the WLLN: Let $\delta_1, \delta_2, \ldots, \delta_n$ be iid Bernoulli random variables with

$$P[\delta_i = 1] = x = 1 - P[\delta_i = 0].$$

Define $S_n = \sum_{i=1}^{n} \delta_i$ so that S_n has a binomial distribution with success probability $p = x$ and

$$E(S_n) = nx, \quad \text{Var}(S_n) = nx(1-x) \leq n.$$

Since f is continuous on $[0, 1]$, f is bounded. Thus, since

$$\frac{S_n}{n} \xrightarrow{P} x,$$

from the WLLN, we get

$$f(\frac{S_n}{n}) \xrightarrow{P} f(x)$$

by continuity of f and by dominated convergence, we get

$$Ef(\frac{S_n}{n}) \to f(x).$$

But

$$Ef(\frac{S_n}{n}) = \sum_{k=0}^{n} f(\frac{k}{n}) \binom{n}{k} x^k (1-x)^{n-k} = B_n(x).$$

We now show convergence is uniform. Since f is continuous on $[0, 1]$, f is uniformly continuous, so define the modulus of continuity as

$$\omega(\delta) = \sup_{\substack{|x-y|\le\delta \\ 0\le x,y\le 1}} |f(x) - f(y)|,$$

and uniform continuity means

$$\lim_{\delta\downarrow 0} w(\delta) = 0.$$

Define

$$\|f\| = \sup\{|f(x)| : 0 \le x \le 1\}.$$

Now we write

$$\begin{aligned}
\sup_{0\le x\le 1} |B_n(x) - f(x)| &= \sup_x |E\big(f(\frac{S_n}{n})\big) - f(x)| \\
&\le \sup_x E\big(|f(\frac{S_n}{n}) - f(x)|\big) \\
&\le \sup_x \Big\{ E\big(|f(\frac{S_n}{n}) - f(x)| 1_{[|\frac{S_n}{n} - x| \le \epsilon]}\big) \\
&\quad + \sup_x E\big(|f(\frac{S_n}{n}) - f(x)| 1_{[|\frac{S_n}{n} - x| > \epsilon]}\big) \Big\} \\
&\le \omega(\epsilon) P[\ ] + 2\|f\| \sup_x P[|\frac{S_n}{n} - x| > \epsilon] \\
&\le \omega(\epsilon) + 2\|f\| \sup_x \frac{\mathrm{Var}(\frac{S_n}{n})}{\epsilon^2} \qquad \text{(by Chebychev)} \\
&\le \omega(\epsilon) + \frac{2\|f\|}{\epsilon^2} \sup_x \frac{nx(1-x)}{n^2} \\
&= \omega(\epsilon) + \frac{2\|f\|}{\epsilon^2} \frac{1}{4} \cdot \frac{1}{n}
\end{aligned}$$

where we have used

$$\sup_{0\le x\le 1} x(1-x) = \frac{1}{4}.$$

So we conclude

$$\sup_{0\le x\le 1} |B_n(x) - f(x)| = \omega(\epsilon) + (\text{const}) \cdot \frac{1}{n},$$

and therefore

$$\limsup_{n\to\infty} \sup_{0\le x\le 1} |B_n(x) - f(x)| \le \omega(\epsilon).$$

Since $\omega(\epsilon) \to 0$ as $\epsilon \to 0$, the proof is complete. □

6.4 Quantile Estimation

A significant statistical application of convergence in probability is to quantile estimation.

Let F be a distribution function. For $0 < p < 1$, the *pth order quantile* of F is $F^{\leftarrow}(p)$. If F is unknown, we may wish to estimate a quantile. Statistical tests, reliability standards, insurance premia, dam construction are all based on estimated quantiles. How can one estimate $F^{\leftarrow}(p)$ based on a random sample?

One non-parametric method uses order statistics. Let $X_1, \dots, X_n$ be a random sample from F; that is, $X_1, \dots, X_n$ are iid with common distribution function F. The order statistics of the sample are

$$X_1^{(n)} \le X_2^{(n)} \le \cdots \le X_n^{(n)},$$

so that $X_1^{(n)}$ is the minimum of the sample and $X_n^{(n)}$ is the maximum. Define the empirical cumulative distribution function (cdf) by

$$F_n(x) = \frac{1}{n}\sum_{1}^{n} 1_{[X_j\le x]}$$

which is the percentage of the sample whose value is no greater than x. Note that if we think of *success* at the jth trial as the random variable X_j having a value $\le x$, then

$$nF_n(x) = \#\text{ successes in } n \text{ trials}$$

is a binomial random variable with success probability $F(x)$. So

$$E(nF_n(x)) = nF(x), \quad \mathrm{Var}(nF_n(x)) = nF(x)(1 - F(x))$$

and the WLLN implies

$$F_n(x) \xrightarrow{P} F(x)$$

for each x. In fact, much more is true as we will see when we study the Glivenko-Cantelli Lemma in the next chapter.

Thus, F_n approximates or estimates F and we hope $F_n^{\leftarrow}$ estimates $F^{\leftarrow}$. But since F_n only jumps at the order statistics, we have

$$\begin{aligned} F_n^{\leftarrow}(p) &= \inf\{y : F_n(y) \geq p\} \\ &= \inf\{X_j^{(n)} : F_n(X_j^{(n)}) \geq p\} \\ &= \inf\{X_j^{(n)} : \frac{j}{n} \geq p\} \qquad (\text{ since } F_n(X_j^{(n)}) = \frac{j}{n}) \\ &= \inf\{X_j^{(n)} : j \geq np\} \\ &= X_{\lceil np \rceil}^{(n)}, \end{aligned}$$

where $\lceil np \rceil$ is the first integer $\geq np$. We will try $X_{\lceil np \rceil}^{(n)}$ as the quantile estimator.

Theorem 6.4.1 *Suppose F is strictly increasing at $F^{\leftarrow}(p)$ which means that for all $\epsilon > 0$*

$$F(F^{\leftarrow}(p) + \epsilon) > p, \quad F(F^{\leftarrow}(p) - \epsilon) < p. \tag{6.4}$$

Then we have $X_{\lceil np \rceil}^{(n)}$ is a weakly consistent quantile estimator,

$$X_{\lceil np \rceil}^{(n)} \xrightarrow{P} F^{\leftarrow}(p).$$

Proof. We begin with a reminder that

$$X_{\alpha}^{(n)} \leq y \text{ iff } nF_n(y) \geq \alpha. \tag{6.5}$$

We must prove for all $\epsilon > 0$,

$$P[|X_{\lceil np \rceil}^{(n)} - F^{\leftarrow}(p)| > \epsilon] \to 0, \tag{6.6}$$

which is equivalent to showing that for all $\epsilon > 0$,

$$P[X_{\lceil np \rceil}^{(n)} > F^{\leftarrow}(p) + \epsilon] \to 0, \tag{6.7}$$

$$P[X_{\lceil np \rceil}^{(n)} \leq F^{\leftarrow}(p) - \epsilon] \to 0. \tag{6.8}$$

From (6.5), this in turn is equivalent to showing that for all $\epsilon > 0$:

$$\begin{aligned} 1 - P[X_{\lceil np \rceil}^{(n)} \leq F^{\leftarrow}(p) + \epsilon] &= 1 - P[nF_n(F^{\leftarrow}(p) + \epsilon) \geq \lceil np \rceil] \\ &= P[nF_n(F^{\leftarrow}(p) + \epsilon) < \lceil np \rceil] \to 0 \end{aligned} \tag{6.9}$$

and

$$P[nF_n(F^{\leftarrow}(p) - \epsilon) \geq \lceil np \rceil] \to 0. \tag{6.10}$$

For (6.10) we have

$$P[F_n(F^{\leftarrow}(p)-\epsilon)\geq \frac{\lceil np\rceil}{n}]$$
$$= P[F_n(F^{\leftarrow}(p)-\epsilon)-F(F^{\leftarrow}(p)-\epsilon)\geq \frac{\lceil np\rceil}{n}-F(F^{\leftarrow}(p)-\epsilon)], \tag{6.11}$$

where we centered the random variable on the left to have zero mean. Now $\frac{\lceil np\rceil}{n}\to p$ and by the WLLN

$$F_n(F^{\leftarrow}(p)-\epsilon)-F(F^{\leftarrow}(p)-\epsilon)\stackrel{P}{\to} 0.$$

Also by (6.4), there exists $\delta>0$ such that

$$\delta := p - F(F^{\leftarrow}(p)-\epsilon)>0.$$

For all large n,

$$\frac{\lceil np\rceil}{n}-F(F^{\leftarrow}(p)-\epsilon)\geq \frac{\delta}{2}>0.$$

So the probability in (6.11) is bounded above by

$$P[|F_n(F^{\leftarrow}(p)-\epsilon)-F(F^{\leftarrow}(p)-\epsilon)|\geq \frac{\delta}{2}]\to 0.$$

Similarly, we can show the convergence in (6.9). □

6.5 L_p Convergence

In this section we examine a form of convergence and a notion of distance that is widely used in statistics and time series analysis. It is based on the L_p metric.

Recall the notation $X\in L_p$ which means $E(|X|^p)<\infty$. For random variables $X, Y\in L_p$, we define the L_p metric for $p\geq 1$ by

$$d(X,Y)=(E|X-Y|^p)^{1/p}.$$

This indeed defines a metric but proving this is not completely trivial. (The triangle inequality is the Minkowski Inequality.) This metric is norm induced because

$$\|X\|_p := (E|X|^p)^{1/p}$$

is a norm on the space L_p.

A sequence $\{X_n\}$ of random variables converges in L_p to X, written

$$X_n \stackrel{L_p}{\to} X,$$

if

$$E(|X_n - X|^p) \to 0$$

as $n \to \infty$.

The most important case is when $p = 2$, in which case L_2 is a Hilbert space with the inner product of X, Y defined by the covariance of X, Y. Here are two simple examples which use the L_2 metric.

1. Define $\{X_n\}$ to be a (2nd order, weakly, covariance) stationary process if $EX_n =: m$ independent of n and

$$\mathrm{Corr}(X_n, X_{n+k}) = \rho(k)$$

for all n. No distributional structure is specified. The *best linear predictor* of X_{n+1} based on $X_1, \dots, X_n$ is the linear combination of $X_1, \dots, X_n$ which achieves *minimum mean square error (MSE)*. Call this predictor $\widehat{X}_{n+1}$. Then $\widehat{X}_{n+1}$ is of the form $\widehat{X}_{n+1} = \sum_1^n \alpha_i X_i$ and $\alpha_1, \dots, \alpha_n$ are chosen so that

$$E(\widehat{X}_{n+1} - X_{n+1})^2 = \min_{\alpha_1, \dots, \alpha_n} E(\sum_{i=1}^n \alpha_i X_i - X_{n+1})^2.$$

2. Suppose $\{X_n\}$ is an iid sequence of random variables with $E(X_n) = \mu$ and $\mathrm{Var}(X_n) = \sigma^2$. Then

$$\bar{X} = \sum_1^n X_i/n \xrightarrow{L_2} \mu,$$

since

$$\begin{aligned} E(\frac{S_n}{n} - \mu)^2 &= \frac{1}{n^2} E(S_n - n\mu)^2 \\ &= \frac{1}{n^2} \mathrm{Var}(S_n) \\ &= \frac{n\sigma^2}{n^2} \to 0. \end{aligned}$$

□

Here are some basic facts about L_p convergence.

(i) L_p convergence implies convergence in probability: For $p > 0$, if $X_n \xrightarrow{L_p} X$ then $X_n \xrightarrow{P} X$.

This follows readily from Chebychev's inequality,

$$P[|X_n - X| \geq \epsilon] \leq \frac{E(|X_n - X|^p)}{\epsilon^p} \to 0. \qquad (6.12)$$

(ii) Convergence in probability does not imply L_p convergence. What can go wrong is that the nth function in the sequence can be huge on a very small set.

Here is a simple example. Let the probability space be $([0,1], \mathcal{B}([0,1]), \lambda)$ where λ is Lebesgue measure and set

$$X_n = 2^n 1_{(0,\frac{1}{n})}.$$

Then

$$P[|X_n| > \epsilon] = P\Big((0, \frac{1}{n})\Big) = \frac{1}{n} \to 0$$

but

$$E\left(|X_n|^p\right) = 2^{np}\frac{1}{n} \to \infty.$$

(iii) L_p convergence does not imply almost sure convergence as shown by the following simple example. Consider the functions $\{X_n\}$ defined on $([0,1], \mathcal{B}([0,1]), \lambda)$ where λ is Lebesgue measure.

$$X_1 = 1_{[0,\frac{1}{2}]}, \quad X_2 = 1_{[\frac{1}{2},1]}$$
$$X_3 = 1_{[0,\frac{1}{3}]}, \quad X_4 = 1_{[\frac{1}{3},\frac{2}{3}]}$$
$$X_5 = 1_{[\frac{2}{3},1]}, \quad X_6 = 1_{[0,\frac{1}{4}]}$$

and so on. Note that for any $p > 0$,

$$E(|X_1|^p) = \frac{1}{2}, \quad E(|X_2|^p) = \frac{1}{2},$$
$$E(|X_3|^p) = \frac{1}{3}, \dots, E(|X_6|^p) = \frac{1}{4}.$$

So $E(|X_n|^p) \to 0$ and

$$X_n \stackrel{L_p}{\to} 0.$$

Observe that $\{X_n\}$ does not converge almost surely to 0. □

Deeper and more useful connections between modes of convergence depend on the notion of uniform integrability (ui) which is discussed next.

6.5.1 Uniform Integrability

Uniform integrability is a property of a family of random variables which says that the first absolute moments are uniformly bounded and the distribution tails of the random variables in the family converge to 0 at a uniform rate. We give the formal definition.

Definition. A family $\{X_t, t \in T\}$ of L_1 random variables indexed by T is *uniformly integrable* (abbreviated ui) if

$$\sup_{t\in T} E\left(|X_t| 1_{[|X_t|>a]}\right) = \sup_{t\in T} \int_{[|X_t|>a]} |X_t| dP \to 0$$

as $a \to \infty$; that is,

$$\int_{[|X_t|>a]} |X_t| dP \to 0$$

as $a \to \infty$, uniformly in $t \in T$.

We next give some simple **criteria** for various families to be uniformly integrable.

(1) If $T = \{1\}$ consists of one element, then

$$\int_{[|X_1|>a]} |X_1| dP \to 0$$

as a consequence of $X_1 \in L_1$ and Exercise 6 of Chapter 5.

(2) **Dominated families.** If there exists a dominating random variable $Y \in L_1$, such that

$$|X_t| \leq Y$$

for all $t \in T$, then $\{X_t\}$ is ui. To see this we merely need to observe that

$$\sup_{t\in T} \int_{[|X_t|>a]} |X_t| dP \leq \int_{[|Y|>a]} |Y| \to 0, \quad a \to \infty.$$

(3) **Finite families.** Suppose $X_i \in L_1$, for $i = 1, \ldots, n$. Then the finite family $\{X_1, X_2, \ldots, X_n\}$ is ui. This follows quickly from the finite family being dominated by an integrable random variable,

$$|X_i| \leq \sum_{i=1}^{n} |X_j| \in L_1$$

and then applying (2).

(4) **More domination.** Suppose for each $t \in T$ that $X_t \in L_1$ and $Y_t \in L_1$ and

$$|X_t| \leq |Y_t|.$$

Then if $\{Y_t\}$ is ui so is $\{X_t\}$ ui.

This is an easy consequence of the definition.

(5) **Crystal Ball Condition.** For $p > 0$, the family $\{|X_n|^p\}$ is ui, if

$$\sup_n E\left(|X_n|^{p+\delta}\right) < \infty, \tag{6.13}$$

for some $\delta > 0$.

For example, suppose $\{X_n\}$ is a sequence of random variables satisfying $E(X_n) = 0$, and $\text{Var}(X_n) = 1$ for all n. Then $\{X_n\}$ is ui.

To verify sufficiency of the crystal ball condition, write

$$\begin{aligned}\sup_n \int_{[|X_n|^p > a]} |X_n|^p dP &= \sup_n \int_{[|\frac{X_n}{a^{1/p}}|>1]} |X_n|^p \cdot 1 dP \\ &= \sup_n \int_{[\frac{|X_n|^\delta}{a^{\delta/p}}>1]} |X_n|^p \cdot 1 dP \\ &\leq \sup_n \int |X_n|^p \frac{|X_n|^\delta}{a^{\delta/p}} dP \\ &\leq a^{-\delta/p} \sup_n E\left(|X_n|^{p+\delta}\right) \\ &\to 0,\end{aligned}$$

as $a \to \infty$. □

We now characterize uniform integrability in terms of uniform absolute continuity and uniform boundedness of the first absolute moments.

Theorem 6.5.1 *Let $\{X_t, t \in T\}$ be L_1 random variables. This family is ui iff*

(A) Uniform absolute continuity: For all $\epsilon > 0$, there exists $\xi = \xi(\epsilon)$, such that

$$\forall A \in \mathcal{B}: \quad \sup_{t \in T} \int_A |X_t| dP < \epsilon \text{ if } P(A) < \xi,$$

and

(B) Uniform bounded first absolute moments:

$$\sup_{t \in T} E\left(|X_t|\right) < \infty.$$

Proof. Suppose $\{X_t\}$ is ui. For any $X \in L_1$ and $a > 0$

$$\begin{aligned}\int_A |X| dP &= \int_{A[|X| \leq a]} |X| dP + \int_{A[|X|>a]} |X| dP \\ &\leq aP(A) + \int_{[|X|>a]} |X| dP.\end{aligned}$$

So

$$\sup_{t\in T}\int_A |X_t|dP \le aP(A) + \sup_{t\in T}\int_{|X_t|>a} |X_t|dP.$$

Insert $A = \Omega$ and we get (B). To get (A) pick "a" so large that

$$\sup_{t\in T}\int_{[|X_t|>a]} |X_t|dP \le \frac{\epsilon}{2}.$$

If

$$P(A) \le \frac{\epsilon/2}{a} = \xi,$$

then

$$\sup_{t\in T}\int_A |X_t|dP \le \frac{\epsilon}{2} + \frac{\epsilon}{2} = \epsilon$$

which is (A).

Conversely: Suppose (A) and (B) hold. Chebychev's inequality implies

$$\sup_{t\in T} P[|X_t| > a] \le \sup_{t\in T} E\left(|X_t|\right)/a = \text{const}/a$$

from (B). Now we apply (A): Given $\epsilon > 0$, there exists ξ such that whenever $P(A) < \xi$, we have

$$\int_A |X_t|dP < \epsilon$$

for all $t \in T$. Pick "a" large enough so that $P[|X_t| > a] < \xi$, for all t. Then for all t we get

$$\int_{[|X_t|>a]} |X_t|dP < \epsilon,$$

which is the ui property. □

Example 6.5.1 Let $\{X_n\}$ be a sequence of random variables with

$$P[X_n = 0] = p, P[X_n = n] = q, \quad p + q = 1.$$

Find a value of $p = p_n$ so that

$$1 = E(X_n) = 0 \cdot p + nq$$

and thus

$$q = \frac{1}{n}, \quad p = 1 - \frac{1}{n}.$$

Since $X_n \ge 0$,

$$\sup_{n\ge1} E\left(|X_n|\right) = 1$$

but the family in not ui since

$$\int_{[|X_n|>a]} |X_n| dP = \begin{cases} 1, & \text{if } a \le n, \\ 0, & \text{if } a > n. \end{cases}$$

This entails

$$\sup_{n\ge1} \int_{[|X_n|>a]} |X_n| dP = 1.$$

□

6.5.2 *Interlude: A Review of Inequalities*

We pause in our discussion of L_p convergence and uniform integrability to discuss some standard moment inequalities. We take these up in turn.

1. Schwartz Inequality: Suppose we have two random variables $X, Y \in L_2$. Then

$$|E(XY)| \le E(|XY|) \le \sqrt{E(X^2)E(Y^2)}.$$

To prove this, note for any $t \in \mathbb{R}$ that

$$\begin{aligned} 0 &\le E(X - tY)^2 = E(X^2) - 2tE(XY) + t^2E(Y^2) \qquad (6.14) \\ &=: q(t) \end{aligned}$$

and that $q(\cdot)$ is differentiable

$$q'(t) = -2E(XY) + 2tE(Y^2).$$

Set $q'(t) = 0$ and solve to get

$$t = E(XY)/EY^2.$$

Substitute this value of t into (6.14) which yields

$$0 \le E(X^2) - 2\frac{E(XY)}{E(Y^2)}E(XY) + \left(\frac{E(XY)}{E(Y^2)}\right)^2 E(Y^2).$$

Multiply through by $E(Y^2)$. □

A minor modification of this procedure shows when equality holds. This is discussed in Problem 9 of Section 6.7

2. Hölder's inequality: Suppose p, q satisfy

$$p > 1,\ q > 1,\ \frac{1}{p} + \frac{1}{q} = 1$$

and that

$$E(|X|^p) < \infty, \quad E(|Y|^q) < \infty.$$

Then

$$|E(XY)| \le E(|XY|) \le (E|X|^p)^{1/p}(E|Y|^q)^{1/q}.$$

In terms of norms this says

$$\|XY\|_1 \le \|X\|_p\|Y\|_q.$$

Note Schwartz's inequality is the special case $p = q = 2$.

The proof of Hölder's inequality is a convexity argument. First note that if $E(|X|^p) = 0$, then $X = 0$ a.s.. Thus, $E(|XY|) = 0$ and the asserted inequality holds. Similarly if $E(|Y|^q) = 0$. So suppose the right side of Hölder's inequality is strictly positive. (This will allow us at the end of the proof to divide by $\|X\|_p\|Y\|_q$.)

Observe for $a > 0,\ b > 0$ there exist $s, t \in \mathbb{R}$ such that

$$a = \exp\{p^{-1}s\}, \quad b = \exp\{q^{-1}t\}. \tag{6.15}$$

Since $\exp\{x\}$ is convex on $\mathbb{R}$ and $p^{-1} + q^{-1} = 1$, we have by convexity

$$\exp\{p^{-1}s + q^{-1}t\} \le p^{-1}\exp\{s\} + q^{-1}\exp\{t\},$$

or from the definition of s, t

$$ab \le p^{-1}a^p + q^{-1}b^q.$$

Now replace a by $|X|/\|X\|_p$ and b by $|Y|/\|Y\|_q$ to get

$$\frac{|XY|}{\|X\|_p\|Y\|_q} \le p^{-1}\left(\frac{|X|}{\|X\|_p}\right)^p + q^{-1}\left(\frac{|Y|}{\|Y\|_q}\right)^q$$

and so, after taking expectations, we get

$$\frac{E(|XY|)}{\|X\|_p\|Y\|_q} \le p^{-1} + q^{-1} = 1.$$

□

3. Minkowski Inequality: For $1 \le p < \infty$, assume $X, Y \in L_p$. Then $X+Y \in L_p$ and

$$\|X+Y\|_p \le \|X\|_p + \|Y\|_p.$$

To see why L_p is closed under addition, note

$$|X+Y|^p \le 2(|X|^p \vee |Y|^p) \le 2(|X|^p + |Y|^p) \in L_p.$$

If $p = 1$, Minkowski's inequality is obvious from the ordinary triangle inequality. So assume $1 < p < \infty$ and choose q conjugate to p so that

$p^{-1}+q^{-1}=1$ and thus $p-1=p/q$. Now we apply Hölder's inequality. We have

$$\begin{aligned}\|X+Y\|_p^p =&E(|X+Y|^p) = E\left(|X+Y||X+Y|^{p/q}\right)\\ \leq&E\left(|X||X+Y|^{p/q}\right)+E\left(|Y||X+Y|^{p/q}\right)\end{aligned}$$

and applying Hölder's inequality, we get the bound

$$\begin{aligned}&\leq(\|X\|_p\||X+Y|^{p/q}\|_q+\|Y\|_p\||X+Y|^{p/q}\|_q\\ &=(\|X\|_p+\|Y\|_p)\||X+Y|^{p/q}\|_q\\ &=(\|X\|_p+\|Y\|_p)\left(E|X+Y|^p\right)^{q^{-1}}\\ &=(\|X\|_p+\|Y\|_p)\|X+Y\|_p^{p/q}\\ &=(\|X\|_p+\|Y\|_p)\|X+Y\|_p^{p-1}.\end{aligned}$$

Assuming the last factor is non-zero, we may divide through to get the result. If the last factor equals zero, both sides of the inequality are zero. □

4. Jensen's inequality: Suppose $u:\mathbb{R}\mapsto\mathbb{R}$ is convex and $E(|X|)<\infty$ and $E(|u(X)|)<\infty$. Then

$$E(u(X))\geq u(E(X)).$$

This is more general than the variance inequality $\text{Var}(X)\geq 0$ implying $E(X^2)\geq(EX)^2$ which is the special case of Jensen's inequality for $u(x)=x^2$.

If u is concave, the inequality reverses.

For the proof of Jensen's inequality, note that u convex means for each $\xi\in\mathbb{R}$, there exists a supporting line L through $(\xi,u(\xi))$ such that graph of u is above the line. So

$$u(x)\geq \text{ line L thru } (\xi,u(\xi))$$

and therefore, parameterizing the line, we have

$$u(x)\geq u(\xi)+\lambda(x-\xi)$$

where λ is the slope of L. Let $\xi=E(X)$. Then for all x

$$u(x)\geq u(E(X))+\lambda(x-E(X)).$$

(Note λ depends on $\xi=E(X)$ but does not depend on x.) Now let $x=X$ and we get

$$u(X)\geq u(E(X))+\lambda(X-E(X)).$$

Taking expectations

$$Eu(X)\geq u(E(X))+\lambda E(X-EX)=u(E(X)).$$ □

Example 6.5.2 (An application of Hölder's Inequality) Let $0 < \alpha < \beta$ and set

$$r = \frac{\beta}{\alpha} > 1, \quad s = \frac{\beta}{\beta - \alpha}.$$

Then

$$\frac{1}{r} + \frac{1}{s} = \frac{\alpha}{\beta} + \frac{\beta - \alpha}{\beta} = \frac{\beta}{\beta} = 1.$$

Set

$$Z = |X|^{\alpha}, \quad Y = 1.$$

With these definitions, we have by Hölder's inequality that

$$E(|ZY|) \leq (E|Z|^r)^{1/r}(E|Y|^s)^{1/s};$$

that is,

$$E(|X|^{\alpha}) \leq (E|X|^{r\alpha})^{1/r} 1 = (E|X|^{\beta})^{\alpha/\beta},$$

so that

$$(E|X|^{\alpha})^{1/\alpha} \leq (E|X|^{\beta})^{1/\beta},$$

and

$$\|X\|_{\alpha} \leq \|X\|_{\beta}.$$

We conclude that $X \in L_{\beta}$ implies $X \in L_{\alpha}$, provided $\alpha < \beta$. Furthermore

$$\|X\|_t = (E|X|^t)^{1/t}$$

is non-decreasing in t.

Also if

$$X_n \xrightarrow{L_p} X$$

and $p' < p$, then

$$X_n \xrightarrow{L_{p'}} X.$$

6.6 More on L_p Convergence

This section utilizes the definitions and basic properties of L_p convergence, uniform integrability and the inequalities discussed in the previous section. We begin with some relatively easy implications of L_p convergence. We work up to an answer to the question: If random variables converge, when do their moments converge?

Assume the random variables $\{X, X_n, n \geq 1\}$ are all defined on $(\Omega, \mathcal{B}, P)$.

1. *A form of Scheffé's lemma*: We have the following equivalence for L_1 convergence: As $n \to \infty$

$$X_n \xrightarrow{L_1} X$$

iff

$$\sup_{A \in \mathcal{B}} \left| \int_A X_n dP - \int_A X dP \right| \to 0. \tag{6.16}$$

Note that if we replace A by Ω in (6.16) that we get

$$|E(X_n) - E(X)| \leq E|X_n - X| \to 0$$

so that first moments converge. This, of course, also follows by the modulus inequality.

To verify (6.16), suppose first that $X_n \xrightarrow{L_1} X$. Then we have

$$\begin{aligned}
\sup_A \left| \int_A X_n dP - \int_A X dP \right| & \\
&= \sup_A \left| \int_A (X_n - X) dP \right| \\
&\leq \sup_A \int_A |X_n - X| dP \\
&\leq \int |X_n - X| dP \\
&= E(|X_n - X|) \to 0.
\end{aligned}$$

For the converse, suppose (6.16) holds. Then

$$\begin{aligned}
E|X_n - X| &= \int_{[X_n > X]} (X_n - X) dP + \int_{[X_n \leq X]} (X - X_n) dP \\
&= \left(\int_{[X_n > X]} X_n - \int_{[X_n > X]} X \right) \\
&\quad + \left(\int_{[X_n \leq X]} X - \int_{[X_n \leq X]} X_n \right) \\
&\leq 2 \sup_A \left| \int_A X_n - \int_A X \right|.
\end{aligned}$$

□

2. If

$$X_n \xrightarrow{L_p} X$$

then

$$E(|X_n|^p) \to E(|X|^p)$$

or equivalently

$$\|X_n\|_p \to \|X\|_p.$$

For this verification, write

$$X = X_n + X - X_n$$

and Minkowski's inequality implies

$$\|X\|_p \leq \|X_n\|_p + \|X - X_n\|_p. \tag{6.17}$$

Interchange the roles of X_n and X in (6.17) to get

$$\|X_n\|_p \leq \|X\|_p + \|X - X_n\|_p. \tag{6.18}$$

So combining (6.17)and (6.18) we get

$$|\|X_n\|_p - \|X\|_p| \leq \|X - X_n\|_p \to 0, \tag{6.19}$$

as was to be proved. □

Towards a resolution of the problem of when moments of a sequence of random variables converge, we present the following result which deals with the case $p = 1$.

Theorem 6.6.1 *Suppose for $n \geq 1$ that $X_n \in L_1$. The following statements are equivalent:*

(a) $\{X_n\}$ is L_1-convergent.

(b) $\{X_n\}$ is L_1-cauchy; that is,

$$E|X_n - X_m| \to 0,$$

as $n, m \to \infty$.

(c) $\{X_n\}$ is uniformly integrable and $\{X_n\}$ converges in probability.

So if $X_n \stackrel{a.s.}{\to} X$ or $X_n \stackrel{P}{\to} X$ and $\{X_n\}$ is ui, then the first moments converge:

$$|E(X_n) - E(X)| \leq E(|X_n - X|) \to 0.$$

Later, we will see that convergence i.p. of $\{X_n\}$ can be replaced by convergence in distribution to X.

Proof. (a)→(b): L_1 convergence implies Cauchy convergence because of the triangle inequality.

(b)→(c): Given (b) we first show that $\{X_n\}$ is ui. Given $\epsilon > 0$, there exists N_ϵ such that if $m, n \geq N_\epsilon$ then

$$\int |X_n - X_m| dP < \epsilon/2. \tag{6.20}$$

To show $\{X_n\}$ is ui, we use Theorem 6.5.1. For any $A \in \mathcal{B}$

$$\int_A |X_n| dP \le \int_A |X_n - X_{N_\epsilon} + X_{N_\epsilon}| dP$$
$$\le \int_A |X_{N_\epsilon}| dP + \int |X_n - X_{N_\epsilon}| dP.$$

For any $n \ge N_\epsilon$

$$\int_A |X_n| dP \le \int_A |X_{N_\epsilon}| dP + \epsilon/2;$$

that is,

$$\sup_{n \ge N_\epsilon} \int_A |X_n| dP \le \int_A |X_{N_\epsilon}| dP + \epsilon/2.$$

and thus

$$\sup_n \int_A |X_n| dP \le \sup_{m \le N_\epsilon} \int_A |X_m| dP + \epsilon/2.$$

If $A = \Omega$, we conclude

$$\sup_n E(|X_n|) \le \sup_{m \le N_\epsilon} E(|X_m|) + \epsilon/2 < \infty.$$

Furthermore, since finite families of L_1 rv's are ui, $\{X_m, m \le N_\epsilon\}$ is ui and given $\epsilon > 0$, there exists $\delta > 0$ such that if $P(A) < \delta$, then

$$\sup_{m \le N_\epsilon} \int_A |X_m| dP < \epsilon/2$$

so we may conclude that whenever $P(A) < \delta$,

$$\sup_n \int_A |X_n| \le \epsilon/2 + \epsilon/2 = \epsilon.$$

Hence $\{X_n\}$ is ui.

To finish the proof that (b) implies (c), we need to check that $\{X_n\}$ converges in probability. But

$$P[|X_n - X_m| > \epsilon] \le E(|X_n - X_m|)/\epsilon \to 0$$

so $\{X_n\}$ is Cauchy i.p. and hence convergent in probability.

(c)$\to$(a): If $X_n \stackrel{P}{\to} X$, then there exists a subsequence $\{n_k\}$ such that

$$X_{n_k} \stackrel{a.s.}{\to} X,$$

and so by Fatou's lemma

$$E(|X|) = E(\liminf_{n_k\to\infty} |X_{n_k}|) \le \liminf_{n_k\to\infty} E(|X_{n_k}|) \le \sup_n E(|X_n|) < \infty$$

since $\{X_n\}$ is ui. So $X \in L_1$. Also, for any $\epsilon > 0$

$$\begin{aligned}\int |X_n - X|dP &\le \int_{[|X_n-X|\le\epsilon]} |X_n - X|dP + \int_{[|X_n-X|>\epsilon]} |X_n|dP \\ &\quad + \int_{[|X_n-X|>\epsilon]} |X|dP \\ &\le \epsilon + A + B.\end{aligned}$$

Since $X_n \xrightarrow{P} X$,

$$P[|X_n - X| > \epsilon] \to 0$$

and hence $B \to 0$ as $n \to \infty$ by Exercise 6 of Chapter 5.

To verify $A \to 0$, note that since $\{X_n\}$ is ui, given $\epsilon > 0$, there exists $\delta > 0$ such that

$$\sup_{k\ge1} \int_A |X_k|dP < \epsilon$$

if $P(A) < \delta$. Choose n so large that

$$P[|X_n - X| > \epsilon] < \delta$$

and then $A < \epsilon$. □

Example. Suppose X_1 and X_2 are iid $N(0, 1)$ random variables and define $Y = X_1/|X_2|$. The variable Y has a Cauchy distribution with density

$$f(y) = \frac{1}{\pi(1+y^2)}, \quad y \in \mathbb{R}.$$

Define

$$Y_n = \frac{X_1}{\frac{1}{n} + |X_2|}.$$

Then

$$Y_n \to Y$$

but $\{Y_n\}$ is NOT ui, since $E(Y_n) = 0$ but $E(|Y|) = \infty$. If $\{Y_n\}$ were ui, then by Theorem 6.6.1 we would have

$$E(Y_n) \to E(Y)$$

but the expectation of Y does not exist.

We give more detail about why $E(Y_n) = 0$. Let $F_1 = F_2$ be standard normal distributions. Then

$$E(Y_n) = \iint_{\mathbb{R}^2} \frac{x_1}{n^{-1} + |x_2|} F_1 \times F_2(dx_1, dx_2).$$

Note that the integrand is in $L_1(F_1 \times F_2)$ since

$$\iint_{\mathbb{R}^2} \left| \frac{x_1}{n^{-1} + |x_2|} \right| F_1 \times F_2(dx_1, dx_2) \leq n \iint_{\mathbb{R}^2} |x_1| F_1 \times F_2(dx_1, dx_2)$$
$$= nE(|X_1|).$$

Thus, from Fubini's theorem

$$E(X_n) = \int_{\mathbb{R}} \frac{1}{n^{-1} + |x_2|} \left[\int_{\mathbb{R}} x_1 F_1(dx_1) \right] F_2(dx_2) = 0,$$

since the inner integral is zero. □

We give the extension of Theorem 6.6.1 to higher moments.

Theorem 6.6.2 *Suppose $p \geq 1$ and $X_n \in L_p$. The following are equivalent.*

(a) *$\{X_n\}$ is L_p convergent.*

(b) *$\{X_n\}$ is L_p-cauchy; that is*

$$\|X_n - X_m\|_p \to 0,$$

as $n, m \to \infty$.

(c) *$\{|X_n|^p\}$ is uniformly integrable and $\{X_n\}$ is convergent in probability.*

Note that this Theorem states that L_p is a complete metric space; that is, every Cauchy sequence has a limit.

Proof. The proof is similar to that of Theorem 6.6.1 and is given briefly.

(a)→(b): Minkowski's inequality means $\|X\|_p$ is a norm satisfying the triangle inequality so

$$\|X_n - X_m\|_p \leq \|X_n - X\|_p + \|X - X_m\|_p \to 0$$

as $n, m \to \infty$.

(b)→(c): If $\{X_n\}$ is L_p Cauchy, then it is Cauchy in probability (see (6.12)) so there exists X such that $X_n \stackrel{P}{\to} X$. To show uniform integrability we verify the conditions of Theorem 6.5.1. Note by (6.19) (with X_m replacing X), we have $\{\|X_n\|_p, n \geq 1\}$ is a Cauchy sequence of real numbers and hence convergent. So $\sup_n \|X_n\|_p < \infty$. This also implies that X, the limit in probability of X_n, is in L_p by Fatou. To finish, note we have

$$\int_A |X_n|^p dP \leq \int_A |X_n - X_m + X_m|^p dP$$

and applying the 2^p inequality of Exercise 3 we get the bound

$$\leq 2^p \int_A |X_n - X_m|^p dP + 2^p \int_A |X_m|^p dP$$
$$\leq 2^p \|X_n - X_m\|_p^p + 2^p \int_A |X_m|^p dP.$$

Given $\epsilon > 0$, there exists m_0 such that $n \geq m_0$ implies

$$\int_A |X_n|^p dP \leq \frac{\epsilon}{2} + 2^p \int_A |X_{m_0}|^p dP.$$

Since $X_{m_0} \in L_p$, we have $2^p \int_A |X_{m_0}|^p dP \to 0$ as $P(A) \to 0$ (see Exercise 6 of Chapter 5). The uniform integrability follows.

(c)→(a): As in Theorem 6.6.1, since $\{X_n\}$ is convergent in probability, there exists X such that along some subsequence $X_{n_k} \stackrel{a.s.}{\to} X$. Since $\{|X_n|^p\}$ is ui

$$E(|X|^p) \leq \liminf_{k\to\infty} E(|X_{n_k}|^p) \leq \bigvee_{n=1}^{\infty} E(|X_n|^p) < \infty,$$

so $X \in L_p$. One may now finish the proof in a manner wholly analogous to the proof of Theorem 6.6.1. □

6.7 Exercises

1. (a) Let $\{X_n\}$ be a monotone sequence of random variables. If

$$X_n \stackrel{P}{\to} X$$

then

$$X_n \stackrel{a.s.}{\to} X.$$

(Think subsequences.)

(b) Let $\{X_n\}$ be any sequence of random variables. Show that

$$X_n \stackrel{a.s.}{\to} X$$

iff

$$\sup_{k\geq n} |X_k - X| \stackrel{P}{\to} 0.$$

(c) Points are chosen at random on the circumference of the unit circle. Y_n is the arc length of the largest arc not containing any points when n points are chosen. Show $Y_n \to 0$ a.s.

(d) Let $\{X_n\}$ be iid with common distribution $F(x)$ which satisfies $F(x_0) = 1$, $F(x) < 1$ for $x < x_0$ with $x_0 < \infty$. Prove

$$\max\{X_1, \ldots, X_n\} \uparrow x_0$$

almost surely.

2. Let $\{X_n\}$ be iid, $EX_n = \mu$, $\text{Var}(X_n) = \sigma^2$. Set $\bar{X} = \sum_{i=1}^n X_i/n$. Show

$$\frac{1}{n}\sum_{i=1}^{n}(X_i - \bar{X})^2 \xrightarrow{P} \to \sigma^2.$$

3. Suppose $X \geq 0$ and $Y \geq 0$ are random variables and that $p \geq 0$.

   (a) Prove
   $$E((X+Y)^p) \leq 2^p \left(E(X^p) + E(Y^p)\right).$$

   (b) If $p > 1$, the factor 2^p may be replaced by 2^{p-1}.

   (c) If $0 \leq p \leq 1$, the factor 2^p may be replaced by 1.

4. Let $\{X_n, n \geq 1\}$ be iid, $EX_n = 0$, $EX_n^2 = \sigma^2$. Let $a_n \in \mathbb{R}$ for $n \geq 1$. Set $S_n = \sum_{i=1}^n a_i X_i$. Prove $\{S_n\}$ is L_2-convergent iff $\sum_{i=1}^\infty a_i^2 < \infty$.

5. Suppose $\{X_n\}$ is iid. Show $\{n^{-1}S_n, n \geq 1\}$ is ui provided $X_i \in L_1$.

6. Let $\{X_n\}$ be ui and let $X \in L_1$. Show $\{X_n - X\}$ is ui.

7. Let X_n be $N(0, \sigma_n^2)$. When is $\{X_n\}$ ui?

8. Suppose $\{X_n\}$ and $\{Y_n\}$ are two families of ui random variables defined on the same probability space. Is $\{X_n + Y_n\}$ ui?

9. When is there equality in the Schwartz Inequality? (Examine the derivation of the Schwartz Inequality.)

10. Suppose $\{X_n\}$ is a sequence for which there exists an increasing function $f : [0, \infty) \mapsto [0, \infty)$ such that $f(x)/x \to \infty$ as $x \to \infty$ and
    $$\sup_{n \geq 1} E\left(f(|X_n|)\right) < \infty.$$
    Show $\{X_n\}$ is ui.

    Specialize to the case where $f(x) = x^p$ for $p > 1$ or $f(x) = x(\log x)^+$.

11. Suppose $\{X_n, n \geq 1\}$ are iid and non-negative and define $M_n = \vee_{i=1}^n X_i$.

    (a) Check that
    $$P[M_n > x] \leq nP[X_1 > x].$$

    (b) If $E(X_1^p) < \infty$, then $M_n/n^{1/p} \xrightarrow{P} 0$.

    (c) If in addition to being iid, the sequence $\{X_n\}$ is non-negative, show $M_n/n \xrightarrow{P} 0$ iff $nP[X_1 > n] \to 0$, as $n \to \infty$.

(d) Review the definition of *rapid variation* in Exercise 27 of Chapter 5. Prove there exists a sequence $b(n) \to \infty$ such that

$$M_n/b(n) \xrightarrow{P} 1, \quad n \to \infty,$$

iff $1 - F(x) := P[X_1 > x]$ is rapidly varying at ∞. In this case, we may take

$$b(n) = \left(\frac{1}{1-F}\right)^{\leftarrow}(n)$$

to be the $1 - \frac{1}{n}$ quantile of F.

(e) Now suppose $\{X_n\}$ is an arbitrary sequence of non-negative random variables. Show that

$$E(M_n 1_{[M_n \geq \delta]}) \leq \sum_{k=1}^{n} E(X_k 1_{[X_k \geq \delta]}).$$

If in addition, $\{X_n\}$ is ui, show $E(M_n)/n \to 0$.

12. Let $\{X_n\}$ be a sequence of random variables.

(a) If $X_n \xrightarrow{P} 0$, then for any $p > 0$

$$\frac{|X_n|^p}{1 + |X_n|^p} \xrightarrow{P} 0 \qquad (6.21)$$

and

$$E\left(\frac{|X_n|^p}{1 + |X_n|^p}\right) \to 0. \qquad (6.22)$$

(b) If for some $p > 0$ (6.21) holds, then $X_n \xrightarrow{P} 0$.

(c) Suppose $p > 0$. Show $X_n \xrightarrow{P} 0$ iff (6.22).

13. Suppose $\{X_n, n \geq 1\}$ are identically distributed with finite variance. Show that

$$nP[|X_1| \geq \epsilon\sqrt{n}] \to 0$$

and

$$\frac{\bigvee_{i=1}^n |X_i|}{\sqrt{n}} \xrightarrow{P} 0.$$

14. Suppose $\{X_k\}$ are independent with

$$P[X_k = k^2] = \frac{1}{k^2} \quad P[X_k = -1] = 1 - \frac{1}{k^2}.$$

Show $\sum_{i=1}^n X_i \to -\infty$ almost surely as $n \to \infty$.

15. Suppose $X_n \geq 0$ for $n \geq 0$ and $X_n \xrightarrow{P} X_0$ and also $E(X_n) \to E(X_0)$. Show $X_n \to X_0$ in L_1. (Hint: Try considering $(X_0 - X_n)^+$.)

16. For any sequence of random variables $\{X_n\}$ set $S_n = \sum_{i=1}^n X_i$.

    (a) Show $X_n \xrightarrow{a.s.} 0$ implies $S_n/n \xrightarrow{a.s.} 0$.

    (b) Show $X_n \xrightarrow{L_p} 0$ implies $S_n/n \xrightarrow{L_p} 0$ for any $p \geq 1$.

    (c) Show $X_n \xrightarrow{P} 0$ does NOT imply $S_n/n \xrightarrow{P} 0$. (Try $X_n = 2^n$ with probability n^{-1} and $= 0$ with probability $1 - n^{-1}$. Alternatively look at functions on $[0, 1]$ which are indicators of $[i/n, (i+1)/n]$.)

    (d) Show $S_n/n \xrightarrow{P} 0$ implies $X_n/n \xrightarrow{P} 0$.

17. In a discrete probability space, convergence in probability is equivalent to almost sure convergence.

18. Suppose $\{X_n\}$ is an uncorrelated sequence, meaning

$$\mathrm{Cov}(X_i, X_j) = 0, \quad i \neq j.$$

    If there exists a constant $c > 0$ such that $\mathrm{Var}(X_n) \leq c$ for all $n \geq 1$, then for any $\alpha > 1/2$ we have

$$\frac{\sum_{i=1}^n X_i}{n^\alpha} \xrightarrow{L_2} 0.$$

19. If $0 \leq X_n \leq Y_n$ and $Y_n \xrightarrow{P} 0$, check $X_n \xrightarrow{P} 0$.

20. Suppose $E(X^2) = 1$ and $E(|X|) \geq a > 0$. Prove for $0 \leq \lambda \leq 1$ that

$$P[|X| \geq \lambda a] \geq (1 - \lambda)^2 a^2.$$

21. Recall the notation $d(A, B) = P(A \Delta B)$ for events A, B. Prove

$$d(A_n, A) \to 0 \text{ iff } 1_{A_n} \xrightarrow{L_2} 1_A.$$

22. Suppose $\{X_n, n \geq 1\}$ are independent non-negative random variables satisfying

$$E(X_n) = \mu_n, \quad \mathrm{Var}(X_n) = \sigma_n^2.$$

    Define for $n \geq 1$, $S_n = \sum_{i=1}^n X_i$ and suppose $\sum_{n=1}^\infty \mu_n = \infty$ and $\sigma_n^2 \leq c\mu_n$ for some $c > 0$ and all n. Show $S_n/E(S_n) \xrightarrow{P} 1$.

23. A classical transform result says the following: Suppose $u_n \geq 0$ and $u_n \to u$ as $n \to \infty$. For $0 < s < 1$, define the generating function

$$U(s) = \sum_{n=0}^{\infty} u_n s^n.$$

Show that

$$\lim_{s \to 1} (1 - s)U(s) = u$$

by the following relatively painless method which uses convergence in probability: Let $T(s)$ be a geometric random variable satisfying

$$P[T(s) = n] = (1 - s)s^n.$$

Then $T(s) \xrightarrow{P} \infty$. What is $E(u_{T(s)})$?

24. Recall a random vector (X_n, Y_n) (that is, a random element of $\mathbb{R}^2$) converges in probability to a limit random vector (X, Y) if

$$d((X_n, Y_n), (X, Y)) \xrightarrow{P} 0$$

where d is the Euclidean metric on $\mathbb{R}^2$.

(a) Prove

$$(X_n, Y_n) \xrightarrow{P} (X, Y) \tag{6.23}$$

iff

$$X_n \xrightarrow{P} X \text{ and } Y_n \xrightarrow{P} Y.$$

(b) If $f : \mathbb{R}^2 \mapsto \mathbb{R}^d$ is continuous ($d \geq 1$), (6.23) implies

$$f(X_n, Y_n) \xrightarrow{P} f(X, Y).$$

(c) If (6.23) holds, then

$$(X_n + Y_n, X_n Y_n) \xrightarrow{P} (X + Y, XY).$$

25. For random variables X, Y define

$$\rho(X, Y) = \inf\{\delta > 0 : P[|X - Y| \geq \delta] \leq \delta\}.$$

(a) Show $\rho(X, Y) = 0$ iff $P[X = Y] = 1$. Form equivalence classes of random variables which are equal almost surely and show that ρ is a metric on the space of such equivalence classes.

(b) This metric metrizes convergence in probability:
$X_n \xrightarrow{P} X$ iff $\rho(X_n, X) \to 0$.

(c) The metric is complete: Every Cauchy sequence is convergent.

(d) It is impossible to metrize almost sure convergence.

26. Let the probability space be the Lebesgue interval; that is, the unit interval with Lebesgue measure.

(a) Define
$$X_n = \frac{n}{\log n} 1_{(0,n^{-1})}, \quad n \geq 3.$$
Then $\{X_n\}$ is ui, $E(X_n) \to 0$ but there is no integrable Y which dominates $\{X_n\}$.

(b) Define
$$X_n = n1_{(0,n^{-1})} - n1_{(n^{-1},2n^{-1})}.$$
Then $\{X_n\}$ is not ui but $X_n \xrightarrow{P} 0$ and $E(X_n) \to 0$.

27. Let X be a random variable in L_1 and consider the map
$$\chi : [1, \infty] \mapsto [0, \infty]$$
defined by $\chi(p) = \|X\|_p$. Let
$$p_0 := \sup\{p \geq 1 : \|X\|_p < \infty\}.$$
Show χ is continuous on $[1, p_0)$. Furthermore on $[1, p_0)$ the continuous function $p \mapsto \log \|X\|_p$ is convex.

28. Suppose u is a continuous and increasing mapping of $[0, \infty]$ onto $[0, \infty]$. Let $u^{\leftarrow}$ be its inverse function. Define for $x \geq 0$
$$U(x) = \int_0^x u(s)ds, \quad V(x) = \int_0^x u^{\leftarrow}(s)ds.$$
Show
$$xy \leq U(x) + V(y), \quad x, y \in [0, \infty].$$
(Draw some pictures.)

Hence, for two random variables X, Y on the same probability space, XY is integrable if $U(|X|) \in L_1$ and $V(|Y|) \in L_1$.

Specialize to the case where $u(x) = x^{p-1}$, for $p > 1$.

29. Suppose the probability space is $((0, 1], \mathcal{B}((0, 1]), \lambda)$ where λ is Lebesgue measure. Define the interval
$$A_n := [2^{-p}q, 2^{-p}(q + 1)],$$
where $2^p + q = n$ is the decomposition of n such that p and q are integers satisfying $p \geq 0, 0 \leq q < 2^p$. Show $1_{A_n} \xrightarrow{P} 0$ but that
$$\limsup_{n\to\infty} 1_{A_n} = 1, \quad \liminf_{n\to\infty} 1_{A_n} = 0.$$

30. **The space L_∞:** For a random variable X define

$$\|X\|_\infty = \sup\{x : P[|X| > x] > 0\}.$$

Let L_∞ be the set of all random variables X for which $\|X\|_\infty < \infty$.

(a) Show that for a random variable X and $1 < p < q < \infty$

$$0 \le \|X\|_1 \le \|X\|_p \le \|X\|_q \le \|X\|_\infty.$$

(b) For $1 < p < q < \infty$, show

$$L_\infty \subset L_q \subset L_p \subset L_1.$$

(c) Show Hölder's inequality holds in the form

$$E(|XY|) \le \|X\|_1 \|Y\|_\infty.$$

(d) Show Minkowski's inequality holds in the form

$$\|X + Y\|_\infty \le \|X\|_\infty + \|Y\|_\infty.$$

31. Recall the definition of *median* from Exercise 31 of Chapter 5.

(a) Let $\{X_n, n \ge 1\}$ be a sequence of random variables such that there exists a sequence of constants $\{c_n\}$ with the property that

$$X_n - c_n \xrightarrow{P} 0.$$

If $m(X_n)$ is a median of X_n, show

$$X_n - m(X_n) \xrightarrow{P} 0$$

and $c_n - m(X_n) \to 0$.

(b) If there exists a random variable X with a unique median such that $X_n \xrightarrow{P} X$, then $m(X_n) \to m(X)$.

32. For a random variable X, let $\gamma(X)$ be the central value defined in Exercise 25 of Chapter 5. For a sequence of random variables $\{X_n, n \ge 0\}$, suppose there exist a sequence of constants $\{c_n\}$ such that $X_n - c_n \to X_0$ almost surely. Show $\lim_{n\to\infty} X_n - \gamma(X_n)$ exists and is finite almost surely, where $\gamma(X_n)$ is the unique root of the equation $E(\arctan(X - \gamma) = 0$. Show $\lim_{n\to\infty}(c_n - \gamma(X_n))$ exists and is finite.

33. Suppose $\{X_{n,k}, 1 \le k \le n, n \ge 1\}$ is a triangular array of non-negative random variables. For $n \ge 1$, set

$$S_n = \sum_{i=1}^{n} X_{n,i}, \quad M_n = \bigvee_{i=1}^{n} X_{n,i}.$$

Show that $M_n \xrightarrow{P} 0$ implies $S_n/n \xrightarrow{P} 0$.

7

Laws of Large Numbers and Sums of Independent Random Variables

This chapter deals with the behavior of sums of independent random variables and with averages of independent random variables. There are various results that say that averages of independent (and approximately independent) random variables are approximated by some population quantity such as the mean. Our goal is to understand these results in detail.

We begin with some remarks on truncation.

7.1 Truncation and Equivalence

We will see that it is easier to deal with random variables that are uniformly bounded or that have moments. Many techniques rely on these desirable properties being present. If these properties are not present, a technique called *truncation* can induce their presence but then a comparison must be made between the original random variables and the truncated ones. For instance, we often want to compare

$$\{X_n\} \text{ with } \{X_n 1_{[|X_n| \le n]}\}$$

where the second sequence is considered the truncated version of the first.

The following is a useful concept, expecially for problems needing almost sure convergence.

Definition. Two sequences $\{X_n\}$ and $\{X'_n\}$ are tail equivalent if

$$\sum_n P[X_n \neq X'_n] < \infty. \tag{7.1}$$

When two sequences are tail equivalent, their sums behave asymptotically the same as shown next.

Proposition 7.1.1 (Equivalence) *Suppose the two sequences $\{X_n\}$ and $\{X'_n\}$ are tail equivalent. Then*

(1) $\sum_n (X_n - X'_n)$ converges a.s.

(2) The two series $\sum_n X_n$ and $\sum_n X'_n$ converge a.s. together or diverge a.s. together; that is

$$\sum_n X_n \text{ converges a.s. iff } \sum_n X'_n \text{ converges a.s.}$$

(3) If there exists a sequence $\{a_n\}$ such that $a_n \uparrow \infty$ and if there exists a random variable X such that

$$\frac{1}{a_n}\sum_{j=1}^{n} X_j \stackrel{a.s.}{\to} X,$$

then also

$$\frac{1}{a_n}\sum_{j=1}^{n} X'_j \stackrel{a.s.}{\to} X.$$

Proof. From the Borel-Cantelli Lemma, we have that (7.1) implies

$$P([X_n \neq X'_n] \text{ i.o.}) = 0,$$

or equivalently

$$P(\liminf_{n\to\infty}[X_n = X'_n]) = 1.$$

So for $\omega \in \liminf_{n\to\infty}[X_n = X'_n]$ we have that $X_n(\omega) = X'_n(\omega)$ from some index onwards, say for $n \geq N(\omega)$. This proves (1).

For (2) note

$$\sum_{n=N}^{\infty} X_n(\omega) = \sum_{n=N}^{\infty} X'_n(\omega).$$

For (3) we need only observe that

$$\frac{1}{a_n}\sum_{j=1}^{n}(X_j - X'_j) \stackrel{a.s.}{\to} 0.$$

□

7.2 A General Weak Law of Large Numbers

Recall that the *weak* law of large numbers refers to averages of random variables converging in the sense of convergence in probability. We first present a fairly general but easily proved result. Before proving the result, we look at several special cases.

Theorem 7.2.1 (General weak law of large numbers) *Suppose $\{X_n, n \geq 1\}$ are independent random variables and define $S_n = \sum_{j=1}^n X_j$. If*

$$(i) \qquad \sum_{j=1}^{n} P[|X_j| > n] \to 0, \tag{7.2}$$

$$(ii) \qquad \frac{1}{n^2} \sum_{j=1}^{n} E X_j^2 1_{[|X_j| \leq n]} \to 0, \tag{7.3}$$

then if we define

$$a_n = \sum_{j=1}^{n} E\left(X_j 1_{[|X_j| \leq n]}\right),$$

we get

$$\frac{S_n - a_n}{n} \xrightarrow{P} 0. \tag{7.4}$$

One of the virtues of this result is that no assumptions about moments need to be made. Also, although this result is presented as conditions which are sufficient for (7.4), the conditions are in fact necessary as well. We will only prove sufficiency, but first we discuss the specialization of this result to the iid case under progressively weaker conditions.

SPECIAL CASES:

(a) **WLLN with variances.** Suppose $\{X_n, n \geq 1\}$ are iid with $E(X_n) = \mu$ and $E(X_n^2) < \infty$. Then as $n \to \infty$,

$$\frac{1}{n} S_n \xrightarrow{P} \mu.$$

The proof is simple since Chebychev's inequality makes it easy to verify (7.2) and (7.3). For instance (7.2) becomes

$$nP[|X_1| > n] \leq nE(X_1)^2/n^2 \to 0$$

and (7.3) becomes

$$\frac{1}{n^2} nE(X_1^2 1_{[|X_1| \leq n]}) \leq \frac{1}{n} E(X_1^2) \to 0.$$

Finally, we observe, as $n \to \infty$

$$\frac{a_n}{n} = E(X_1 1_{[|X_1| \leq n]}) \to E(X_1) = \mu.$$

(b) **Khintchin's WLLN under the first moment hypothesis.** Suppose that $\{X_n, n \geq 1\}$ are iid with $E(|X_1|) < \infty$ and $E(X_n) = \mu$. (No assumption is made about variances.) Then

$$S_n/n \xrightarrow{P} \mu.$$

To prove this by means of Theorem 7.2.1, we show that (7.2) and (7.3) hold. For (7.3) observe that

$$\begin{aligned} nP[|X_1| > n] =& E(n1_{[|X_1|>n]}) \\ \le& E(|X_1|1_{[|X_1|>n]}) \to 0, \end{aligned}$$

since $E(|X_1|) < \infty$.

Next for (7.3), we use a divide and conquer argument. We have for any $\epsilon > 0$

$$\begin{aligned} \frac{1}{n}EX_1^2 1_{[|X_1|\le n]} &\le \frac{1}{n}\left(E(X_1^2 1_{[|X_1|\le \epsilon\sqrt{n}]}) + E(X_1^2 1_{[\epsilon\sqrt{n}\le|X_1|\le n]})\right) \\ &\le \frac{\epsilon^2 n}{n} + \frac{1}{n}E(n|X_1|1_{[\epsilon\sqrt{n}\le|X_1|\le n]}) \\ &\le \epsilon^2 + E(|X_1|1_{\epsilon\sqrt{n}\le|X_1|}) \\ &\to \epsilon^2, \end{aligned}$$

as $n \to \infty$, since $E(|X_1|) < \infty$. So applying Theorem 7.2.1 we conclude

$$\frac{S_n - nE(X_1 1_{[|X_1|\le n]})}{n} \xrightarrow{P} 0.$$

Since

$$\left|\frac{nE(X_1 1_{[|X_1|\le n]})}{n} - E(X_1)\right| \le E(|X_1|1_{[|X_1|>n]}) \to 0,$$

the result follows.

(c) **Feller's WLLN without a first moment assumption:** Suppose that $\{X_n, n \ge 1\}$ are iid with

$$\lim_{x\to\infty} xP[|X_1| > x] = 0. \tag{7.5}$$

Then

$$\frac{S_n}{n} - E(X_1 1_{[|X_1|\le n]}) \xrightarrow{P} 0.$$

The converse is true although we will not prove it. Note that this result makes no assumption about a finite first moment.

As before, we show (7.2) and (7.3) hold. The condition (7.2) becomes in the iid case $nP[|X_1| > n] \to 0$ which is covered by (7.5). To show (7.3) holds, we need to write it in terms of

$$\tau(x) := xP[|X_1| > x].$$

Let $P[X_1 \le x] = F(x)$ and because of the iid assumption (7.3) becomes

$$
\begin{aligned}
\frac{1}{n}\int_\Omega |X_1|^2 1_{[|X_1|\le n]} dP &= \frac{1}{n}\int_{\{x:|x|\le n\}} x^2 F(dx)\\
&= \frac{1}{n}\int_{|x|\le n} \Big(\int_{s=0}^{|x|} 2s ds\Big) F(dx)\\
&= \frac{1}{n}\int_{s=0}^{n} 2s\Big[\int_{s<|x|\le n} F(dx)\Big] ds \quad (\text{ by Fubini })\\
&= \frac{1}{n}\int_0^n 2s(P[|X_1|>s] - P[|X_1|>n]) ds\\
&= \frac{1}{n}\int_0^n 2\tau(s) ds - \frac{1}{n}\int_0^n 2s ds P[|X_1|>n]\\
&= \frac{2}{n}\int_0^n \tau(s) ds - \underbrace{nP[|X_1|>n]}_{\tau(n)} \to 0
\end{aligned}
$$

since if $\tau(s) \to 0$ so does its average. □

Proof of Theorem 7.2.1. Define

$$X'_{nj} = X_j 1_{[|X_j|\le n]} \text{ and } S'_n = \sum_{j=1}^n X'_{nj}.$$

Then

$$\sum_{j=1}^n P[X'_{nj} \ne X_j] = \sum_{j=1}^n P[|X_j| > n] \to 0.$$

So

$$
\begin{aligned}
P[|S_n - S'_n| > \epsilon] &\le P[S_n \ne S'_n]\\
&\le P\{\bigcup_{j=1}[X'_{nj} \ne X_j]\}\\
&\le \sum_{j=1}^n P[X'_{nj} \ne X_j] \to 0
\end{aligned}
$$

and therefore

$$S_n - S'_n \xrightarrow{P} 0. \tag{7.6}$$

The variance of a random variable is always bounded above by the second moment about 0 since

$$\text{Var}(X) = E(X^2) - (E(X))^2 \le E(X^2).$$

So by Chebychev's inequality

$$\begin{aligned}P\left[\left|\frac{S'_n - ES'_n}{n}\right| > \epsilon\right] &\le \frac{\text{Var}(S'_n)}{n^2\epsilon^2} \\ &\le \frac{1}{n^2\epsilon^2}\sum_{j=1}^{n} E(X'_{nj})^2 \\ &= \frac{1}{n^2\epsilon^2}\sum_{j=1}^{n} E\left(X_j^2 1_{[|X_j|\le n]}\right) \to 0\end{aligned}$$

where that convergence is due to (7.3).

Note $a_n = ES'_n = \sum_{j=1}^{n} EX_j 1_{[|X_j|\le n]}$, and thus

$$\frac{S'_n - a_n}{n} \xrightarrow{P} 0. \tag{7.7}$$

We therefore get

$$\frac{S_n - a_n}{n} = \frac{S_n - S'_n}{n} + \frac{S'_n - a_n}{n} \xrightarrow{P} 0$$

where the first difference goes to zero because of (7.6) and the second difference goes to zero because of (7.7). □

Example: Let F be a symmetric distribution on $\mathbb{R}$ and suppose

$$1 - F(x) = \frac{e}{2x\log x}, \quad x \ge e$$

Suppose $\{X_n, n \ge 1\}$ is an iid sequence of random variables with common distribution function F. Note that

$$EX^+ = \int_e^\infty \frac{e}{2x\log x}dx = \frac{e}{2}\int_1^\infty \frac{dy}{y} = \infty.$$

so because the distribution is symmetric

$$E(X^+) = E(X^-) = \infty$$

and $E(X)$ does not exist. However,

$$\tau(x) = xP[|X_1| > x] = x \cdot \frac{e}{x\log x} = \frac{e}{\log x} \to 0$$

and $a_n = 0$ because F is symmetric, so

$$\frac{S_n}{n} \xrightarrow{P} 0.$$

Thus, without a mean existing, the WLLN still holds. When we consider the strong law of large numbers, we will see that a.s. convergence fails. □

7.3 Almost Sure Convergence of Sums of Independent Random Variables

This section considers the basic facts about when sums of independent random variables converge. We begin with an inequality about tail probabilities of maxima of sums.

Proposition 7.3.1 (Skorohod's inequality) *Suppose $\{X_n, n \geq 1\}$ is an independent sequence of random variables and suppose $\alpha > 0$ is fixed. For $n \geq 1$, define $S_n = \sum_{i=1}^n X_i$, and set*

$$c := \sup_{j \leq N} P[|S_N - S_j| > \alpha].$$

Also, suppose $c < 1$. Then

$$P[\sup_{j \leq N} |S_j| > 2\alpha] \leq \frac{1}{1-c} P[|S_N| > \alpha]. \tag{7.8}$$

There are several similar inequalities which bound the tail probabilities of the maximum of the partial sums, the most famous of which is Kolmogorov's inequality. One advantage of Skorohod's inequality is that it does not require moment assumptions.

Note that in the iid case, the constant c somewhat simplifies since we can express it as

$$c = \bigvee_{j \leq N} P[|S_j| > \alpha] = \bigvee_{j \leq N} P[|S_N - S_j| > \alpha]$$

due to the fact that

$$(X_1, \ldots, X_N) \stackrel{d}{=} (X_N, \ldots, X_1).$$

This means that the sums of the variables written in reverse order ($S_N - S_j$, $j = N-1, N-2, \ldots, 1$) have the same distribution as $(S_1, \ldots, S_N)$.

Proof of Proposition 7.3.1. Define

$$J := \inf\{j : |S_j| > 2\alpha\},$$

with the convention that $\inf \emptyset = \infty$. Note that

$$[\sup_{j \leq N} |S_j| > 2\alpha] = [J \leq N] = \sum_{j=1}^{N} [J = j],$$

where the last union is a union of disjoint sets. Now we write

$$\begin{aligned} P[|S_N| > \alpha] &\geq P[|S_N| > \alpha, J \leq N] \\ &= \sum_{j=1}^{N} P[|S_N| > \alpha, J = j] \\ &\geq \sum_{j=1}^{N} P[|S_N - S_j| \leq \alpha, J = j]. \end{aligned} \tag{7.9}$$

To justify this last step, suppose

$$|S_N(\omega) - S_j(\omega)| \le \alpha, \text{ and } J(\omega) = j$$

so that $|S_j(\omega)| > 2\alpha$. If it were the case that $|S_N(\omega)| \le \alpha$, then it would follow that $|S_N(\omega) - S_j(\omega)| > \alpha$ which is impossible, so $|S_N(\omega)| > \alpha$. We have checked that

$$[|S_N - S_j| \le \alpha, J = j] \subset [|S_N| > \alpha, J = j]$$

which justifies (7.9). Thus,

$$P[|S_N| > \alpha] \ge \sum_{j=1}^{N} P[|S_N - S_j| \le \alpha, J = j].$$

It is also true that

$$S_N - S_j = \sum_{i=j+1}^{N} X_j \in \mathcal{B}(X_{j+1}, \dots, X_N)$$

and

$$[J = j] = [\sup_{i<j} |S_i| \le 2\alpha, |S_j| > 2\alpha] \in \mathcal{B}(X_1 \dots X_j).$$

Since

$$\mathcal{B}(X_{j+1}, \dots, X_N) \perp\!\!\!\perp \mathcal{B}(X_1 \dots X_j)$$

we have

$$\begin{aligned} P[|S_N| > \alpha] &\ge \sum_{j=1}^{N} P[|S_N - S_j| \le \alpha] P[J = j] \\ &\ge \sum_{j=1}^{N} (1 - c) P[J = j] \quad (\text{from the definition of } c) \\ &= (1 - c) P[J \le N] \\ &= (1 - c) P[\sup_{j \le N} |S_j| > 2\alpha]. \end{aligned}$$

□

Based on Skorohod's inequality, we may now present a rather remarkable result due to Lévy which shows the equivalence of convergence in probability to almost sure convergence for sums of independent random variables.

Reminder: If $\{\xi_n\}$ is a monotone sequence of random variables, then

$$\xi_n \xrightarrow{P} \xi$$

implies (and hence is equivalent to)

$$\xi_n \xrightarrow{a.s.} \xi.$$

Theorem 7.3.2 (Lévy's theorem) *If $\{X_n, n \geq 1\}$ is an independent sequence of random variables,*

$$\sum_n X_n \text{ converges i.p. iff } \sum_n X_n \text{ converges a.s.}$$

This means that if $S_n = \sum_{i=1}^n X_i$, then the following are equivalent:

1. *$\{S_n\}$ is Cauchy in probability.*
2. *$\{S_n\}$ converges in probability.*
3. *$\{S_n\}$ converges almost surely.*
4. *$\{S_n\}$ is almost surely Cauchy.*

Proof. Assume $\{S_n\}$ is convergent in probability, so that $\{S_n\}$ is Cauchy in probability. We show $\{S_n\}$ is almost surely convergent by showing $\{S_n\}$ is almost surely Cauchy. To show that $\{S_n\}$ is almost surely Cauchy, we need to show

$$\xi_N = \sup_{m,n \geq N} |S_m - S_n| \to 0 \text{ a.s.},$$

as $N \to \infty$. But $\{\xi_N, N \geq 1\}$ is a decreasing sequence so from the reminder it suffices to show $\xi_N \xrightarrow{P} 0$ as $N \to \infty$. Since

$$\begin{aligned}
\xi_N &= \sup_{m,n \geq N} |S_m - S_N + S_N - S_n| \\
&\leq \sup_{m \geq N} |S_m - S_N| + \sup_{n \geq N} |S_n - S_N| \\
&= 2 \sup_{n \geq N} |S_n - S_N| \\
&= 2 \sup_{j \geq 0} |S_{N+j} - S_N|,
\end{aligned}$$

it suffices to show that

$$\sup_{j \geq 0} |S_{N+j} - S_N| \xrightarrow{P} 0. \tag{7.10}$$

For any $\epsilon > 0$, and $0 < \delta < \frac{1}{2}$, the assumption that $\{S_n\}$ is cauchy i.p. implies that there exists $N_{\epsilon,\delta}$ such that

$$P[|S_m - S_{m'}| > \frac{\epsilon}{2}] \leq \delta \tag{7.11}$$

if $m, m' \geq N_{\epsilon,\delta}$, and hence

$$P[|S_{N+j} - S_N| > \frac{\epsilon}{2}] \leq \delta, \quad \forall j \geq 0, \tag{7.12}$$

if $N \geq N_{\epsilon,\delta}$.

Now write

$$P[\sup_{j\geq 0} |S_{N+j} - S_N| > \epsilon] = P\{\lim_{N'\to\infty} [\sup_{N'\geq j\geq 0} |S_{N+j} - S_N| > \epsilon]\}$$
$$= \lim_{N'\to\infty} P[\sup_{N'\geq j\geq 0} |S_{N+j} - S_N| > \epsilon].$$

Now we seek to apply Skorohod's inequality. Let $X'_i = X_{N+i}$ and

$$S'_j = \sum_{i=1}^{j} X'_i = \sum_{i=1}^{j} X_{N+i} = S_{N+j} - S_N.$$

With this notation we have

$$\begin{aligned} P[\sup_{N'\geq j\geq 0} &|S_{N+j} - S_N| > \epsilon] \\ &= P[\sup_{N'\geq j\geq 0} |S'_j| > \epsilon] \\ &\leq \left(\frac{1}{1 - \vee_{j\leq N'} P[|S'_{N'} - S'_j| > \frac{1}{2}\epsilon]}\right) P[|S'_{N'}| > \frac{1}{2}\epsilon] \\ &\leq \frac{1}{1-\delta} \cdot \delta \leq 2\delta \end{aligned}$$

from the choice of δ. Note that from (7.11)

$$\bigvee_{j\leq N'} P[|S'_{N'} - S'_j| > \frac{1}{2}\epsilon] = \bigvee_{j\leq N'} P[|S_{N+N'} - S_{N+j}| > \frac{1}{2}\epsilon] \leq \delta.$$

Since δ can be chosen arbitrarily small, this proves the result. □

Lévy's theorem gives us an easy proof of the Kolmogorov convergence criterion which provides the easiest method of testing when a series of independent random variables with finite variances converges; namely by checking convergence of the sum of the variances.

Theorem 7.3.3 (Kolmogorov Convergence Criterion) *Suppose $\{X_n, n \geq 1\}$ is a sequence of independent random variables. If*

$$\sum_{j=1}^{\infty} Var(X_j) < \infty,$$

then

$$\sum_{j=1}^{\infty} (X_j - E(X_j)) \text{ converges almost surely.}$$

Proof. Without loss of generality, we may suppose for convenience that $E(X_j) = 0$. The sum of the variances then becomes

$$\sum_{j=1}^{\infty} EX_j^2 < \infty.$$

This implies that $\{S_n\}$ is L_2 Cauchy since $(m < n)$

$$\|S_n - S_m\|_2^2 = \mathrm{Var}(S_n - S_m) = \sum_{j=m+1}^{n} EX_j^2 \to 0,$$

as $m, n \to \infty$ since $\sum_j E(X_j^2) < \infty$. So $\{S_n\}$, being L_2–Cauchy, is also Cauchy in probability since

$$P[|S_n - S_m| > \epsilon] \le \epsilon^{-2}\mathrm{Var}(S_n - S_m) = \epsilon^{-2}\sum_{j=m}^{n} \mathrm{Var}(X_j) \to 0$$

as $n, m \to \infty$. By Lévy's theorem $\{S_n\}$ is almost surely convergent. □

Remark. The series in Theorem 7.3.3 is L_2-convergent. Call the L_2 limit $\sum_{j=1}^{\infty}(X_j - E(X_j))$. Because of L_2 convergence, the first and second moments converge; that is,

$$0 = E(\sum_{j=1}^{n}(X_j - EX_j)) \to E\left(\sum_{j=1}^{\infty}(X_j - EX_j)\right)$$

and

$$\sum_{j=1}^{n}\mathrm{Var}(X_j - E(X_j)) = \mathrm{Var}\left(\sum_{j=1}^{n}(X_j - E(X_j))\right) \to \mathrm{Var}\left(\sum_{j=1}^{\infty}(X_j - E(X_j))\right)$$

so we may conclude

$$E\left(\sum_{1}^{\infty}(X_j - EX_j)\right) = 0,$$

$$\mathrm{Var}\left(\sum_{j=1}^{\infty}(X_j - E(X_j))\right) = \sum_{j=1}^{\infty}\mathrm{Var}(X_j - E(X_j)).$$

7.4 Strong Laws of Large Numbers

This section considers the problem of when sums of independent random variables properly scaled and centered converge almost surely. We will prove that sample averages converge to mathematical expectations when the sequence is iid and a mean exists.

We begin with a number theory result which is traditionally used in the development of the theory.

Lemma 7.4.1 (Kronecker's lemma) *Suppose we have two sequences* $\{x_k\}$ *and* $\{a_n\}$ *such that* $x_k \in \mathbb{R}$ *and* $0 < a_n \uparrow \infty$. *If*

$$\sum_{k=1}^{\infty} \frac{x_k}{a_k} \text{ converges,}$$

then

$$\lim_{n\to\infty} a_n^{-1} \sum_{k=1}^{n} x_k = 0.$$

Proof. Let $r_n = \sum_{k=n+1}^{\infty} x_k/a_k$ so that $r_n \to 0$ as $n \to \infty$. Given $\epsilon > 0$, there exists $N_0 = N_0(\epsilon)$ such that for $n \geq N_0$, we have $|r_n| \leq \epsilon$. Now

$$\frac{x_n}{a_n} = r_{n-1} - r_n$$

so

$$x_n = a_n(r_{n-1} - r_n), \quad n \geq 1,$$

and

$$\begin{aligned}\sum_{k=1}^{n} x_k &= \sum_{k=1}^{n} (r_{k-1} - r_k)a_k \\ &= \sum_{j=1}^{n-1} (a_{j+1} - a_j)r_j + a_1 r_0 - a_n r_n.\end{aligned}$$

Then for $n \geq N_0$,

$$\begin{aligned}\left|\frac{\sum_{k=1}^{n} x_k}{a_n}\right| &\leq \sum_{j=1}^{N_0-1} \frac{(a_{j+1} - a_j)}{a_n}|r_j| + \sum_{j=N_0}^{n-1} \frac{(a_{j+1} - a_j)}{a_n}|r_j| \\ &\quad + |\frac{a_1 r_0}{a_n}| + |\frac{a_n r_n}{a_n}| \\ &= \frac{const}{a_n} + \frac{\epsilon}{a_n}(a_{N_0+1} - a_{N_0} + a_{N_0+2} - a_{N_0+1} \\ &\quad + a_{N_0+3} - a_{N_0+2} + \cdots + a_n - a_{n-1}) + r_n \\ &\leq o(1) + \frac{\epsilon(a_n - a_{N_0})}{a_n} + \epsilon \\ &\leq 2\epsilon + o(1).\end{aligned}$$

This shows the result. □

The Kronecker lemma quickly gives the following strong law.

Corollary 7.4.1 *Let* $\{X_n, n \geq 1\}$ *be an independent sequence of random variables satisfying* $E(X_n^2) < \infty$. *Suppose we have a monotone sequence* $b_n \uparrow \infty$. *If*

$$\sum_k Var(\frac{X_k}{b_k}) < \infty,$$

then

$$\frac{S_n - E(S_n)}{b_n} \xrightarrow{a.s.} 0.$$

Proof. Because the sum of the variances is finite, the Kolmogorov convergence criterion yields

$$\sum_n \left(\frac{X_n - E(X_n)}{b_n} \right) \text{ converges a.s.}$$

and the Kronecker Lemma implies

$$\sum_{k=1}^{n} (X_k - E(X_k))/b_n \to 0$$

almost surely. □

7.4.1 *Two Examples*

This section discusses two interesting results that require knowledge of sums of independent random variables. The first shows that the number of records in an iid sequence grows logarithmically and the second discusses explosions in a pure birth process.

Example 1: Record counts. Suppose $\{X_n, n \geq 1\}$ is an iid sequence with common continuous distribution function F. Define

$$\mu_N = \sum_{j=1}^{N} 1_{[X_j \text{ is a record }]} = \sum_{j=1}^{N} 1_j$$

where

$$1_j = 1_{[X_j \text{ is a record }]}.$$

So μ_N is the number of records in the first N observations.

Proposition 7.4.1 (Logarithmic growth rate) *The number of records in an iid sequence grows logarithmically and we have the almost sure limit*

$$\lim_{N \to \infty} \frac{\mu_N}{\log N} \to 1.$$

Proof of Proposition 7.4.1. We need the following fact taken from analysis: There is a constant c (Euler's constant) such that

$$\log n - \sum_{j=1}^{n} \frac{1}{j} \to c,$$

as $n \to \infty$.

Recall that $\{1_j, j \geq 1\}$ are independent. The basic facts about $\{1_j\}$ are that

$$P[1_j = 1] = \frac{1}{j}, \quad E(1_j) = \frac{1}{j},$$
$$\operatorname{Var}(1_j) = E(1_j)^2 - (E1_j)^2 = \frac{1}{j} - \frac{1}{j^2} = \frac{j-1}{j^2}.$$

This implies that

$$\sum_{j=2}^{\infty} \operatorname{Var}\left(\frac{1_j}{\log j}\right) = \sum_{2}^{\infty} \frac{1}{(\log j)^2} \operatorname{Var}(1_j) = \sum_{2}^{\infty} \frac{j-1}{j^2(\log j)^2} < \infty$$

since by the integral test

$$\sum_{j=2}^{\infty} \frac{j-1}{j^2(\log j)^2} \approx \sum_{j=2}^{\infty} \frac{j}{j^2(\log j)^2} = \sum_{j=2}^{\infty} \frac{1}{j(\log j)^2} \approx \int_e^{\infty} \frac{dx}{x(\log x)^2} = \int_1^{\infty} \frac{dy}{y^2} < \infty.$$

The Kolmogorov convergence criterion implies that

$$\sum_{j=2}^{\infty} \left(\frac{1_j - E(1_j)}{\log j}\right) = \sum_{j=2}^{\infty} \frac{(1_j - \frac{1}{j})}{\log j} \text{ converges}$$

and Kronecker's lemma yields

$$0 \overset{a.s.}{\leftarrow} \frac{\sum_{j=1}^{n}(1_j - j^{-1})}{\log n} = \frac{\sum_{j=1}^{n} 1_j - \sum_{j=1}^{n} j^{-1}}{\log n} = \frac{\mu_n - \sum_{j=1}^{n} j^{-1}}{\log n}.$$

Thus

$$\frac{\mu_n}{\log n} - 1 = \frac{\mu_n - \sum_{j=1}^{n} j^{-1}}{\log n} + \frac{\sum_{j=1}^{n} j^{-1} - \log n}{\log n} \to 0.$$

This completes the derivation. □

Example 2: Explosions in the Pure Birth Process. Next we present a stochastic processes example where convergence of a series of independent random variables is a crucial issue and where the necessary and sufficient criterion for convergence can be decided on the basis of first principles.

Let $\{X_j, j \geq 1\}$ be non-negative independent random variables and suppose

$$P[X_n > x] = e^{-\lambda_n x}, x > 0$$

where $\lambda_n \geq 0, n \geq 1$ are called the *birth* parameters. Define the birth time process $S_n = \sum_{i=1}^n X_i$ and the population size process $\{X(t), t \geq 0\}$ of the *pure birth process* by

$$X(t) = \begin{cases} 1, & \text{if } 0 \leq t < S_1, \\ 2, & \text{if } S_1 \leq t < S_2, \\ 3, & \text{if } S_2 \leq t < S_3, \\ \vdots \end{cases}.$$

Next define the event *explosion* by

$$\begin{aligned} [\text{ explosion }] &= [\sum_{n=1}^{\infty} X_n < \infty] \\ &= [X(t) = \infty \text{ for some finite } t]. \end{aligned}$$

Here is the basic fact about the probability of *explosion* in the pure birth process.

Proposition 7.4.2 *For the probability of explosion we have*

$$P[\textit{ explosion }] = \begin{cases} 1, & \textit{if } \sum_n \lambda_n^{-1} < 0, \\ 0, & \textit{if } \sum_n \lambda_n^{-1} = \infty. \end{cases}$$

Recall that we know that $P[\sum_n X_n < \infty] = 0$ or 1 by the Kolmogorov Zero-One Law.

Proof. If $\sum_n \lambda_n^{-1} < \infty$, then by the series version of the monotone convergence theorem

$$E(\sum_{n=1}^{\infty} X_n) = \sum_{n=1}^{\infty} E(X_n) = \sum_{n=1}^{\infty} \lambda_n^{-1} < \infty,$$

and so $P[\sum_{n=1}^{\infty} X_n < \infty] = 1$. (Otherwise, recall that $E(\sum_n X_n) = \infty$.)

Conversely, suppose $P[\sum_n X_n < \infty] = 1$. Then $\exp\{-\sum_{n=1}^{\infty} X_n\} > 0$ a.s., which implies that $E(\exp\{-\sum_n X_n\}) > 0$. But

$$\begin{aligned} 0 < E(e^{-\sum_{n=1}^{\infty} X_n}) &= E(\prod_{n=1}^{\infty} e^{-X_n}) \\ &= E(\lim_{N\to\infty} \prod_{k=1}^{N} e^{-X_n}) \\ &= \lim_{N\to\infty} E\left(\prod_{n=1}^{N} e^{-X_n}\right) \quad \text{(by Monotone Convergence)} \\ &= \lim_{N\to\infty} \prod_{n=1}^{N} E(e^{-X_n}) \quad \text{(by independence)} \\ &= \lim_{N\to\infty} \prod_{n=1}^{N} \int_0^{\infty} e^{-x} \lambda_n e^{-\lambda_n x}\, dx \\ &= \lim_{N\to\infty} \prod_{n=1}^{N} \frac{\lambda_n}{1+\lambda_n}. \end{aligned}$$

Now

$$\begin{aligned} E(\exp\{-\sum_n X_n\}) > 0 \text{ iff } &-\log E(e^{-\sum_n X_n}) < \infty \\ &\text{iff } \sum_{n=1}^{\infty} -\log\left(\frac{\lambda_n}{1+\lambda_n}\right) < \infty \\ &\text{iff } \sum_{n=1}^{\infty} \log(1+\lambda_n^{-1}) < \infty. \end{aligned}$$

If $\sum_{n=1}^{\infty} \log(1+\lambda_n^{-1}) < \infty$, then $\log(1+\lambda_n^{-1}) \to 0$ implies $\lambda_n^{-1} \to 0$. Since

$$\lim_{x\downarrow 0} \frac{\log(1+x)}{x} = 1,$$

by L'Hôpital's rule, we have

$$\log(1+\lambda_n^{-1}) \sim \lambda_n^{-1} \text{ as } n \to \infty$$

and thus

$$\sum_{n=1}^{\infty} \log(1+\lambda_n^{-1}) < \infty \text{ iff } \sum_{n=1}^{\infty} \lambda_n^{-1} < \infty.$$

□

7.5 The Strong Law of Large Numbers for IID Sequences

This section proves that the sample mean is almost surely approximated by the mathematical expectation. We begin with a preparatory lemma.

Lemma 7.5.1 *Let $\{X_n, n \geq 1\}$ be an iid sequence of random variables. The following are equivalent:*

(a) $E|X_1| < \infty$.

(b) $\lim_{n\to\infty} |\frac{X_n}{n}| = 0$ *almost surely.*

(c) For every $\epsilon > 0$

$$\sum_{n=1}^{\infty} P[|X_1| \geq \epsilon n] < \infty.$$

Proof. (a) $\leftrightarrow$ (c): Observe that:

$$\begin{aligned} E(|X_1|) &= \int_0^{\infty} P[|X_1| \geq x]dx \\ &= \sum_{n=0}^{\infty} \int_n^{n+1} P[|X_1| \geq x]dx \\ &\geq \sum_{n=0}^{\infty} P[|X_1| \geq n+1] \\ &\leq \sum_{n=0}^{\infty} P[|X_1| \geq n]. \end{aligned}$$

Thus $E(|X_1|) < \infty$ iff $\sum_{n=0}^{\infty} P[|X_1| \geq n] < \infty$. Set $Y = \frac{X_1}{\epsilon}$ and we get the following chain of equivalences:

$$\begin{aligned} E(|X_1|) < \infty \text{ iff } & E(|Y|) < \infty \\ \text{iff } & \sum_{n=0}^{\infty} P[|Y| \geq n] < \infty \\ \text{iff } & \sum_{n=0}^{\infty} P[|X_1| \geq \epsilon n] < \infty. \end{aligned}$$

(c) $\leftrightarrow$ (b): Given $\epsilon > 0$,

$$\sum_n P[|X_1| \geq \epsilon n] = \sum_n P[|X_n| \geq \epsilon n] < \infty$$

is equivalent, by the Borel zero-one law, to

$$P\left\{\left[\frac{|X_n|}{n} > \epsilon\right] \text{ i.o.}\right\} = 0,$$

which is in turn equivalent to

$$\limsup_{n\to\infty} \frac{|X_n|}{n} \le \epsilon$$

almost surely. Since $\limsup_{n\to\infty} \frac{|X_n|}{n}$ is a tail function, it is a.s. constant. If this constant is bounded by ϵ for any $\epsilon > 0$, then

$$\limsup_{n\to\infty} \frac{|X_n|}{n} = 0.$$

This gives (b) and the converse is similar. □

We are now prepared to consider Kolmogorov's version of the strong law of large numbers (SLLN).

Theorem 7.5.1 (Kolmogorov's SLLN) *Let $\{X_n, n \ge 1\}$ be an iid sequence of random variables and set $S_n = \sum_{i=1}^n X_i$. There exists $c \in \mathbb{R}$ such that*

$$\bar{X}_n = S_n/n \stackrel{a.s.}{\to} c$$

iff $E(|X_1|) < \infty$ in which case $c = E(X_1)$.

Corollary 7.5.1 *If $\{X_n\}$ is iid, then*

$$E(|X_1|) < \infty \textit{ implies } \bar{X}_n \stackrel{a.s.}{\to} \mu = E(X_1)$$

and

$$EX_1^2 < \infty \textit{ implies } S_n := \frac{1}{n}\sum_1^n (X_i - \bar{X})^2 \stackrel{a.s.}{\to} \sigma^2 =: Var(X_1).$$

Proof of Kolmogorov's SLLN (a) We show first that

$$\frac{S_n}{n} \stackrel{a.s.}{\to} c$$

implies $E(|X_1|) < \infty$. We have

$$\begin{aligned}
\frac{X_n}{n} &= \frac{S_n - S_{n-1}}{n} \\
&= \frac{S_n}{n} - \left(\frac{n-1}{n}\right)\frac{S_{n-1}}{n-1} \\
&\stackrel{a.s.}{\to} c - c = 0.
\end{aligned}$$

Since

$$X_n/n \stackrel{a.s.}{\to} 0,$$

Lemma 7.5.1 yields $E(|X_1|) < \infty$.

(b) Now we show that $E(|X_1|) < \infty$ implies $S_n/n \stackrel{a.s.}{\to} E(X_1)$. To do this, we use a truncation argument. Define

$$X'_n = X_n 1_{[|X_n| \le n]}, \quad n \ge 1.$$

Then

$$\sum_n P[X_n \ne X'_n] = \sum_n P[|X_n| > n] < \infty$$

(since $E|X_1| < \infty$) and hence $\{X_n\}$ and $\{X'_n\}$ are tail equivalent. Therefore by Proposition 7.1.1

$$S_n/n \stackrel{a.s.}{\to} E(X_1) \text{ iff } S'_n/n = \sum_{j=1}^n X'_j/n \stackrel{a.s.}{\to} E(X_1).$$

So it suffices to consider the truncated sequence.

Next observe that

$$\begin{aligned}\left|\frac{S'_n - E(S'_n)}{n} - \frac{S'_n - E(S_n)}{n}\right| &= \left|\frac{nE(X_1) - \sum_{j=1}^n E(X_1 1_{[|X_1| \le j]})}{n}\right| \\ &= \left|E(X_1) - \sum_{j=1}^n \frac{E(X_1 1_{[|X_1| \le j]})}{n}\right| \\ &\to 0.\end{aligned}$$

This last step follows from the fact that

$$|E(X_1) - E(X_1 1_{[|X_1| \le n]})| \le E(|X_1| 1_{[|X_1| > n]}) \to 0,$$

and hence the Cesaro averages converge. We thus conclude that

$$\frac{S'_n}{n} - E(X_1) \stackrel{a.s.}{\to} 0 \text{ iff } \frac{\sum_1^n (X'_j - E(X'_j))}{n} \stackrel{a.s.}{\to} 0.$$

To prove the last statement, it is enough by Kronecker's lemma to prove

$$\sum_j \mathrm{Var}\left(\frac{X'_j}{j}\right) < \infty.$$

However,

$$\begin{aligned}\sum_j \mathrm{Var}\left(\frac{X'_j}{j}\right) &= \sum_j \frac{1}{j^2}\mathrm{Var}(X'_j) \le \sum_j \frac{E(X'_j)^2}{j^2} \\ &= \sum_{j=1}^{\infty} \frac{1}{j^2} E(X_1^2 1_{[|X_1|\le j]}) \\ &= \sum_{j=1}^{\infty}\sum_{k=1}^{j} \frac{1}{j^2} E(X_1^2 1_{[k-1<|X_1|\le k]}) \\ &= \sum_{k=1}^{\infty}\Big(\sum_{j=k}^{\infty}\frac{1}{j^2}\Big) E(X_1^2 1_{[k-1<|X_1|\le k]}).\end{aligned}$$

Now

$$\frac{1}{j^2} \le \int_{j-1}^{j} \frac{1}{x^2}dx,$$

and therefore

$$\begin{aligned}\sum_{j=k}^{\infty}\frac{1}{j^2} \le \sum_{j=k}^{\infty}\int_{j-1}^{j}\frac{1}{x^2}dx &= \int_{k-1}^{\infty}\frac{1}{x^2}dx \\ &= \frac{1}{k-1} \le \frac{2}{k}\end{aligned}$$

provided $k \ge 2$. So

$$\begin{aligned}\sum_{k=2}^{\infty}\Big(\sum_{j=k}^{\infty}\frac{1}{j^2}\Big)E(X_1^2 1_{[k-1<|X_1|\le k]}) &\le \sum_{k=2}^{\infty}\frac{2}{k}E\big(|X_1|^2 1_{[k-1<|X_1|\le k]}\big) \\ &\le \sum_{k=2}^{\infty}\frac{2}{k}\cdot kE\big(|X_1| 1_{[k-1<|X_1|\le k]}\big) \\ &= 2E(|X_1|) < \infty.\end{aligned}$$

□

7.5.1 *Two Applications of the SLLN*

Now we present two standard applications of the Kolmogorov SLLN to *renewal theory* and to the *Glivenko–Cantelli Lemma*.

Renewal Theory. Suppose $\{X_n, n \ge 1\}$ is an iid sequence of non-negative random variables. Assume $E(X_n) = \mu$ and that $0 < \mu < \infty$. Then

$$\frac{S_n}{n} \stackrel{a.s.}{\to} \mu > 0$$

so that $S_n \stackrel{a.s.}{\to} \infty$. Let $S_0 = 0$ and define

$$N(t) := \sum_{j=0}^{\infty} 1_{[S_j \le t]}.$$

We call $N(t)$ the number of renewals in $[0, t]$. Then

$$[N(t) \le n] = [S_n > t] \tag{7.13}$$

and

$$S_{N(t)-1} \le t < S_{N(t)}. \tag{7.14}$$

Note that since we assume $S_0 = 0$, we have $N(t) \ge 1$. Also observe $\{N(t), t \ge 0\}$ is non-decreasing in t. We need to show $N(t) \to \infty$ a.s. as $t \to \infty$ and because of the monotonicity, it suffices to show $N(t) \stackrel{P}{\to} \infty$. Since for any m

$$\lim_{t\to\infty} P[N(t) \le m] = \lim_{t\to\infty} P[S_m > t] \to 0,$$

we get the desired result that $N(t) \stackrel{P}{\to} \infty$. Now define the sets

$$\Lambda_1 = \{\omega : \frac{S_n(\omega)}{n} \to \mu\},$$
$$\Lambda_2 = \{\omega : N(t, \omega) \to \infty\},$$

so that

$$P(\Lambda_1) = P(\Lambda_2) = 1.$$

Then $\Lambda := \Lambda_1 \cap \Lambda_2$ has $P(\Lambda) = 1$. For $\omega \in \Lambda$, as $t \to \infty$

$$\frac{S_{N(t,\omega)}(\omega)}{N(t, \omega)} \to \mu,$$

and so

$$S_{N(t)}/N(t) \stackrel{a.s.}{\to} \mu,$$

as $t \to \infty$. From (7.14)

$$\frac{t}{N(t)} \le \frac{S_{N(t)}}{N(t)} \to \mu$$

and

$$\frac{t}{N(t)} \ge \frac{S_{N(t)-1}}{N(t)} = \frac{S_{N(t)-1}}{N(t)-1} \cdot \frac{N(t)-1}{N(t)} \to \mu \cdot 1,$$

so we conclude that $t/N(t) \stackrel{a.s.}{\to} \mu$ and thus $N(t)/t \to \mu^{-1}$. Thus the long run rate of renewals is μ^{-1}. □

Glivenko–Cantelli Theorem. The Glivenko–Cantelli theorem says that the empirical distribution function is a uniform approximation for the true distribution function.

Let $\{X_n, n \geq 1\}$ be iid random variables with common distribution F. We imagine F is unknown and on the basis of a sample $X_1, \dots, X_n$ we seek to estimate F. The estimator will be the *empirical distribution function* (edf) defined by

$$\hat{F}_n(x, \omega) = \frac{1}{n} \sum_{j=1}^{n} 1_{[X_j \leq x]}(\omega).$$

By the SLLN we get that for each fixed x, $\hat{F}_n(x) \to F(x)$ a.s. as $n \to \infty$. In fact the convergence is uniform in x.

Theorem 7.5.2 (Glivenko–Cantelli Theorem) *Define*

$$D_n := \sup_x |\hat{F}_n(x) - F(x)|.$$

Then

$$D_n \to 0 \text{ a.s.}$$

as $n \to \infty$.

Proof. Define

$$x_{\nu,k} := F^{\leftarrow}(\nu/k), \quad \nu = 1, \dots, k,$$

where $F^{\leftarrow}(x) = \inf\{u : F(u) \geq x\}$. Recall

$$F^{\leftarrow}(u) \leq t \text{ iff } u \leq F(t) \tag{7.15}$$

and

$$F(F^{\leftarrow}(u)) \geq u, \quad F(F^{\leftarrow}(u)-) \leq u, \tag{7.16}$$

since for any $\epsilon > 0$ $F(F^{\leftarrow}(u) - \epsilon) < u$. If $x_{\nu,k} \leq x < x_{\nu+1,k}$, then monotonicity implies

$$F(x_{\nu,k}) \leq F(x) \leq F(x_{\nu+1,k}-), \quad \hat{F}_n(x_{\nu k}) \leq \hat{F}_n(x) \leq \hat{F}_n(x_{\nu+1,k}-),$$

and for such x

$$\begin{aligned} \hat{F}_n(x_{\nu,k}) - F(x_{\nu+1,k}-) &\leq \hat{F}_n(x) - F(x) \\ &\leq \hat{F}_n(x_{\nu+1,k}-) - F(x_{\nu,k}). \end{aligned} \tag{7.17}$$

Since

$$F(x_{\nu+1,k}-) - F(x_{\nu k}) \leq \frac{\nu+1}{k} - \frac{\nu}{k} = \frac{1}{k},$$

we modify (7.17) to get

$$\hat{F}_n(X_{\nu_k}) - F(x_{\nu_k}) - \frac{1}{k} \le \hat{F}_n(x) - F(x)$$
$$\le \hat{F}_n(x_{\nu+1,k}-) - F(x_{\nu+1,k}-) + \frac{1}{k}. \tag{7.18}$$

Therefore

$$\sup_{x\in[x_{\nu_k},x_{\nu+1,k})} |\hat{F}_n(x) - F(x)|$$
$$\le (|\hat{F}_n(x_{\nu,k}) - F(x_{\nu,k})| \vee |\hat{F}_n(x_{\nu+1,k}-) - F(x_{\nu+1,k}-)|) + \frac{1}{k},$$

which is valid for $\nu = 1, \ldots, k-1$, and taking the supremum over ν gives

$$\sup_{x\in[x_{1k},x_{k,k})} |\hat{F}_n(x) - F(x)|$$
$$\le \frac{1}{k} + \bigvee_{\nu=1}^{k} |\hat{F}_n(x_{\nu,k}) - F(x_{\nu,k})| \vee |\hat{F}_n(x_{\nu,k}-) - F(x_{\nu,k}-)|$$
$$= \text{RHS}.$$

We now show that this inequality also holds for $x < x_{1,k}$ and $x \ge x_{k,k}$. If $x \ge x_{k,k}$, then $F(x) = \hat{F}_n(x) = 1$ so $\hat{F}_n(x) - F(x) = 0$ and RHS is still an upper bound. If $x < x_{1,k}$, either

(i) $F(x) \ge \hat{F}_n(x)$ in which case

$$|\hat{F}_n(x,\omega) - F(x)| = F(x) - F_n(x,\omega)$$
$$\le F(x) \le F(x_{1,k}-)$$
$$\le \frac{1}{k}$$

so RHS is still the correct upper bound,

or

(ii) $\hat{F}_n(x) > F(x)$ in which case

$$|\hat{F}_n(x,\omega) - F(x)| = \hat{F}_n(x,\omega) - F(x)$$
$$\le \hat{F}_n(x_{1,k}-,\omega) - F(x_{1,k}-) + F(x_{1,k}-) - F(x)$$
$$\le |\hat{F}_n(x_{1,k}-,\omega) - F(x_{1,k}-)| + |F(x_{1k}-) - F(x)|$$

and since the last term is bounded by $1/k$ we have the bound

$$\le \frac{1}{k} + |\hat{F}_n(x_{1k}-,\omega) - F(x_{1k}-)|$$
$$\le \text{RHS}.$$

We therefore conclude that

$$D_n \leq \text{RHS}.$$

The SLLN implies that there exist sets $\Lambda_{\nu,k}$, and $\tilde{\Lambda}_{\nu,k}$, such that

$$P(\Lambda_{\nu,k}) = P(\tilde{\Lambda}_{\nu,k}) = 1,$$

and such that

$$\hat{F}_n(x_{\nu,k}) \to F(x_{\nu k}), \quad n \to \infty$$

and

$$\hat{F}_n(x_{\nu,k}-) = \frac{1}{n}\sum_1^n 1_{[X_j < x_{\nu,k}]} \to P[X_1 < x_{\nu,k}] = F(x_{\nu,k}-)$$

provided $\omega \in \Lambda_{\nu k}$ and $\tilde{\Lambda}_{\nu k}$ respectively. Let

$$\Lambda_k = \bigcap_\nu \Lambda_{\nu,k} \bigcap \bigcap_\nu \tilde{\Lambda}_{\nu,k},$$

so $P(\Lambda_k) = 1$. Then for $\omega \in \Lambda_k$

$$\limsup_{n\to\infty} D_n(\omega) \leq \frac{1}{k}.$$

For $\omega \in \bigcap_k \Lambda_k$

$$\lim_{n\to\infty} D_n(\omega) = 0,$$

and $P(\bigcap_k \Lambda_k) = 1$. □

7.6 The Kolmogorov Three Series Theorem

The Kolmogorov three series theorem provides necessary and sufficient conditions for a series of independent random variables to converge. The result is especially useful when the Kolmogorov convergence criterion may not be applicable, for example, when existence of variances is not guaranteed.

Theorem 7.6.1 *Let $\{X_n, n \geq 1\}$ be an independent sequence of random variables. In order for $\sum_n X_n$ to converge a.s., it is necessary and sufficient that there exist $c > 0$ such that*

(i) $\sum_n P[|X_n| > c] < \infty$.

(ii) $\sum_n Var(X_n 1_{[|X_n| \leq c]}) < \infty$.

(iii) $\sum_n E(X_n 1_{[|X_n| \leq c]})$ *converges.*

If $\sum_n X_n$ converges a.s., then (i), (ii), (iii) hold for any $c > 0$. Thus if the three series converge for one value of $c > 0$, they converge for all $c > 0$.

In this section, we consider the proof of sufficiency and an example. The proof of necessity is given in Section 7.6.1.

Proof of Sufficiency. Suppose the three series converge. Define

$$X'_n = X_n 1_{[|X_n| \leq c]}.$$

Then

$$\sum_n P[X'_n \neq X_n] = \sum_n P[|X_n| > c] < \infty$$

by (i) so $\{X'_n\}$ and $\{X_n\}$ are tail equivalent. Thus $\sum_n X_n$ converges almost surely iff $\sum_n X'_n$ converges almost surely.

From (ii)

$$\sum_n \mathrm{Var}(X'_n) < \infty,$$

so by the Kolmogorov convergence criterion

$$\sum_j (X'_j - E(X'_j)) \text{ converges a.s.}$$

But (iii) implies

$$\sum_n E(X'_n) \text{ converges}$$

and thus $\sum_j X'_j$ converges, as desired. □

Remark 7.6.1 In the exercises there awaits an interesting fact for you to verify. When the independent random variables in the three series theorem are non-negative, it is only necessary to check convergence of two series; the third series involving the truncated variances is redundant.

Example. Heavy tailed time series models: It is increasingly common to encounter data sets which must be modeled by heavy-tailed times series. Many time series are defined by means of recursions; for example, pth order autoregressions are defined as

$$X_n = \sum_{i=1}^{p} \phi_i X_{n-i} + Z_n, \quad n = 0, 1, \ldots \tag{7.19}$$

where $\{Z_n\}$ is an iid sequence. When does there exist a stationary process $\{X_n\}$ satisfying (7.19)? It is usually possible to iterate the recursion and guess that the solution is some infinite series. Provided the infinite series converges, it is relatively simple to show that this infinite sum satisfies the recursion. For instance, (7.19) for the case $p = 1$ is (set $\phi_1 = \phi$)

$$\begin{aligned} X_n &= \phi X_{n-1} + Z_n = \phi(\phi X_{n-2} + Z_{n-1}) + Z_n \\ &= \phi^2 X_{n-2} + Z_n + \phi Z_{n-1}. \end{aligned}$$

Continuing the iteration backward we get

$$X_n = \phi^m X_{n-m} + \sum_{i=0}^{m-1} \phi^i Z_{n-i}.$$

This leads to the suspicion that $\sum_{i=0}^{\infty} \phi^i Z_{n-i}$ is a solution to (7.19) when $p = 1$. Of course, this depends on $\sum_{i=0}^{\infty} \phi^i Z_{n-i}$ being an almost surely convergent series.

Showing the infinite sum converges can sometimes be tricky, especially when there is little information about existence of moments which is usually the case with heavy-tailed time series. Kolmogorov's three series theorem helps in this regard.

Suppose a time series is defined by

$$X_n = \sum_{j=0}^{\infty} \rho_j Z_{n-j}, \quad n = 0, 1, \ldots \tag{7.20}$$

where $\{\rho_n\}$ is a sequence of real constants and $\{Z_n\}$ is an iid sequence with Pareto tails satisfying

$$\bar{F}(x) := P[|Z_1| > x] \sim kx^{-\alpha}, \quad x \to \infty, \tag{7.21}$$

for some $\alpha > 0$, and $k > 0$. (Tail conditions somewhat more general than (7.21) such as regular variation could easily be assumed at the expense of slightly extra labor in the verifications to follow.) A sufficient condition for existence of a process satisfying (7.20) is that

$$\sum_{j=1}^{\infty} |\rho_j Z_j| < \infty, \tag{7.22}$$

almost surely. Condition (7.22) is often verified under the condition

$$\sum_j |\rho_j|^\delta < \infty, \quad 0 < \delta < \alpha \wedge 1, \tag{7.23}$$

(cf. Brockwell and Davis (1991)) especially when (7.21) is replaced by a condition of regular variation.

We now prove that (7.23) is sufficient for (7.22) and by Remark 7.6.1 we need to verify that (7.23) implies convergence of the two series

$$\sum_{j=1}^{\infty} P[|\rho_j Z_j| > 1] = \sum_{j=1}^{\infty} \bar{F}(1/|\rho_j|) < \infty, \tag{7.24}$$

$$\sum_{j=1}^{\infty} E\big(|\rho_j Z_j| 1_{[|\rho_j Z_j| \le 1]}\big) =: \sum_{j=1}^{\infty} |\rho_j| m(1/|\rho_j|) < \infty, \tag{7.25}$$

where

$$m(t) := E\big(|Z_1| 1_{[|Z_1| \le t]}\big).$$

Verifying (7.24) is relatively easy since, as $j \to \infty$, we have

$$P[|\rho_j Z_j| > 1] \sim k|\rho_j|^{\alpha}$$

which is summable due to (7.23). To verify (7.25), we observe that by Fubini's theorem

$$\begin{aligned} m(t) &= \int_0^t x F(dx) = \int_{x=0}^t \left[\int_{u=0}^x du\right] F(dx) \\ &= \int_{u=0}^t \left[\int_{x=u}^t F(dx)\right] du = \int_0^t \bar{F}(u)du - t\bar{F}(t) \\ &\le \int_0^t \bar{F}(u)du. \end{aligned} \tag{7.26}$$

From (7.21), given $\theta > 0$, there exists x_0 such that $x \ge x_0$ implies

$$\bar{F}(x) \le (1+\theta)kx^{-\alpha} =: k_1 x^{-\alpha}. \tag{7.27}$$

Thus from (7.26)

$$m(t) \le \int_0^{x_0} + \int_{x_0}^t \le c + k_1 \int_{x_0}^t u^{-\alpha} du, \quad t \ge x_0. \tag{7.28}$$

Now for $\alpha > 1$, $E(|Z_1|) < \infty$ so that

$$\sum_j |\rho_j| m(c/|\rho_j|) \le \sum_j |\rho_j| E(|Z_1|) < \infty$$

by (7.23). For $\alpha = 1$, we find from (7.28) that

$$m(t) \le c' + k_2 \log t, \quad t \ge x_0$$

for positive constants c', k_2. Now choose $\eta > 0$ so small that $1 - \eta > \delta$, and for another constant $c'' > 0$

$$\begin{aligned} \sum_j |\rho_j| m(c/|\rho_j|) &\le c'' \sum_j |\rho_j| + k_2 \sum_j |\rho_j| \log\left(\frac{1}{|\rho_j|}\right) \\ &\le c'' \sum_j |\rho_j| + k_3 \sum_j |\rho_j|^{1-\eta} < \infty \end{aligned}$$

where we have used (7.23) and

$$x^{-1}\log x \le x^{-1+\eta}, \quad x \ge x'.$$

Finally for $\alpha < 1, t > x_0$

$$m(t) \le c_1 + k_1 t^{1-\alpha}$$

so that

$$\sum_j |\rho_j| m(1/|\rho_j|) \le c_2 \sum_j |\rho_j| + k_4 \sum_j |\rho_j|^{1+\alpha-1} < \infty$$

from (7.23). □

7.6.1 *Necessity of the Kolmogorov Three Series Theorem*

We now examine a proof of the necessity of the Kolmogorov three series theorem. Two lemmas pave the way. The first is a partial converse to the Kolmogorov convergence criterion.

Lemma 7.6.1 *Suppose $\{X_n, n \ge 1\}$ are independent random variables which are uniformly bounded, so that for some $\alpha > 0$ and all $\omega \in \Omega$ we have $|X_n(\omega)| \le \alpha$. If $\sum_n (X_n - E(X_n))$ converges almost surely, then $\sum_{n=1}^{\infty} Var(X_n) < \infty$.*

Proof. Without loss of generality, we suppose $E(X_n) = 0$ for all n and we prove the statement: If $\{X_n, n \ge 1\}$ are independent, $E(X_n) = 0$, $|X_n| \le \alpha$, then $\sum_n X_n$ almost surely convergent implies $\sum_n E(X_n^2) < \infty$.

We set $S_n = \sum_{i=1}^n X_i$, $n \ge 1$ and begin by estimating $\mathrm{Var}(S_N) = \sum_{i=1}^N E(X_i^2)$ for a positive integer N. To help with this, fix a constant $\lambda > 0$, and define the first passage time out of $[-\lambda, \lambda]$

$$\tau := \inf\{n \ge 1 : |S_n| > \lambda\}.$$

Set $\tau = \infty$ on the set $[\vee_{n=1}^{\infty} |S_n| \le \lambda]$. We then have

$$\begin{aligned}\sum_{i=1}^N E(X_i^2) &= E(S_N^2) = E(S_N^2 1_{[\tau \le N]}) + E(S_N^2 1_{[\tau > N]}) \\ &= I + II.\end{aligned} \tag{7.29}$$

Note on $\tau > N$, we have $\vee_{i=1}^N |S_i| \le \lambda$, so that in particular, $S_N^2 \le \lambda^2$. Hence,

$$II \le \lambda^2 P[\tau > N] \le (\lambda + \alpha)^2 P[\tau > N]. \tag{7.30}$$

For I we have

$$I = \sum_{j=1}^N E(S_N^2 1_{[\tau = j]}).$$

For $j < N$

$$E(S_N^2 1_{[\tau=j]}) = E((S_j + \sum_{i=j+1}^{N} X_i)^2) 1_{[\tau=j]}).$$

Note

$$[\tau = j] = [\bigvee_{i=1}^{j-1} |S_i| \leq \lambda, |S_j| > \lambda] \in \sigma(X_1, \ldots, X_j)$$

while

$$\sum_{i=j+1}^{N} X_i \in \sigma(X_{j+1}, \ldots, X_N),$$

and thus

$$1_{[\tau=j]} \perp\!\!\!\perp \sum_{i=j+1}^{N} X_i.$$

Hence, for $j < N$,

$$\begin{aligned}
E(S_N^2 1_{[\tau=j]}) &= E\Big((S_j^2 + 2S_j \sum_{i=j+1}^{N} X_i + (\sum_{i=j+1}^{N} X_i)^2) 1_{[\tau=j]} \Big) \\
&= E(S_j^2 1_{[\tau=j]}) + 2E(S_j 1_{[\tau=j]}) E(\sum_{i=j+1}^{N} X_i) \\
&\quad + E(\sum_{i=j+1}^{N} X_i)^2 P[\tau = j] \\
&= E(S_j^2 1_{[\tau=j]}) + 0 + E(\sum_{i=j+1}^{N} X_i)^2 P[\tau = j] \\
&\leq E\big((|S_{j-1}| + |X_j|)^2 1_{[\tau=j]}\big) + \sum_{i=j+1}^{N} E(X_i)^2 P[\tau = j] \\
&\leq (\lambda + \alpha)^2 P[\tau = j] + \sum_{i=j+1}^{N} E(X_i)^2 P[\tau = j].
\end{aligned}$$

Summarizing, we conclude for $j < N$ that

$$E(S_N^2 1_{[\tau=j]}) \leq \Big((\lambda + \alpha)^2 + \sum_{i=j+1}^{N} E(X_i)^2 \Big) P[\tau = j], \tag{7.31}$$

and defining $\sum_{i=N+1}^{N} E(X_i)^2 = 0$, we find that (7.31) holds for $1 \leq j \leq N$. Adding over j yields

$$I = E(S_N^2 1_{[\tau \leq N]}) \leq \left((\lambda+\alpha)^2 + \sum_{i=1}^{N} E(X_i^2)\right) P[\tau \leq N]$$
$$= \left((\lambda+\alpha)^2 + E(S_N^2)\right) P[\tau \leq N]. \qquad (7.32)$$

Thus combining (7.30) and (7.32)

$$\begin{aligned} E(S_N^2) &= \sum_{i=1}^{N} E(X_i^2) = I + II \\ &\leq \left((\lambda+\alpha)^2 + E(S_N^2)\right) P[\tau \leq N] + (\lambda+\alpha)^2 P[\tau > N] \\ &\leq (\lambda+\alpha)^2 + E(S_N^2) P[\tau \leq N] \end{aligned}$$

and solving for $E(S_N^2)$ yields

$$E(S_N^2) \leq \frac{(\lambda+\alpha)^2}{P[\tau > N]}.$$

Let $N \to \infty$. We get

$$\sum_{i=1}^{\infty} E(X_i^2) \leq \frac{(\lambda+\alpha)^2}{P[\tau = \infty]},$$

which is helpful and gives the desired result only if $P[\tau = \infty] > 0$. However, note that we assume $\sum_n X_n$ is almost surely convergent, and hence for almost all ω, we have $\{S_n(\omega), n \geq 1\}$ is a bounded sequence of numbers. So $\vee_n |S_n|$ is almost surely a finite random variable, and there exists $\lambda > 0$ such that

$$P[\tau = \infty] = P[\bigvee_{n=1}^{\infty} |S_n| \leq \lambda] > 0,$$

else $P[\vee_{n=1}^{\infty} |S_n| < \infty] = 0$, which is a contradiction. This completes the proof of the lemma. □

Lemma 7.6.2 *Suppose $\{X_n, n \geq 1\}$ are independent random variables which are uniformly bounded in the sense that there exists $\alpha > 0$ such that $|X_n(\omega)| \leq \alpha$ for all $n \geq 1$ and $\omega \in \Omega$. Then $\sum_n X_n$ convergent almost surely implies that $\sum_n E(X_n)$ converges.*

Proof. The proof uses a technique called *symmetrization*. Define an independent sequence $\{Y_n, n \geq 1\}$ which is independent of $\{X_n, n \geq 1\}$ satisfying $Y_n \stackrel{d}{=} X_n$. Let

$$Z_n = X_n - Y_n, \quad n \geq 1.$$

Then $\{Z_n, n \geq 1\}$ are independent random variables, $E(Z_n) = 0$, and the distribution of each Z_n is symmetric which amounts to the statement that

$$Z_n \stackrel{d}{=} -Z_n.$$

Further,

$$\mathrm{Var}(Z_n) = \mathrm{Var}(X_n) + \mathrm{Var}(Y_n) = 2\mathrm{Var}(X_n)$$

and $|Z_n| \leq |X_n| + |Y_n| \leq 2\alpha$.

Since $\{X_n, n \geq 1\} \stackrel{d}{=} \{Y_n, n \geq 1\}$ as random elements of $\mathbb{R}^\infty$, the convergence properties of the two sequences are identical and since $\sum_n X_n$ is assumed almost surely convergent, the same is true of $\sum_n Y_n$. Hence also $\sum_n Z_n$ is almost surely convergent. Since $\{Z_n\}$ is also uniformly bounded, we conclude from Lemma 7.6.1 that

$$\sum_n \mathrm{Var}(Z_n) = \sum_n 2\mathrm{Var}(X_n) < \infty.$$

From the Kolmogorov convergence criterion we get $\sum_n (X_n - E(X_n))$ almost surely convergent. Since we also assume $\sum_n X_n$ is almost surely convergent, it can only be the case that $\sum_n E(X_n)$ converges. □

We now turn to the proof of necessity of the Kolmogorov three series theorem.

Re-statement: Given independent random variables $\{X_n, n \geq 1\}$ such that $\sum_n X_n$ converges almost surely, it follows that the following three series converge for any $c > 0$:

(i) $\sum_n P[|X_n| > c]$,

(ii) $\sum_n \mathrm{Var}(X_n 1_{[|X_n| \leq c]})$,

(iii) $\sum_n E(X_n 1_{[|X_n| \leq c]})$.

Proof of necessity. Since $\sum_n X_n$ converges almost surely, we have $X_n \stackrel{a.s.}{\to} 0$ and thus

$$P([|X_n| > c] \text{ i.o.}) = 0.$$

By the Borel zero-one law, it follows that

$$\sum_n P[|X_n| > c] < \infty. \tag{7.33}$$

If (7.33) holds, then $\{X_n\}$ and $\{X_n 1_{[|X_n| \leq c]}\}$ are tail equivalent and one converges iff the other does. So we get that the uniformly bounded sequence $\{X_n 1_{[|X_n| \leq c]}\}$ satisfies $\sum_n X_n 1_{[|X_n| \leq c]}$ converges almost surely. By Lemma 7.6.2, $\sum_n E(X_n 1_{[|X_n| \leq c]})$ (the series in (iii)) is convergent. Thus the infinite series of uniformly bounded summands

$$\sum_n (X_n 1_{[|X_n| \leq c]} - E(X_n 1_{[|X_n| \leq c]}))$$

is almost surely convergent and by Lemma 7.6.1

$$\sum_n \mathrm{Var}(X_n 1_{[|X_n|\le c]}) < \infty$$

which is (ii). □

7.7 Exercises

1. **Random signs.** Does $\sum_n 1/n$ converge? Does $\sum_n (-1)^n 1/n$ converge? Let $\{X_n\}$ be iid with
$$P[X_n = \pm 1] = \frac{1}{2}.$$
Does $\sum_n X_n/n$ converge?

2. Let $\{X_n\}$ be iid, $EX_n = \mu$, $\mathrm{Var}(X_n) = \sigma^2$. Set $\bar{X} = \sum_{i=1}^n X_i/n$. Show that
$$\frac{1}{n}\sum_{i=1}^n (X_i - \bar{X})^2 \xrightarrow{P} \sigma^2.$$

3. **Occupancy problems.** Randomly distribute r balls in n boxes so that the sample space Ω consists of n^r equally likely elements. Write
$$N_n = \sum_{i=1}^n 1_{[i\text{th box is empty}]}$$
for the number of empty boxes. Check
$$P[i\text{th box is empty}] = (1 - \frac{1}{n})^r$$
so that $E(N_n) = n(1 - n^{-1})^r$. Check that as $r/n \to c$
$$E(N_n)/n \to e^{-c} \tag{7.34}$$
$$N_n/n \xrightarrow{P} e^{-c}. \tag{7.35}$$
For the second result, compute $\mathrm{Var}(N_n)$ and show $\mathrm{Var}(N_n/n) \to 0$.

4. Suppose $g : [0, 1] \mapsto \mathbb{R}$ is measurable and Lebesgue integrable. Let $\{U_n, n \ge 1\}$ be iid uniform random variables and define $X_i = g(U_i)$. In what sense does $\sum_{i=1}^n X_i/n$ approximate $\int_0^1 g(x)dx$? (This offers a way to approximate the integral by Monte Carlo methods.) How would one guarantee a desired degree of precision?

5. (a) Let $\{X_n, n \geq 1\}$ be iid, $EX_n = 0$, $EX_n^2 = \sigma^2$. Let $a_n \in \mathbb{R}$ for $n \geq 1$. Set $S_n = \sum_{i=1}^n a_i X_i$. Prove $\{S_n\}$ is L_2-convergent iff $\sum_{i=1}^\infty a_i^2 < \infty$. If $\sum_{i=1}^\infty a_i^2 < \infty$, then $\{S_n, n \geq 1\}$ is almost surely convergent.

(b) Suppose $\{X_n, n \geq 1\}$ are arbitrary random variables such that $\sum_n \pm X_n$ converges almost surely for all choices ± 1. Show $\sum_n X_n^2 < \infty$ almost surely. (Hint: Consider $\sum_n B_n(t) X_n(\omega)$ where the random variables $\{B_n, n \geq 1\}$ are coin tossing or Bernoulli random variables. Apply Fubini on the space of (t, ω).)

(c) Suppose $\{B_n, n \geq 1\}$ are iid with possible values $\{1, -1\}$ and

$$P[B_n = \pm 1] = \frac{1}{2}.$$

Show for constants a_n that

$$\sum_n a_n B_n \text{ converges iff } \sum_n a_n^2 < \infty.$$

(d) Show $\sum_n B_n n^{-\theta}$ converges a.s. iff $\theta > 1/2$.

6. Suppose $\{X_k, k \geq 1\}$ are independent random variables and suppose X_k has a gamma density $f_k(x)$

$$f_k(x) = \frac{x^{\gamma_k - 1} e^{-x}}{\Gamma(\gamma_k)}, \quad x > 0, \gamma_k > 0.$$

Give necessary and sufficient conditions for $\sum_{k=1}^\infty X_k$ to converge almost surely. (Compare with the treatment of sums of exponentially distributed random variables.)

7. Let $\{E_n\}$ be events.

(a) Verify

$$\sum_{k=1}^n 1_{E_k} = 1_{\cup_{k=1}^n E_k} \sum_{k=1}^n 1_{E_k}$$

and then, using the Schwartz inequality, prove that

$$P(\cup_{k=1}^n E_k) \geq \frac{\left(E(\sum_{k=1}^n 1_{E_k})\right)^2}{E(\sum_{k=1}^n 1_{E_k})^2}.$$

(b) Deduce from this that if

(i) $\sum_n P(E_n) = \infty$, and

(ii) there exists $c > 0$ such that for all $m < n$, we have

$$P(E_m E_n) \leq c P(E_m) P(E_{n-m})$$

then

(iii) $P(\limsup_{n\to\infty} E_n) > 0$.

Thus, we get a partial converse to Borel-Cantelli: if (iii) fails, that is, if $P(\limsup_{n\to\infty} E_n) = 0$, and (ii) holds, then (i) fails so that $\sum_n P(E_n) < \infty$.

(c) Let $\{Y_n, n \geq 1\}$ be iid positive random variables with common distribution function G and suppose $\{X_n, n \geq 1\}$ is a sequence of iid positive random variables with common distribution F. Suppose $\{X_n\}$ and $\{Y_n\}$ are independent. Prove using Borel-Cantelli that if for all $\epsilon > 0$

$$\int_0^\infty \frac{G(dy)}{1 - F(\epsilon y)} < \infty,$$

then as $n \to \infty$

$$\frac{Y_n}{\bigvee_{i=1}^n X_i} \to 0,$$

almost surely. Prove the converse using the converse to Borel-Cantelli proved in (b) above.

8. The SLLN says that if $\{X_n, n \geq 1\}$ are iid with $E|X_1| < \infty$, then

$$S_n/n \stackrel{a.s.}{\to} E(X_1).$$

Show also that

$$S_n/n \stackrel{L_1}{\to} E(X_1).$$

(Think of uniform integrability.)

9. Suppose $\{X_n\}$ are iid, $E|X_1| < \infty$, $EX_1 = 0$ and suppose that $\{c_n\}$ is a bounded sequence of real numbers. Prove

$$\frac{1}{n}\sum_{i=1}^n c_j X_j \to 0 \text{ a.s.}$$

(If necessary, examine the proof of the SLLN.)

10. (a) Suppose that $\{X_n\}$ are m-dependent in the sense that random variables more than m apart in the sequence are independent. More precisely, let

$$\mathcal{B}_j^k = \mathcal{B}(X_j, \dots, X_k),$$

and assume that $\mathcal{B}_{j_1}^{k_1}, \dots, \mathcal{B}_{j_l}^{k_l}$ are independent if $k_{i-1} + m < j_i$ for $i = 2, \dots, l$. (Independent random variables are 0-dependent.) Suppose that the $\{X_n\}$ have this property and are uniformly bounded and that $EX_n = 0$. Show that $n^{-1}S_n \to 0$ a.s.

Hint: Consider the subsequences $X_i, X_{i+m+1}, X_{i+2(m+1)}, \dots$ for $1 \leq i \leq m+1$.

(b) Suppose that $\{X_n\}$ are iid and each X_i has finite range $x_1, \dots, x_l$ and

$$P[X_1 = x_i] = p(x_i), \quad i = 1, \dots, l.$$

For $u_1, \dots, u_k$, a k-tuple of the x_i's, let $N_n(u_1, \dots, u_k)$ be the frequency of the k-tuple in the first $n+k-1$ trials; that is, the number of m such that $1 \le m \le n$ and

$$X_m = u_1, \dots, X_{m+k-1} = u_k.$$

Show that with probability 1, all asymptotic relative frequencies are what they should be—that is, with probability 1,

$$n^{-1} N_n(u_1, \dots, u_k) \to p(u_1) \cdots p(u_k)$$

for every k and every k-tuple $u_1, \dots, u_k$.

11. Suppose $\{X_n, n \ge 1\}$ are iid with a symmetric distribution. Then $\sum_n X_n/n$ converges almost surely iff $E(|X_1|) < \infty$.

12. Suppose $\{X_n\}$ is defined iteratively in the following way: Let X_0 have a uniform distribution on $[0, 1]$ and for $n \ge 1$, X_{n+1} has a uniform distribution on $[0, X_n]$. Show that

$$\frac{1}{n} \log X_n \text{ converges a.s.}$$

and find the almost sure limit.

13. Use the three series theorem to derive necessary and sufficient conditions for $\sum_n X_n$ to converge a.s. when $\{X_n\}$ are independent and exponentially distributed.

14. Suppose $\{X_n, n \ge 1\}$ are independent, normally distributed with

$$E(X_n) = \mu_n, \quad \mathrm{Var}(X_n) = \sigma_n^2.$$

Show that $\sum_n X_n$ converges almost surely iff $\sum_n \mu_n$ converges and $\sum_n \sigma_n^2 < \infty$.

15. Prove the three series theorem reduces to a two series theorem when the random variables are positive. If $V_n \ge 0$ are independent, then $\sum_n V_n < \infty$ a.s. iff for any $c > 0$, we have

$$\sum_n P[V_n > c] < \infty, \tag{i}$$

$$\sum_n E(V_n 1_{[V_n \le c]}) < \infty. \tag{ii}$$

16. If $\{X_n\}$ are iid with $E|X_1| < \infty$, $EX_1 \ne 0$, show that

$$\frac{\bigvee_{i=1}^n |X_i|}{|S_n|} \xrightarrow{a.s.} 0.$$

17. Suppose $\{X_n\}$ are independent with

$$P[X_k = k^2] = \frac{1}{k^2}, \quad P[X_k = -1] = 1 - \frac{1}{k^2}.$$

Prove

$$\lim_{n\to\infty} \sum_{i=1}^{n} X_k$$

exists a.s. and find the limit.

18. Supppse $\{X_n, n \geq 1\}$ are iid with $E(X_n) = 0$, and $E(X_n^2) = 1$. Set $S_n = \sum_{i=1}^{n} X_i$. Show

$$\frac{S_n}{n^{1/2}\log n} \to 0$$

almost surely.

19. Suppose $\{X_n, n \geq 1\}$ are non-negative independent random variables. Then $\sum_n X_n < \infty$ almost surely iff

$$\sum_n E\left(\frac{X_n}{1+X_n}\right) < \infty$$

iff

$$\sum_n E(X_n \wedge 1) < \infty.$$

20. Suppose $\{X_n, n \geq 1\}$ are independent random variables with $E(X_n) = 0$ for all n. If

$$\sum_n E\left(X_n^2 1_{[|X_n|\leq 1]} + |X_n| 1_{[|X_n|>1]}\right) < \infty,$$

then $\sum_n X_n$ converges almost surely.

Hint: We have

$$0 = E(X_n) = E(X_n 1_{[|X_n|\leq 1]}) + E(X_n 1_{[|X_n|>1]}).$$

21. Suppose $\{X_n(\theta), n \geq 1\}$ are iid with common exponential distribution with mean θ. Then

$$\sum_{i=1}^{n} X_i(\theta)/n \xrightarrow{P} \theta.$$

Show for any $u : \mathbb{R}_+ \mapsto \mathbb{R}_+$ which is bounded and continuous that

$$\int_0^\infty u(y) e^{-ny\theta^{-1}} \frac{(ny\theta^{-1})^{n-1}}{(n-1)!} n\theta^{-1} dy \to u(\theta).$$

Show this convergence is uniform on finite θ-intervals. (Compare this with the Bernstein polynomials of item 6.3 on page 176.)

22. **The WLLN and inversion of Laplace transforms.** Let $X > 0$ be a non-negative random variable with distribution function F and Laplace transform

$$\widehat{F}(\lambda) = E(e^{-\lambda X}) = \int_{[0,\infty)} e^{-\lambda x} F(dx).$$

Show that $\widehat{F}$ determines F using the weak law of large numbers and the following steps.

(a) Show

$$\widehat{F}^{(k)}(\lambda) = \int_{[0,\infty)} (-1)^k y^k e^{-\lambda x} F(dx).$$

(b) Suppose $\{\xi_n(\theta), n \geq 1\}$ are iid non-negative random variables with a Poisson distribution, parameter $\theta > 0$. Use the weak law of large numbers to prove

$$\lim_{n\to\infty} P[\sum_{i=1}^{n} \xi_i(\theta)/n \leq x] = \begin{cases} 1, & \text{if } x > \theta, \\ 0, & \text{if } x < \theta, \end{cases}$$

and therefore

$$\lim_{n\to\infty} \sum_{j \leq nx} e^{n\theta} \frac{(n\theta)^j}{j!} = \begin{cases} 1, & \text{if } x > \theta, \\ 0, & \text{if } x < \theta. \end{cases}$$

(c) Conclude for any $x \geq 0$ which is a point of continuity of F that

$$\sum_{j \leq nx} \frac{(-1)^j}{j!} n^j \hat{F}^{(j)}(n) \to F(x).$$

23. Suppose $\{X_n, n \geq 1\}$ are iid and uniformly distributed on $(-1, 1)$. What is $E(X_1^2)$? Verify

$$\sum_{i=1}^{n} X_i^2/n \xrightarrow{P} \frac{1}{3}$$

so that if we define

$$\|\mathbf{X}_n\|_n = (\sum_{i=1}^{n} X_i^2)^{1/2},$$

then

$$\|\mathbf{X}_n\|_n/\sqrt{n} \xrightarrow{P} \sqrt{\frac{1}{3}}.$$

Now define the n-dimensional annulus

$$B_{n,\delta} := \{\mathbf{x} \in \mathbb{R}^n : \sqrt{\frac{1}{3}} - \delta < \frac{\|\mathbf{x}\|_n}{\sqrt{n}} < \sqrt{\frac{1}{3}} + \delta\}.$$

Further define the n-dimensional cube

$$I_n = (-1, 1)^n := \{\mathbf{x} \in \mathbb{R}^n : \bigvee_{i=1}^{n} |x_i| < 1\}.$$

If λ_n is n-dimensional Lebesgue measure, show that

$$2^{-n}\lambda_n(B_{n,\delta} \cap I_n) \to 1.$$

This gives the curious result that for large n, the cube is well approximated by the annulus.

24. **Relative stability of sums.** Prove the following are equivalent for iid non-negative random variables $\{X_n, n \geq 1\}$.

(a) There exist constants $a_n > 0$ such that

$$a_n^{-1} \sum_{i=1}^{n} X_i \xrightarrow{P} 1.$$

(b) As $n \to \infty$

$$\frac{\bigvee_{i=1}^{n} X_i}{\sum_{i=1}^{n} X_i} \xrightarrow{P} 0.$$

(c) We have

$$\lim_{x\to\infty} \frac{E(X_1 1_{[X_1 \leq x]})}{xP[X_1 > x]} = \infty.$$

(d) We have that the function $\mu(x) := E(X_1 1_{[X_1 \leq x]})$ is slowly varying; that is,

$$\lim_{t\to\infty} \frac{\mu(tx)}{\mu(t)} = 1, \quad \forall x > 0.$$

This is equivalent to

$$U(x) = \int_0^x P[X_1 > s]ds$$

being slowly varying.

(e) In this case, show we may select a_n as follows. Set $H(x) = x/U(x)$ and then set

$$a_n = H^{\leftarrow}(n)$$

where $H^{\leftarrow}$ is the inverse function of H satisfying $H(H^{\leftarrow}(x)) \sim x$.

(f) Now apply this to the *St. Petersburg paradox*. Let $\{X_n, n \geq 1\}$ be iid with

$$P[X_1 = 2^k] = 2^{-k}, \quad k \geq 1.$$

What is $E(X_1)$? Set $S_n = \sum_{i=1}^{n} X_i$. The goal is to show

$$\frac{S_n}{(n \log n)/\log 2} \xrightarrow{P} 1.$$

Proceed as follows:

i. Check $P[X > 2^n] = 2^{-n}$, $n \geq 1$.

ii. Evaluate

$$\int_0^{2^n} P[X_1 > s]ds = \sum_{j=1}^{n} \int_{2^{j-1}}^{2^j} P[X_1 > s]ds$$

to get

$$U(x) \sim \frac{\log x}{\log 2}.$$

So $H(x) = x/U(x) \sim \log 2(x/\log x)$ and

$$a_n \sim (n \log n)/\log 2.$$

25. Formulate and prove a generalization of Theorem 7.2.1 applicable to triangular arrays $\{X_{n,k}, 1 \leq k \leq n; n \geq 1\}$ where $\{X_{n,k}, 1 \leq k \leq n\}$ is independent and where n is replaced by a general sequence of constants $\{b_n\}$. For iid random variables, the conditions should reduce to

$$nP[|X_{n,1}| > b_n] \to 0, \tag{7.36}$$

$$\frac{n}{b_n^2} E(X_{n,1}^2 1_{[|X_{n,1}| \leq b_n]}) \to 0. \tag{7.37}$$

If $S_n = \sum_{i=1}^n X_{n,i}$, the conclusion for the row iid case is

$$\frac{S_n - nE(X_{n,1} 1_{[|X_{n,1}| \leq b_n]})}{b_n} \xrightarrow{P} 0.$$

Apply this to the *St. Petersburg paradox* of Problem 24.

26. If $\{X_n, n \geq 1\}$ is defined as in the St. Petersburg paradox of Problem 24, show almost surely that

$$\limsup_{n\to\infty} \frac{X_n}{n \log_2 n} = \infty,$$

so that for the partial sums we also have almost surely

$$\limsup_{n\to\infty} \frac{S_n}{n \log_2 n} = \infty.$$

Thus we have another example of a sequence converging in probability but not almost surely.

27. **More applications of WLLN.**

(a) Suppose $u(x, y)$ is continuous on the triangle

$$\text{TRI} := \{(x, y) : x \geq 0, y \geq 0, x + y = 1\}.$$

Show that uniformly on TRI

$$\sum u(\frac{j}{n}, \frac{k}{n}) \frac{n!}{j!k!(n-j-k)!} x^j y^k (1-x-y)^{n-j-k} \to u(x, y).$$

(b) Suppose $u : [0, \infty) \mapsto \mathbb{R}$ is continuous with

$$\lim_{x\to\infty} u(x) =: u(\infty)$$

existing finite. Show u can be approximated uniformly by linear combinations of e^{nx}.

28. Suppose $\{X_n, n \geq 1\}$ are uncorrelated random variables satisfying

$$E(X_n) = \mu, \quad \mathrm{Var}(X_n) \leq C, \quad \mathrm{Cov}(X_i, X_j) = 0,\ i \neq j.$$

Show that as $n \to \infty$ that $\sum_{i=1}^n X_i/n \to \mu$ in probability and in L_2.

Now suppose $E(X_n) = 0$ and $E(X_i X_j) \leq \rho(i-j)$ for $i > j$ and $\rho(n) \to 0$ as $n \to \infty$. Show $\sum_{i=1}^n X_i/n \xrightarrow{P} 0$.

29. Suppose $\{X_n, n \geq 1\}$ is iid with common distribution described as follows. Define

$$p_k = \frac{1}{2^k k(k+1)}, \quad k \geq 1,$$

and $p_0 = 1 - \sum_{k=1}^\infty p_k$. Suppose

$$P[X_n = 2^k - 1] = p_k, \quad k \geq 1$$

and $P[X_n = -1] = p_0$. Observe that

$$\sum_{k=1}^\infty 2^k p_k = \sum_{k=1}^\infty \left(\frac{1}{k} - \frac{1}{k+1}\right) = 1,$$

and that $E(X_n) = 0$. For $S_n = \sum_{i=1}^n X_i,\ n \geq 1$, prove

$$\frac{S_n}{n/\log_2 n} \xrightarrow{P} -1.$$

30. **Classical coupon collecting.** Suppose $\{X_k, k \geq 1\}$ is iid and uniformly distributed on $\{1, \dots, n\}$. Define

$$T_n = \inf\{m : \{X_1, \dots, X_m\} = \{1, \dots, n\}\}$$

to be the first time all values are sampled.

The problem name stems from the game of collecting coupons. There are n different coupons and one samples with replacement repeatedly from the population $\{1, \dots, n\}$ until all coupons are collected. The random variable T_n is the number of samples necessary to obtain all coupons.

Show

$$\frac{T_n}{n \log n} \xrightarrow{P} 1.$$

Hints: Define

$$\tau_k(n) = \inf\{m : \text{card}\{X_1, \dots, X_m\} = k\}$$

to be the number of samples necessary to draw k different coupons. Verify that $\tau_1(n) = 1$ and $\{\tau_k(n) - \tau_{k-1}(n), 2 \le k \le n\}$ are independent and geometrically distributed. Verify that

$$E(T_n) = n \sum_{i=1}^{n} \frac{1}{i} \sim n \log n,$$

$$\text{Var}(T_n) \le n^2 \sum_{i=1}^{n} i^{-2}.$$

Check

$$\text{Var}(T_n / E(T_n)) \to 0.$$

31. Suppose $\{X_n, n \ge 1\}$ are independent Poisson distributed random variables with $E(X_n) = \lambda_n$. Suppose $\{S_n = \sum_{i=1}^{n} X_i, n \ge 1\}$ and that $\sum_n \lambda_n = \infty$. Show $S_n / E(S_n) \to 1$ almost surely.

32. Suppose $\{X_n, n \ge 1\}$ are iid with $P[X_i > x] = e^{-x}$, $x > 0$. Show as $n \to \infty$

$$\bigvee_{i=1}^{n} X_i / \log n \to 1,$$

almost surely. Hint: You already know from Example 4.5.2 of Chapter 4 that

$$\limsup_{n \to \infty} \frac{X_n}{\log n} = 1,$$

almost surely.

33. Suppose $\{X_j, j \ge 1\}$ are independent with

$$P[X_n = n^{-\alpha}] = P[X_n = -n^{-\alpha}] = \frac{1}{2}.$$

Use the Kolmogorov convergence criterion to verify that if $\alpha > 1/2$, then $\sum_n X_n$ converges almost surely. Use the Kolmogorov three series theorem to verify that $\alpha > 1/2$ is necessary for convergence. Verify that $\sum_n E(|X_n|) < \infty$ iff $\alpha > 1$.

34. Let $\{N_n, n \ge 1\}$ be iid $N(0, 1)$ random variables. Use the Kolmogorov convergence criterion to verify quickly that $\sum_{n=1}^{\infty} \frac{N_n}{n} \sin(n\pi t)$ converges almost surely.

35. Suppose $\{X_n, n \geq 1\}$ is iid and

$$E(X_n^+) < \infty, \quad E(X_n^-) = \infty.$$

Show $S_n/n \to -\infty$ almost surely. (Try truncation of X_n^- and use the classical SLLN.)

36. Suppose $\{X_n, n \geq 1\}$ is independent and $X_k \geq 0$. If for some $\delta \in (0, 1)$ there exists x such that for all k

$$\int_{[X_k > x]} X_k dP \leq \delta E(X_k),$$

then almost sure convergence of $\sum_n X_n$ implies $\sum_n E(X_n) < \infty$ as well.

37. Use only the three series theorem to come up with a necessary and sufficient condition for sums of independent exponentially distributed random variables to converge.

38. Suppose $\{X_n, n \geq 1\}$ are iid with

$$P[X_n = 0] = P[X_n = 2] = \frac{1}{2}.$$

Show $\sum_{n=1}^{\infty} X_n/3^n$ converges almost surely. The limit has the Cantor distribution.

39. If $\{A_n, n \geq 1\}$ are independent events, show that

$$\frac{1}{n}\sum_{i=1}^{n} 1_{A_i} - \frac{1}{n}\sum_{i=1}^{n} P(A_i) \xrightarrow{P} 0.$$

40. Suppose $\{X_n, n \geq 1\}$ are independent and set $S_n = \sum_{i=1}^{n} X_i$. Then $S_n/n \to 0$ almost surely iff the following two conditions hold:

(a) $S_n/n \xrightarrow{P} 0$,

(b) $S_{2^n}/2^n \to 0$ almost surely.

41. Suppose $\{X_n, Y_n, n \geq 1\}$ are independent random variables such that $X_n \stackrel{d}{=} Y_n$ for all $n \geq 1$. Suppose further that, for all $n \geq 1$, there is a constant K such that

$$|X_n| \vee |Y_n| \leq K.$$

Then $\sum_n (X_n - Y_n)$ converges almost surely iff $\sum_n \mathrm{Var}(X_n) < \infty$.

42. Suppose $\{X_n, n \geq 1\}$ is an arbitrary sequence of random variables that have finite means and variances and satisfying

(a) $\lim_{n\to\infty} E(X_n) = c$, for some finite constant c, and

(b) $\sum_{n=1}^{n} \mathrm{Var}(X_n) < \infty$.

Show $X_n \to c$ almost surely. (Hint: Define $\xi_n = X_n - E(X_n)$ and show $\sum_n \xi_n^2 < \infty$ almost surely. Alternatively, try to proceed using Chebychev's inequality.)

If (b) is replaced by the hypothesis $\mathrm{Var}(X_n) \to 0$, show $X_n \stackrel{P}{\to} c$.

43. (a) Fix a real number $p \in (0, 1)$ and let $\{B_n, n \geq 1\}$ be iid Bernoulli random variables with

$$P[B_n = 1] = p = 1 - P[B_n = 0].$$

Define $Y = \sum_{n=1}^{\infty} B_n/2^n$. Verify that the series converges. What is the range of Y? What is the mean and variance of Y? Let Q_p be the distribution of Y. Where does Q_p concentrate?

Use the SLLN to show that if $p \neq p'$, then Q_p and $Q_{p'}$ are mutually singular; that is, there exists a set A such that $Q_p(A) = 1$ and $Q_{p'}(A) = 0$.

(b) Let $F_p(x)$ be the distribution function corresponding to Q_p. Show $F_p(x)$ is continuous and strictly increasing on $[0, 1]$, $F_p(0) = 0$, $F_p(1) = 1$ and satisfies

$$F_p(x) = \begin{cases} (1-p)F_p(2x), & \text{if } 0 \leq x \leq 1/2, \\ 1 - p + pF_p(2x - 1), & \text{if } 1/2 \leq x \leq 1. \end{cases}$$

44. Let $\{X_n, n \geq 1\}$ be iid with values in the set $S = \{1, \ldots, 17\}$. Define the (discrete) density

$$f_0(y) = P[X_1 = y], \quad y \in S.$$

Let $f_1 \neq f_0$ be another probability mass function on S so that for $y \in S$, we have $F_1(y) \geq 0$ and $\sum_{j \in S} f_1(j) = 1$. Set

$$Z_n = \prod_{i=1}^{n} \frac{f_1(X_i)}{f_0(X_i)}, \quad n \geq 1.$$

Prove that $Z_n \stackrel{a.s.}{\to} 0$. (Consider $Y_n = \log Z_n$.)

45. Suppose $\{X_n, n \geq 1\}$ are iid random variables taking values in the *alphabet* $S = \{1, \ldots, r\}$ with positive probabilities $p_1, \ldots, p_r$. Define

$$p_n(i_1, \ldots, i_n) = P[X_1 = i_1, \ldots, X_n = i_n],$$

and set

$$\chi_n(\omega) := p_n(X_1(\omega), \ldots, X_n(\omega)).$$

Then $\chi_n(\omega)$ is the probability that in a new sample of n observations, what is observed matches the original observations. Show that

$$-\frac{1}{n} \log \chi_n(\omega) \stackrel{a.s.}{\to} H := -\sum_{i=1}^{r} p_i \log p_i.$$

46. Suppose $\{X_n, n \geq 1\}$ are iid with Cauchy density. Show that $\{S_n/n, n \geq 1\}$ does not converge almost surely but $\vee_{i=1}^{n} X_i/n$ converges in distribution. To what?

8

Convergence in Distribution

This chapter discusses the basic notions of convergence in distribution. Given a sequence of random variables, when do their distributions converge in a useful way to a limit?

In statisticians' language, given a random sample $X_1, \ldots, X_n$, the sample mean $\bar{X}_n$ is CAN; that is, consistent and asymptotically normal. This means that $\bar{X}$ has an approximately normal distribution as the sample size grows. What exactly does this mean?

8.1 Basic Definitions

Recall our notation that *df* stands for *distribution function*. For the time being, we will understand this to correspond to a probability measure on $\mathbb{R}$.

Recall that F is a df if

(i) $0 \le F(x) \le 1$;

(ii) F is non-decreasing;

(iii) $F(x+) = F(x)\ \forall x \in \mathbb{R}$, where

$$F(x+) = \lim_{\substack{\epsilon>0 \\ \epsilon\downarrow 0}} F(x+\epsilon);$$

that is, F is right continuous.

Also, remember the shorthand notation

$$F(\infty) := \lim_{y\uparrow\infty} F(y)$$
$$F(-\infty) := \lim_{y\downarrow\infty} F(y).$$

F is a *probability distribution function* if

$$F(-\infty) = 0, \qquad F(+\infty) = 1.$$

In this case, F is *proper* or *non-defective.*

If $F(x)$ is a df, set

$$\mathcal{C}(F) = \{x \in \mathbb{R} : F \text{ is continuous at } x\}.$$

A finite interval I with endpoints $a < b$ is called an *interval of continuity* for F if both $a, b \in \mathcal{C}(F)$. We know that

$$(\mathcal{C}(F))^c = \{x : F \text{ is discontinuous at } x\}$$

is at most countable, since

$$\Lambda_n = \{x : F(\{x\}) = F(x) - F(x-) > \frac{1}{n}\}$$

has at most n elements (otherwise (i) is violated) and therefore

$$(\mathcal{C}(F))^c = \bigcup_n \Lambda_n$$

is at most countable.

For an interval $I = (a, b]$, we write, as usual, $F(I) = F(b) - F(a)$. If $a, b \in \mathcal{C}(F)$, then $F((a, b)) = F((a, b])$.

Lemma 8.1.1 *A distribution function $F(x)$ is determined on a dense set. Let D be dense in $\mathbb{R}$. Suppose $F_D(\cdot)$ is defined on D and satisfies the following:*

(a) *$F_D(\cdot)$ is non-decreasing on D.*

(b) $0 \le F_D(x) \le 1$, *for all $x \in D$.*

(c) $\lim_{x\in D, x\to+\infty} F_D(x) = 1, \quad \lim_{x\in D, x\to-\infty} F_D(x) = 0.$

Define for all $x \in \mathbb{R}$

$$F(x) := \inf_{\substack{y>x \\ y\in D}} F_D(y) = \lim_{\substack{y\downarrow x \\ y\in D}} F_D(y). \tag{8.1}$$

Then F is a right continuous probability df. Thus, any two right continuous df's agreeing on a dense set will agree everywhere.

Remark 8.1.1 The proof of Lemma 8.1.1 below shows the following: We let $g : \mathbb{R} \mapsto \mathbb{R}$ have the property that for all $x \in \mathbb{R}$

$$g(x+) = \lim_{y \downarrow x} g(y)$$

exists. Set $h(x) = g(x+)$. Then h is right continuous.

Proof of Lemma 8.1.1. We check that F, defined by (8.1), is right continuous. The plan is to fix $x \in \mathbb{R}$ and show that F is right continuous at x. Given $\epsilon > 0$, there exists $x' \in D$, $x' > x$ such that

$$F(x) + \epsilon \geq F_D(x'). \tag{8.2}$$

From the definition of F, for $y \in (x, x')$,

$$F_D(x') \geq F(y) \tag{8.3}$$

so combining inequalities (8.2) and (8.3) yields

$$F(x) + \epsilon \geq F(y), \quad \forall y \in (x, x').$$

Now F is monotone, so let $y \downarrow x$ to get

$$F(x) + \epsilon \geq F(x+).$$

This is true for all small $\epsilon > 0$, so let $\epsilon \downarrow 0$ and we get

$$F(x) \geq F(x+).$$

Since monotonicity of F implies $F(x+) \geq F(x)$, we get $F(x) = F(x+)$ as desired. □

Four definitions. We now consider four definitions related to weak convergence of probability measures. Let $\{F_n, n \geq 1\}$ be probability distribution functions and let F be a distribution function which is not necessarily proper.

(1) *Vague convergence.* The sequence $\{F_n\}$ converges vaguely to F, written $F_n \stackrel{v}{\to} F$, if for every finite interval of continuity I of F, we have

$$F_n(I) \to F(I).$$

(See Chung (1968), Feller (1971).)

(2) *Proper convergence.* The sequence $\{F_n\}$ converges properly to F, written $F_n \to F$ if $F_n \stackrel{v}{\to} F$ and F is a proper df; that is $F(\mathbb{R}) = 1$. (See Feller (1971).)

(3) *Weak convergence.* The sequence $\{F_n\}$ converges weakly to F, written $F_n \stackrel{w}{\to} F$, if

$$F_n(x) \to F(x),$$

for all $x \in C(F)$. (See Billingsley (1968, 1995).)

(4) *Complete convergence*. The sequence $\{F_n\}$ converges completely to F, written $F_n \xrightarrow{c} F$, if $F_n \xrightarrow{w} F$ and F is proper. (See Loève (1977).)

Example. Define

$$F_n(x) := F(x + (-1)^n n).$$

Then

$$F_{2n}(x) = F(x + 2n) \to 1$$
$$F_{2n+1}(x) = F(x - (2n+1)) \to 0.$$

Thus $\{F_n(x)\}$ does not converge for any x. Thus weak convergence fails. However, for any $I = (a, b]$

$$F_{2n}(a, b] = F_{2n}(b) - F_{2n}(a) \to 1 - 1 = 0$$
$$F_{2n+1}(a, b] = F_{2n+1}(b) - F_{2n+1}(a) \to 0 - 0 = 0.$$

So $F_n(I) \to 0$ and vague convergence holds: $F_n \xrightarrow{v} G$ where $G(\mathbb{R}) = 0$. So the limit is not proper.

Theorem 8.1.1 (Equivalence of the Four Definitions) *If F is proper, then the four definitions (1), (2), (3), (4) are equivalent.*

Proof. If F is proper, then (1) and (2) are the same and also (3) and (4) are the same.

We check that (4) implies (2). If

$$F_n(x) \to F(x), \quad \forall x \in \mathcal{C}(F),$$

then

$$F_n(a, b] = F_n(b) - F_n(a) \to F(b) - F(a) = F(a, b]$$

if $(a, b]$ is an interval of continuity.

Next we show (2) implies (4): Assume

$$F_n(I) \to F(I),$$

for all intervals of continuity I. Let $a, b \in \mathcal{C}(F)$. Then

$$F_n(b) \geq F_n(a, b] \to F(a, b],$$

so

$$\liminf_{n\to\infty} F_n(b) \geq F(a, b], \quad \forall a < b, a \in \mathcal{C}(F).$$

Let $a \downarrow -\infty$, $a \in \mathcal{C}(F)$ to get

$$\liminf_{n\to\infty} F_n(b) \geq F(b).$$

For the reverse inequality, suppose $l < b < r$, $l, r \in \mathcal{C}(F)$, and l chosen so small and r chosen so large that

$$F((l, r]^c) < \epsilon.$$

Then since $F_n(l, r] \to F(l, r]$, we have

$$F_n((l, r]^c) \to F((l, r]^c).$$

So given $\epsilon > 0$, there exists $n_0 = n_0(\epsilon)$ such that $n \geq n_0$ implies

$$F_n((l, r]^c) \leq 2\epsilon.$$

For $n \geq n_0$,

$$\begin{aligned} F_n(b) &= F_n(b) - F_n(l) + F_n(l) \\ &= F_n(l, b] + F_n(l) \\ &\leq F_n(l, b] + 2\epsilon, \end{aligned}$$

since $F_n(l) \leq F_n((l, b]^c)$. So

$$\begin{aligned} \limsup_{n\to\infty} F_n(b) &\leq F(l, b] + 2\epsilon \\ &\leq F(b) + 2\epsilon. \end{aligned}$$

Since $\epsilon > 0$ is arbitrary

$$\limsup_{n\to\infty} F_n(b) \leq F(b).$$

□

Notation: If $\{F, F_n, n \geq 1\}$ are probability distributions, write $F_n \Rightarrow F$ to mean any of the equivalent notions given by (1)–(4). If X_n is a random variable with distribution F_n and X is a random variable with distribution F, we write $X_n \Rightarrow X$ to mean $F_n \Rightarrow F$. This is read "X_n converges in distribution to X" or "F_n converges weakly to F." Notice that unlike almost sure, in probability, or L_p convergence, convergence in distribution says nothing about the behavior of the random variables themselves and only comments on the behavior of the distribution functions of the random variables.

Example 8.1.1 Let N be an $N(0, 1)$ random variable so that the distribution function is symmetric. Define for $n \geq 1$

$$X_n = (-1)^n N.$$

Then $X_n \stackrel{d}{=} N$, so automatically

$$X_n \Rightarrow N.$$

But of course $\{X_n\}$ neither converges almost surely nor in probability.

Remark 8.1.2 Weak limits are unique. If $F_n \xrightarrow{w} F$, and also $F_n \xrightarrow{w} G$, then $F = G$. There is a simple reason for this. The set $(\mathcal{C}(F))^c \cup (\mathcal{C}(G))^c$ is countable so

$$\begin{aligned}\text{INT} &= \mathcal{C}(F) \cap \mathcal{C}(G) \\ &= \mathbb{R} \setminus \text{a countable set}\end{aligned}$$

and hence is dense. For $x \in \text{INT}$,

$$F_n(x) \to F(x), \quad F_n(x) \to G(x),$$

so $F(x) = G(x)$ for $x \in \text{INT}$, and hence by Lemma 8.1.1, we have $F = G$.

Here is a simple example of weak convergence.

Example 8.1.2 Let $\{X_n, n \geq 1\}$ be iid with common unit exponential distribution

$$P[X_n > x] = e^{-x}, \quad x > 0.$$

Set $M_n = \vee_{i=1}^n X_i$ for $n \geq 1$. Then

$$M_n - \log n \Rightarrow Y, \tag{8.4}$$

where

$$P[Y \leq x] = \exp\{-e^{-x}\}, \quad x \in \mathbb{R}.$$

To prove (8.4), note that for $x \in \mathbb{R}$,

$$\begin{aligned}P[M_n - \log n \leq x] &= P\Big(\bigcap_{i=1}^n [X_i \leq x + \log n]\Big) \\ &= \big(1 - e^{-(x+\log n)}\big)^n \\ &= \Big(1 - \frac{e^{-x}}{n}\Big)^n \to \exp\{-e^{-x}\}.\end{aligned}$$

□

8.2 Scheffé's lemma

Consider the following modes of convergence that are stronger than weak convergence.

(a) $F_n(A) \to F(A), \ \forall A \in \mathcal{B}(\mathbb{R})$.

(b) $\sup_{A \in \mathcal{B}(\mathbb{R})} |F_n(A) - F(A)| \to 0$.

Definition (a) (and hence (b)) would rule out many circumstances we would like to fall under weak convergence. Two examples illustrate the point of this remark.

Example 8.2.1 (i). Suppose F_n puts mass $\frac{1}{n}$ at points $\{\frac{1}{n}, \frac{2}{n}, \dots, \frac{n}{n}\}$. If

$$F(x) = x, \quad 0 \le x \le 1$$

is the uniform distribution on [0, 1], then for $x \in (0, 1)$

$$F_n(x) = \frac{[nx]}{n} \to x = F(x).$$

Thus we have weak convergence $F_n \Rightarrow F$. However if $\mathbb{Q}$ is the set of rationals in [0, 1],

$$F_n(\mathbb{Q}) = 1, \quad F(\mathbb{Q}) = 0,$$

so convergence in the sense of (a) fails even though it seems natural that the discrete uniform distribution should be converging to the continuous uniform distribution.

(ii) *DeMoivre–Laplace central limit theorem*: This is a situation similar to what was observed in (a). Suppose $\{X_n, n \ge 1\}$ are iid, with

$$P[X_n = 1] = p = 1 - P[X_n = 0].$$

Set $S_n = \sum_{i=1}^n X_i$, which has a binomial distribution with parameters n, p. Then the DeMoivre–Laplace central limit theorem states that

$$\begin{aligned} P[\frac{S_n - np}{\sqrt{npq}} \le x] \to N(x) &= \int_{-\infty}^{x} n(u)du \\ &= \int_{-\infty}^{x} \frac{1}{\sqrt{2\pi}} e^{-u^2/2} du. \end{aligned}$$

But if

$$A = \{\frac{k - np}{\sqrt{npq}} : k \ge 0, n \ge 0\},$$

we have

$$P[\frac{S_n - np}{\sqrt{npq}} \in A] = 1 \ne N(A) = 0.$$

Weak convergence, because of its connection to continuous functions (see Theorem 8.4.1) is more useful than the convergence notions (a) or (b). The convergence definition (b) is called total variation convergence and has connections to density convergence through Scheffé's lemma.

Lemma 8.2.1 (Scheffé's lemma) *Suppose $\{F, F_n, n \ge 1\}$ are probability distributions with densities $\{f, f_n, n \ge 1\}$. Then*

$$\sup_{B \in \mathcal{B}(\mathbb{R})} |F_n(B) - F(B)| = \frac{1}{2} \int |f_n(x) - f(x)| dx. \tag{8.5}$$

If $f_n(x) \to f(x)$ almost everywhere (that is, for all x except a set of Lebesgue measure 0), then

$$\int |f_n(x) - f(x)|dx \to 0.$$

and thus $F_n \to F$ in total variation (and hence weakly).

Remarks.

- If $F_n \overset{w}{\to} F$ and F_n and F have densities f_n, f, it does **not** necessarily follow that $f_n(x) \to f(x)$. See Exercise 12.
- Although Scheffé's lemma as presented above looks like it deals with densities with respect to Lebesgue measure, in fact it works with densities with respect to any measure. This is a very useful observation and sometimes a density with respect to counting measure is employed to deal with convergence of sums. See, for instance, Exercise 4.

Proof of Scheffé's lemma. Let $B \in \mathcal{B}(\mathbb{R})$. Then

$$1 - 1 = \int (f_n(x) - f(x))dx = 0,$$

so

$$0 = \int_B (f_n(x) - f(x))dx + \int_{B^c} (f_n(x) - f(x))dx,$$

which implies

$$|\int_B (f_n(x) - f(x))dx| = |\int_{B^c} (f_n(x) - f(x))dx|. \tag{8.6}$$

This leads to

$$\begin{aligned}
2|F_n(B) - F(B)| &= 2|\int_B (f_n(x) - f(x))dx| \\
&= |\int_B (f_n(x) - f(x))dx| + |\int_{B^c} (f_n(x) - f(x))dx| \\
&\le \int_B |f_n(x) - f(x)|dx + \int_{B^c} |f_n(x) - f(x)|dx \\
&= \int |f_n(x) - f(x)|dx.
\end{aligned}$$

To summarize:

$$\sup |F_n(B) - F(B)| \le \frac{1}{2} \int |f_n(x) - f(x)|dx.$$

If we find some set B for which equality actually holds, then we will have shown (8.5). Set $B = [f_n \geq f]$. Then from (8.6)

$$2|F_n(B) - F(B)| = |\int_B (f_n(x) - f(x))dx| + |\int_{B^c} (f_n(x) - f(x))dx|,$$

and because the first integrand on the right is non-negative and the second is non-positive, we have the equality

$$= \int_B |f_n(x) - f(x)|dx + \int_{B^c} |f_n(x) - f(x)|dx$$
$$= \int |f_n(x) - f(x)|dx.$$

So equality holds in (8.5).

Now suppose $f_n(x) \to f(x)$ almost everywhere. So $f - f_n \to 0$ a.e., and therefore $(f - f_n)^+ \to 0$ almost everywhere. Also

$$(f - f_n)^+ \leq f,$$

and f is integrable on $\mathbb{R}$ with respect to Lebesgue measure. Since

$$0 = \int (f(x) - f_n(x))dx = \int (f(x) - f_n(x))^+ dx - \int (f(x) - f_n(x))^- dx,$$

it follows that

$$\int |f(x) - f_n(x)|dx = \int (f(x) - f_n(x))^+ dx + \int (f(x) - f_n(x))^- dx$$
$$= 2\int (f(x) - f_n(x))^+ dx.$$

Thus

$$(f - f_n)^+ \leq f \in L_1,$$

and

$$(f - f_n)^+ \to 0,$$

a.e. and dominated convergence implies

$$\int |f(x) - f_n(x)|dx \to 0.$$

□

8.2.1 *Scheffé's lemma and Order Statistics*

As an example of one use for Scheffé's lemma, we consider the following limit result for order statistics.

Proposition 8.2.1 *Suppose $\{U_n, n \geq 1\}$ are iid $U(0,1)$ random variables so that*

$$P[U_j \leq x] = x, \quad 0 \leq x \leq 1$$

and suppose

$$U_{(1,n)} \leq U_{(2,n)} \leq \cdots \leq U_{(n,n)}$$

are the order statistics so that $U_{(1,n)} = \min\{U_1, \ldots, U_n\}$, $U_{(2,n)}$ is the second smallest and $U_{(n,n)}$ is the largest. Assume $k = k(n)$ is a function of n satisfying $k(n) \to \infty$ and $k/n \to 0$. as $n \to \infty$. Let

$$\xi_n = \frac{U_{(k,n)} - \frac{k}{n}}{\sqrt{\frac{k}{n}(1 - \frac{k}{n})\frac{1}{n}}}.$$

Then the density of ξ_n converges to a standard normal density and hence by Scheffé's lemma

$$\sup_{B \in \mathcal{B}(\mathbb{R})} \left|P[\xi_n \in B] - \int_B \frac{1}{\sqrt{2\pi}} e^{-u^2/2} du\right| \to 0$$

as $n \to \infty$.

Proof. The distribution of $U_{(k,n)}$ can be obtained from a binomial probability since for $0 < x < 1$, $P[U_{(k,n)} \leq x]$ is the binomial probability of at least k successes in n trials when the success probability is x. Differentiating, we get the density $f_n(x)$ of $U_{(k,n)}$ to be

$$f_n(x) = \frac{n!}{(k-1)!(n-k)!} x^{k-1}(1-x)^{n-k}, \quad 0 < x < 1.$$

(This density can also be obtained directly by a multimomial argument where there are 3 cells with proabilities x, dx and $(1-x)$ and cell frequencies $k-1$, 1 and $n-k$.) Since

$$\sqrt{\frac{k}{n}(1 - \frac{k}{n})\frac{1}{n}} \sim \frac{\sqrt{k}}{n},$$

as $n \to \infty$, the convergence to types Theorem 8.7.1 discussed below assures us we can replace the square root in the expression for ξ_n by $\sqrt{k}/n$ and thus we consider the density

$$\frac{\sqrt{k}}{n} f_n(\frac{\sqrt{k}}{n}x + \frac{k}{n}).$$

By Stirling's formula (see Exercise 8 of Chapter 9), as $n \to \infty$,

$$\frac{n!}{(k-1)!(n-k)!} \sim \frac{\sqrt{n}}{\sqrt{2\pi}\left(\frac{k}{n}\right)^{k-1/2}\left(1 - \frac{k}{n}\right)^{n-k+1/2}}.$$

Neglecting the factorials in the expression for the density, we have two factors of the form

$$(\frac{\sqrt{k}}{n}x+\frac{k}{n})^{k-1}(1-\frac{k}{n}-\frac{\sqrt{k}}{n}x)^{n-k}$$
$$=(\frac{k}{n})^{k-1}(1+\frac{x}{\sqrt{k}})^{k-1}(1-\frac{k}{n})^{n-k}(1-\frac{x}{(n-k)/\sqrt{k}})^{n-k}.$$

Thus we get for the density of ξ_n, the following asymptotic expression

$$\frac{1}{\sqrt{2\pi}}(1+\frac{x}{\sqrt{k}})^{k-1}(1-\frac{x}{(n-k)/\sqrt{k}})^{n-k}.$$

It suffices to prove that

$$(1+\frac{x}{\sqrt{k}})^{k-1}(1-\frac{x}{(n-k)/\sqrt{k}})^{n-k}\to e^{-x^2/2}$$

or equivalently,

$$(k-1)\log\big(1+\frac{x}{\sqrt{k}}\big)+(n-k)\log\big(1-\frac{x}{(n-k)/\sqrt{k}}\big)\to\frac{x^2}{2}. \tag{8.7}$$

Observe that, for $|t|<1$,

$$-\log(1-t)=\sum_{n=1}^{\infty}\frac{t^n}{n},$$

and therefore

$$\delta(t):=|-\log(1-t)-(t+\frac{t^2}{2})|$$
$$\leq\sum_{n=3}^{\infty}|t|^n=\frac{|t|^3}{1-|t|}\leq 2|t|^3, \tag{8.8}$$

if $|t|<1/2$. So the left side of (8.7) is of the form

$$(k-1)[\frac{x}{\sqrt{k}}-\frac{x^2}{2k}]-(n-k)[\frac{x}{(n-k)/\sqrt{k}}+\frac{x^2}{2(n-k)^2/k}]+o(1)$$

where

$$o(1)=(k-1)\delta(\frac{x}{\sqrt{k}})+(n-k)\delta(\frac{x}{(n-k)/\sqrt{k}})\to 0.$$

Neglecting $o(1)$, (8.7) simplifies to

$$-\frac{x}{\sqrt{k}}-\frac{x^2}{2}(1-\frac{1}{k}+\frac{1}{\frac{n}{k}-1})\to -x^2/2.$$

□

8.3 The Baby Skorohod Theorem

Skorohod's theorem is a conceptual aid which makes certain weak convergence results easy to prove by continuity arguments. The theorem is true in great generality. We only consider the result for real valued random variables and hence the name *Baby* Skorohod Theorem.

We begin with a brief discussion of the relationship of almost sure convergence and weak convergence.

Proposition 8.3.1 *Suppose $\{X, X_n, n \geq 1\}$ are random variables. If*

$$X_n \stackrel{a.s.}{\to} X,$$

then

$$X_n \Rightarrow X.$$

Proof. Suppose $X_n \stackrel{\text{a.s.}}{\to} X$ and let F be the distribution function of X. Set

$$N = [X_n \to X]^c$$

so that $P(N) = 0$. For any $h > 0$ and $x \in C(F)$, we have the following set containments:

$$\begin{aligned} N^c \cap [X \leq x - h] &\subset \liminf_{n\to\infty}[X_n \leq x] \cap N^c \\ &\subset \limsup_{n\to\infty}[X_n \leq x] \cap N^c \\ &\subset [X \leq x] \cap N^c, \end{aligned}$$

and hence, taking probabilities

$$\begin{aligned} F(x-h) &\leq P(\liminf_{n\to\infty}[X_n \leq x]) \\ &\leq \liminf_{n\to\infty} P[X_n \leq x] \qquad \text{(from Fatou's lemma)} \\ &\leq \limsup_{n\to\infty} P[X_n \leq x] \\ &\leq P(\limsup_{n\to\infty}[X_n \leq x]) \quad \text{(from Fatou's lemma)} \\ &\leq F(x). \end{aligned}$$

Since $x \in C(F)$, let $h \downarrow 0$ to get

$$F(x) \leq \liminf_{n\to\infty} F_n(x) \leq \limsup_{n\to\infty} F_n(x) \leq F(x).$$ □

The converse if false: Recall Example 8.1.1.

Despite the fact that convergence in distribution does not imply almost sure convergence, Skorohod's theorem provides a partial converse.

Theorem 8.3.2 (Baby Skorohod Theorem) *Suppose $\{X_n, n \geq 0\}$ are random variables defined on the probability space $(\Omega, \mathcal{B}, P)$ such that*

$$X_n \Rightarrow X_0.$$

Then there exist random variables $\{X_n^\#, n \geq 0\}$ defined on the Lebesgue probability space $([0,1], \mathcal{B}([0,1]), \lambda =$ Lebesgue measure $)$ such that for each fixed $n \geq 0$,

$$X_n \stackrel{d}{=} X_n^\#,$$

and

$$X_n^\# \stackrel{a.s.}{\to} X_0^\#$$

where a.s. means almost surely with respect to λ.

Note that Skorohod's theorem ignores dependencies in the original $\{X_n\}$ sequence. It produces a sequence $\{X_n^\#\}$ whose one dimensional distributions match those of the original sequence but makes no attempt to match the finite dimensional distributions.

The proof of Skorohod's theorem requires the following result.

Lemma 8.3.1 *Suppose F_n is the distribution function of X_n so that $F_n \Rightarrow F_0$. If*

$$t \in (0,1) \cap \mathcal{C}(F_0^{\leftarrow}),$$

then

$$F_n^{\leftarrow}(t) \to F_0^{\leftarrow}(t).$$

Proof of Lemma 8.3.1. Since $\mathcal{C}(F_0)^c$ is at most countable, given $\epsilon > 0$, there exists $x \in \mathcal{C}(F_0)$ such that

$$F_0^{\leftarrow}(t) - \epsilon < x < F_0^{\leftarrow}(t).$$

From the definition of the inverse function, $x < F_0^{\leftarrow}(t)$ implies that $F_0(x) < t$. Also, $x \in \mathcal{C}(F_0)$ implies $F_n(x) \to F_0(x)$. So for large n, we have $F_n(x) < t$. Again, using the definition of the inverse function, we get $x \leq F_n^{\leftarrow}(t)$. Thus

$$F_0^{\leftarrow}(t) - \epsilon < x \leq F_n^{\leftarrow}(t)$$

for all large n and since $\epsilon > 0$ is arbitrary, we conclude

$$F_0^{\leftarrow}(t) \leq \liminf_{n \to \infty} F_n^{\leftarrow}(t). \tag{8.9}$$

Note that we have not yet used the assumption that t is a continuity point of $F_0^{\leftarrow}$ and this is used for the reverse inequality.

Whenever $t' > t$, we may find $y \in \mathcal{C}(F_0)$ such that

$$F_0^{\leftarrow}(t') < y < F_0^{\leftarrow}(t') + \epsilon.$$

This gives

$$F_0(y) \geq t' > t.$$

Since $y \in \mathcal{C}(F_0)$, $F_n(y) \to F_0(y)$ and for large n, $F_n(y) \geq t$, therefore $y \geq F_n^{\leftarrow}(t)$, and thus

$$F_0^{\leftarrow}(t') + \epsilon > y \geq F_n^{\leftarrow}(t)$$

for all large n. Moreover, since $\epsilon > 0$ is arbitrary,

$$\limsup_{n\to\infty} F_n^{\leftarrow}(t) \leq F_0^{\leftarrow}(t').$$

Let $t' \downarrow t$ and use continuity of $F_0^{\leftarrow}$ at t to conclude that

$$\limsup_{n\to\infty} F_n^{\leftarrow}(t) \leq F_0^{\leftarrow}(t). \tag{8.10}$$

The two inequalities (8.9) and (8.10) combine to yield the result. □

This lemma only guarantees convergence of $F_n^{\leftarrow}$ to $F_0^{\leftarrow}$ at continuity points of the limit. However, convergence could take place on more points. For instance, if $F_n = F_0$ for all n, $F_n^{\leftarrow} = F_0^{\leftarrow}$ and convergence would be everywhere.

Lemma 8.3.1 allows a rapid proof of the Baby Skorohod theorem.

Proof of the Baby Skorohod Theorem. On the sample space $[0, 1]$, define the random variable $U(t) = t$ so that U is uniformly distributed, since for $0 \leq x \leq 1$

$$\lambda[U \leq x] = \lambda\{t \in [0,1] : U(t) \leq x\} = \lambda[0, x] = x.$$

For $n \geq 0$ define $X_n^\#$ on $[0, 1]$ by

$$X_n^\# = F_n^{\leftarrow}(U).$$

Then for $y \in \mathbb{R}$

$$\lambda[X_n^\# \leq y] = \lambda\{t \in [0,1] : F_n^{\leftarrow}(t) \leq y\} = \lambda\{t \in [0,1] : t \leq F_n(y)\} = F_n(y).$$

So we conclude that $X_n^\# \stackrel{d}{=} X_n$, for each $n \geq 0$.

Next, we write

$$\begin{aligned}\lambda\{t \in [0,1] : &X_n^\#(t) \not\to X_0^\#(t)\}\\ &= \lambda\{t \in [0,1] : F_n^{\leftarrow}(t) \not\to F_0^{\leftarrow}(t)\},\end{aligned}$$

and using Lemma 8.3.1, this is bounded by

$$\begin{aligned}&\leq \lambda\{t \in [0,1] : F_0^{\leftarrow} \text{ is not continuous at } t\ \}\\ &= \lambda\{\text{ a countable set }\} = 0.\end{aligned}$$

□

The next corollary looks tame when restricted to $\mathbb{R}$, but its multidimensional generalizations have profound consequences. For a map $h : \mathbb{R} \mapsto \mathbb{R}$, define

$$\operatorname{Disc}(h) = \{x : h \text{ is not continuous at } x\ \} = (\mathcal{C}(h))^c.$$

Corollary 8.3.1 (Continuous Mapping Theorem) *Let $\{X_n, n \geq 0\}$ be a sequence of random variables such that*

$$X_n \Rightarrow X_0.$$

For $n \geq 0$, assume F_n is the distribution function of X_n. Let $h : \mathbb{R} \mapsto \mathbb{R}$ satisfy

$$P[X_0 \in Disc(h)] = 0.$$

Then

$$h(X_n) \Rightarrow h(X_0),$$

and if h is bounded, dominated convergence implies

$$Eh(X_n) = \int h(x)F_n(dx) \to Eh(x) = \int h(x)F_0(dx).$$

Remark. Disc(h) is always measurable even if h is not.

As a quick example, if $X_n \Rightarrow X_0$, then $X_n^2 \to X_0^2$ which is checked by applying the continuous function (so Disc(h) is empty) $h(x) = x^2$.

Proof. The proof of the corollary uses the Baby Skorohod Theorem which identifies new random variables $X_n^\# \stackrel{d}{=} X_n, n \geq 0$, with $X_n^\#$ defined on [0, 1]. Also $X_n^\#(t) \to X_0^\#(t)$ for a.a. t. If $X_0^\#(t) \in C(h)$, then $h(X_n^\#(t)) \to h(X_0^\#(t))$. Thus,

$$\begin{aligned}\lambda\{t \in [0,1] : h(X_n^\#(t)) \to h(X_0^\#(t))\} \\ &\geq \lambda\{t \in [0,1] : X_0^\#(t) \in (\text{Disc}(h))^c\} \\ &= P([X_0 \in \text{Disc}(h)]^c) = 1.\end{aligned}$$

So $h(X_n^\#) \to h(X_0^\#)$ almost surely with respect to λ, and since almost sure convergence implies convergence in distribution, we have

$$h(X_n) \stackrel{d}{=} h(X_n^\#) \Rightarrow h(X_0^\#) \stackrel{d}{=} h(X_0)$$

so that $h(X_n) \Rightarrow h(X_0)$. □

8.3.1 The Delta Method

The delta method allows us to take a basic convergence, for instance to a limiting normal distribution, and apply smooth functions and conclude that the functions are asymptotically normal as well.

In statistical estimation we try to estimate a parameter θ from a parameter set Θ based on a random sample of size n with a statistic

$$T_n = T_n(X_1, \ldots, X_n).$$

This means we have a family of probability models

$$\{(\Omega, \mathcal{B}, P_\theta), \theta \in \Theta\},$$

and we are trying to choose the correct model. The estimator T_n is *consistent* if

$$T_n \stackrel{P_\theta}{\to} \theta,$$

for every θ. The estimator T_n is CAN, or consistent and asymptotically normal, if for all $\theta \in \Theta$

$$\lim_{n\to\infty} P_\theta[\sigma_n(T_n - \theta) \le x] = N(0, 1, x),$$

for some $\sigma_n \to \infty$.

Suppose we have a CAN estimator of θ, but we need to estimate a smooth function $g(\theta)$. For example, in a family of exponential densities, θ may represent the mean but we are interested in the variance θ^2. We see from the delta method that $g(T_n)$ is also CAN for $g(\theta)$.

We illustrate the method using the central limit theorem (CLT) to be proved in the next chapter. Let $\{X_j, j \ge 1\}$ be iid with $E(X_n) = \mu$ and $\mathrm{Var}(X_n) = \sigma^2$. From the CLT we get

$$\frac{S_n - n\mu}{\sigma\sqrt{n}} \Rightarrow N(0, 1),$$

where $N(0, 1)$ is a normal random variable with mean 0 and variance 1. Equivalently, we can express this in terms of $\bar{X} = \sum_{i=1}^n X_i/n$ as

$$\sqrt{n}\left(\frac{\bar{X} - \mu}{\sigma}\right) \Rightarrow N(0, 1).$$

So $\bar{X}$ is consistent and an asymptotically normal estimator of μ. The *delta method* asserts that if $g(x)$ has a non-zero derivative $g'(\mu)$ at μ, then

$$\sqrt{n}\left(\frac{g(\bar{X}) - g(\mu)}{\sigma g'(\mu)}\right) \Rightarrow N(0, 1). \tag{8.11}$$

So $g(\bar{X})$ is CAN for $g(\mu)$.

Remark. The proof does not depend on the limiting random variable being $N(0, 1)$ and would work equally well if $N(0, 1)$ were replaced by any random variable Y.

Proof of (8.11). By the Baby Skorohod Theorem there exist random variables $Z_n^\#$ and $N^\#$ on the probability space $((0, 1), \mathcal{B}((0, 1)), \lambda)$ such that

$$Z_n^\# \stackrel{d}{=} \sqrt{n}\left(\frac{\bar{X} - \mu}{\sigma}\right), \quad N^\# \stackrel{d}{=} N$$

and

$$Z_n^\# \to N^\# \quad \text{a.s. } (\lambda).$$

Define

$$\bar{X}^\# = \mu + \sigma Z_n^\#/\sqrt{n},$$

so that $\bar{X} \stackrel{d}{=} \bar{X}^{\#}$. Then using the definition of derivative

$$\begin{aligned}\sqrt{n}\left(\frac{g(\bar{X})-g(\mu)}{\sigma g'(\mu)}\right) &\stackrel{d}{=} \sqrt{n}\left(\frac{g(\mu+\sigma Z_n^{\#}/\sqrt{n})-g(\mu)}{\sigma g'(\mu)}\right)\\ &= \frac{g(\mu+\sigma Z_n^{\#}/\sqrt{n})-g(\mu)}{\sigma Z_n^{\#}/\sqrt{n}} \cdot \frac{\sigma Z_n^{\#}}{\sigma g'(\mu)}\\ &\stackrel{a.s.(\lambda)}{\to} g'(\mu) \cdot \frac{\sigma N^{\#}}{\sigma g'(\mu)} = N^{\#} \stackrel{d}{=} N,\end{aligned}$$

since $\sigma Z_n^{\#}/\sqrt{n} \to 0$ almost surely. This completes the proof. □

Remark. Suppose $\{X_n, n \geq 0\}$ is a sequence of random variables such that

$$X_n \Rightarrow X_0.$$

Suppose further that

$$h : \mathbb{R} \mapsto \mathbb{S},$$

where $\mathbb{S}$ is some nice metric space, for example, $\mathbb{S} = \mathbb{R}^2$. Then if

$$P[X_0 \in \text{Disc}(h)] = 0,$$

Skorohod's theorem suggests that it should be the case that

$$h(X_n) \Rightarrow h(X)$$

in $\mathbb{S}$. But what does weak convergence in $\mathbb{S}$ mean? Read on.

8.4 Weak Convergence Equivalences; Portmanteau Theorem

In this section we discuss several conditions which are equivalent to weak convergence of probability distributions. Some of these are of theoretical use and some allow easy generalization of the notion of weak convergence to higher dimensions and even to function spaces. The definition of weak convergence of distribution functions on $\mathbb{R}$ is notable for not allowing easy generalization to more sophisticated spaces. The modern theory of weak convergence of stochastic processes rests on the equivalences to be discussed next.

We nead the following definition. For $A \in \mathcal{B}(\mathbb{R})$, let

$$\begin{aligned}\partial(A) &= \text{ the boundary of } A\\ &= A^{-} \setminus A^{0} = \text{ the closure of } A \text{ minus the interior of } A\\ &= \{x : \exists\, y_n \in A,\ y_n \to x \text{ and } \exists\, z_n \in A^c, z_n \to x\}\\ &= \text{ points reachable from both outside and inside } A.\end{aligned}$$

Theorem 8.4.1 (Portmanteau Theorem) *Let $\{F_n, n \geq 0\}$ be a family of proper distributions. The following are equivalent.*

(1) $F_n \Rightarrow F_0$.

(2) For all $f : \mathbb{R} \mapsto \mathbb{R}$ *which are bounded and continuous,*

$$\int f dF_n \to \int f dF_0.$$

Equivalently, if X_n is a random variable with distribution F_n ($n \geq 0$), then for f bounded and continuous

$$Ef(X_n) \to Ef(X_0).$$

(3) If $A \in \mathcal{B}(\mathbb{R})$ satisfies $F_0(\partial(A)) = 0$, then

$$F_n(A) \to F_0(A).$$

Remarks. (i) Item (2) allows for the easy generalization of the notion of weak convergence of random elements $\{\xi_n, n \geq 0\}$ whose range $\mathbb{S}$ is a metric space. The definition is

$$\xi_n \Rightarrow \xi_0$$

iff

$$E(f(\xi_n)) \to E(f(\xi_0))$$

as $n \to \infty$ for all test functions $f : \mathbb{S} \mapsto \mathbb{R}$ which are bounded and continuous. (The notion of continuity is natural since $\mathbb{S}$ is a metric space.)

(ii) The following clarification is necessary. *Portmanteau* is not the name of the inventor of this theorem. A *portmanteau* is a large leather suitcase that opens into two hinged compartments. Billingsley (1968) may be the first to call this result and its generalizations by the name portmanteau theorem. He dates the result back to 1940 and attributes it to Alexandrov.

Proof. (1) → (2): This follows from Corollary 8.3.1 of the continuous mapping theorem.

(1) → (3): Let $f(x) = 1_A(x)$. We claim that

$$\partial(A) = \text{Disc}(1_A). \tag{8.12}$$

To verify (8.12), we proceed with verifications of two set inclusions.

(i) $\partial(A) \subset \text{Disc}(1_A)$. This is checked as follows. If $x \in \partial(A)$, then there exists

$$y_n \in A, \ \text{ and } y_n \to x,$$
$$z_n \in A^c, \text{ and } z_n \to x.$$

So

$$1 = 1_A(y_n) \to 1, \quad 0 = 1_A(z_n) \to 0$$

implies $x \in \text{Disc}(1_A)$.

(ii) $\text{Disc}(1_A) \subset \partial(A)$: This is verified as follows. Let $x \in \text{Disc}(1_A)$. Then there exists $x_n \to x$, such that

$$1_A(x_n) \not\to 1_A(x).$$

Now there are two cases to consider.

Case (i) $1_A(x) = 1$. Then there exists n' such that $1_A(x_{n'}) \to 0$. So for all large n', $1_A(x_{n'}) = 0$ and $x_{n'} \in A^c$. Thus $x_{n'} \in A^c$, and $x_{n'} \to x$. Also let $y_n = x \in A$ and then $y_n \to x$, so $x \in \partial(A)$.

Case (ii) $1_A(x) = 0$. This is handled similarly.

Given $A \in \mathcal{B}(\mathbb{R})$ such that $F_0(\partial(A)) = 0$, we have that $F_0(\{x : x \in \text{Disc}(1_A)\} = 0$ and by the continuous mapping theorem

$$\int 1_A dF_n = F_n(A) \to \int 1_A dF_0 = F_0(A).$$

(3)$\to$(1): Let $x \in \mathcal{C}(F_0)$. We must show $F_n(x) \to F(x)$. But if $A = (-\infty, x]$, then $\partial(A) = \{x\}$ and $F_0(\partial(A)) = 0$ since $F_0(\{x\}) = 0$ because $x \in \mathcal{C}(F_0)$. So

$$F_n(A) = F_n(x) \to F_0(A) = F_0(x).$$

(Recall, we are using both F_n and F_0 in two ways, once as a measure and once as a distribution function.)

(2) $\to$ (1). This is the last implication needed to show the equivalence of (1), (2) and (3). Let $a, b \in \mathcal{C}(F)$. Given (2), we show $F_n(a, b] \to F_0(a, b]$.

Define the bounded continuous function g_k whose graph is the trapezoid of height 1 obtained by taking a rectangle of height 1 with base $[a, b]$ and extending the base symmetrically to $[a - k^{-1}, b + k^{-1}]$. Then $g_k \downarrow 1_{[a,b]}$ as $k \to \infty$ and for all k,

$$F_n(a, b] = \int_{\mathbb{R}} 1_{(a,b]} dF_n \le \int g_k dF_n \to \int g_k dF_0$$

as $n \to \infty$ due to (2). Since $g_k \le 1$, and $g_k \downarrow 1_{[a,b]}$ we have

$$\int g_k dF_0 \downarrow F_0([a, b]) = F_0((a, b]).$$

We conclude that

$$\limsup_{n\to\infty} F_n(a, b] \le F_0(a, b].$$

Next, define new functions h_k whose graphs are trapezoids of height 1 obtained by taking a rectangle of height 1 with base $[a + k^{-1}, b - k^{-1}]$ and stretching the base symmetrically to obtain $[a, b]$. Then $h_k \uparrow 1_{(a,b)}$ and

$$F_n(a, b] \ge \int h_k dF_n \to \int h_k dF_0,$$

for all k. By monotone convergence

$$\int h_k dF_0 \uparrow F_0((a,b)) = F_0((a,b])$$

as $k \to \infty$, so that

$$\liminf_{n\to\infty} F_n((a,b]) \geq F_0((a,b]). \qquad \square$$

Sometimes one of the characterizations of Theorem 8.4.1 is much easier to verify than the definition.

Example 8.4.1 The discrete uniform distribution is close to the continuous uniform distribution. Suppose F_n has atoms at $i/n, 1 \leq i \leq n$ of size $1/n$. Let F_0 be the uniform distribution on [0, 1]; that is

$$F(x) = x, \quad 0 \leq x \leq 1.$$

Then

$$F_n \Rightarrow F_0.$$

To verify this, it is easiest to proceed by showing that integrals of arbitrary bounded continuous test functions converge. Let f be real valued, bounded and continuous with domain [0, 1]. Observe that

$$\begin{aligned}\int f dF_n &= \sum_{i=1}^{n} f(i/n)\frac{1}{n}\\ &= \text{Riemann approximating sum}\\ &\to \int_0^1 f(x)dx \qquad (n \to \infty)\\ &= \int f dF_0\end{aligned}$$

where F_0 is the uniform distribution on [0, 1]. $\square$

It is possible to restrict the test functions in the portmanteau theorem to be uniformly continuous and not just continuous.

Corollary 8.4.1 *Let $\{F_n, n \geq 0\}$ be a family of proper distributions. The following are equivalent.*

(1) $F_n \Rightarrow F_0$.

(2) For all $f : \mathbb{R} \mapsto \mathbb{R}$ which are bounded and uniformly continuous,

$$\int f dF_n \to \int f dF_0.$$

Equivalently, if X_n is a random variable with distribution F_n $(n \geq 0)$, then for f bounded and uniformly continuous

$$Ef(X_n) \to Ef(X_0).$$

Proof. In the proof of (2) → (1) in the portmanteau theorem, the trapezoid functions are each bounded, continuous, vanish off a compact set, and are hence uniformly continuous. This observation suffices. □

8.5 More Relations Among Modes of Convergence

We summarize three relations among the modes of convergence in the next proposition.

Proposition 8.5.1 *Let $\{X, X_n, n \geq 1\}$ be random variables on the probability space $(\Omega, \mathcal{B}, P)$.*

(i) If

$$X_n \stackrel{a.s.}{\to} X,$$

then

$$X_n \stackrel{P}{\to} X.$$

(ii) If

$$X_n \stackrel{P}{\to} X,$$

then

$$X_n \Rightarrow X.$$

All the converses are false.

Proof. The statement (i) is just Theorem 6.2.1 of Chapter 6. To verify (ii), suppose $X_n \stackrel{P}{\to} X$ and f is a bounded and continuous function. Then

$$f(X_n) \stackrel{P}{\to} f(X)$$

by Corollary 6.3.1 of Chapter 6. Dominated convergence implies

$$E(f(X_n)) \to E(f(X))$$

(see Corollary 6.3.2 of Chapter 6) so

$$X_n \Rightarrow X$$

by the portmanteau theorem. □

There is one special case where convergence in probability and convergence in distribution are the same.

Proposition 8.5.2 *Suppose $\{X_n, n \geq 1\}$ are random variables. If c is a constant such that*

$$X_n \stackrel{P}{\to} c,$$

then

$$X_n \Rightarrow c$$

and conversely.

Proof. It is always true that convergence in probability implies convergence in distribution, so we focus on the converse. If

$$X_n \Rightarrow c$$

then

$$P[X_n \leq x] \to \begin{cases} 0, & \text{if } x < c, \\ 1, & \text{if } x > c, \end{cases}$$

and

$$X_n \stackrel{P}{\to} c$$

means $P[|X_n - c| > \epsilon] \to 0$ which happens iff

$$P[X_n < c - \epsilon] \to 0 \text{ and } P[X_n < c + \epsilon] \to 1.$$

□

8.6 New Convergences from Old

We now present two results that express the following fact. If X_n converges in distribution to X and Y_n is close to X_n, then Y_n converges in distribution to X as well.

Theorem 8.6.1 (Slutsky's theorem) *Suppose $\{X, X_n, Y_n, \xi_n, n \geq 1\}$ are random variables.*

(a) If $X_n \Rightarrow X$, and

$$X_n - Y_n \stackrel{P}{\to} 0,$$

then

$$Y_n \Rightarrow X.$$

(b) Equivalently, if $X_n \Rightarrow X$, and $\xi_n \stackrel{P}{\to} 0$, then

$$X_n + \xi_n \Rightarrow X.$$

Proof. It suffices to prove (b). Let f be real valued, bounded and uniformly continuous. Define the modulus of continuity

$$\omega_\delta(f) = \sup_{|x-y|\leq\delta} |f(x) - f(y)|.$$

Because f is uniformly continuous,

$$\omega_\delta(f) \to 0, \quad \delta \to 0. \tag{8.13}$$

From Corollary 8.4.1 if suffices to show $Ef(X_n + \xi_n) \to Ef(X)$. To do this, observe

$$\begin{aligned}
&|Ef(X_n + \xi_n) - Ef(X)| \\
&\quad \le |Ef(X_n + \xi_n) - Ef(X_n)| + |Ef(X_n) - Ef(X)| \\
&\quad = E|f(X_n + \xi_n) - f(X_n)| 1_{[|\xi_n| \le \delta]} + 2 \sup_x |f(x)| P[|\xi_n| > \delta] + o(1)
\end{aligned}$$

(since $X_n \Rightarrow X$)

$$= o(1) + \omega_\delta(f) + (\text{const}) P[|\xi_n| > \delta].$$

The last probability goes to 0 by assumption. Let $\delta \to 0$ and use (8.13). □

Slutsky's theorem is sometimes called the converging together lemma. Here is a generalization which is useful for truncation arguments and analyzing time series models.

Theorem 8.6.2 (Second Converging Together Theorem) *Let us suppose that $\{X_{un}, X_u, Y_n, X; n \ge 1, u \ge 1\}$ are random variables such that for each n, $Y_n, X_{un}, u \ge 1$ are defined on a common domain. Assume for each u, as $n \to \infty$,*

$$X_{un} \Rightarrow X_u,$$

and as $u \to \infty$

$$X_u \Rightarrow X.$$

Suppose further that for all $\epsilon > 0$,

$$\lim_{u \to \infty} \limsup_{n \to \infty} P[|X_{un} - Y_n| > \epsilon] = 0.$$

Then we have

$$Y_n \Rightarrow X$$

as $n \to \infty$.

Proof. For any bounded, uniformly continuous function f, we must show

$$\lim_{n \to \infty} Ef(Y_n) = Ef(X).$$

Without loss of generality, we may, for neatness sake, suppose that

$$\sup_{x \in \mathbb{R}} |f(x)| \le 1.$$

Now write

$$|Ef(Y_n) - Ef(X)| \le E|f(Y_n) - f(X_{un})| + E|f(X_{un}) - f(X_u)| + E|f(X_u) - f(X)|$$

so that

$$\begin{aligned}
&\limsup_{n\to\infty} |Ef(Y_n) - Ef(X)| \\
&\quad \le \lim_{u\to\infty} \limsup_{n\to\infty} E|f(Y_n) - f(X_{un})| + 0 + 0 \\
&\quad \le \lim_{u\to\infty} \limsup_{n\to\infty} E|f(Y_n) - f(X_{un})| 1_{[|Y_n - X_{un}| \le \epsilon]} \\
&\qquad + \lim_{u\to\infty} \limsup_{n\to\infty} E|f(Y_n) - f(X_{un})| 1_{[|Y_n - X_{un}| > \epsilon]} \\
&\quad \le \sup\{|f(x) - f(y)| : |x - y| \le \epsilon\} \\
&\qquad + \lim_{u\to\infty} \limsup_{n\to\infty} P[|Y_n - X_{un}| > \epsilon] \\
&\quad \to 0
\end{aligned}$$

as $\epsilon \to 0$. □

8.6.1 *Example: The Central Limit Theorem for m-dependent random variables*

This section discusses a significant application of the second converging together theorem which is often used in time series analysis. In the next chapter, we will state and prove the central limit theorem (CLT) for iid summands: Let $\{X_n, n \ge 1\}$ be iid with $\mu = E(X_1)$, $\sigma^2 = \text{Var}(X_1)$. Then with $S_n = \sum_{i=1}^n X_i$, we have partial sums being asymptotically normally distributed

$$\frac{S_n - n\mu}{\sigma\sqrt{n}} = \frac{S_n - E(S_n)}{\sqrt{\text{Var}(S_n)}} \Rightarrow N(0,1). \tag{8.14}$$

In this section, based on (8.14), we will prove the CLT for stationary, m-dependent summands.

Call a sequence $\{X_n, n \ge 1\}$ *strictly stationary* if, for every k, the joint distribution of

$$(X_{n+1}, \dots, X_{n+k})$$

is independent of n for $n = 0, 1, \dots$. Call the sequence *m-dependent* if for any integer t, the σ-fields $\sigma(X_j, j \le t)$ and $\sigma(X_j, j \ge t + m + 1)$ are independent. Thus, variables which are lagged sufficiently far apart are independent.

The most common example of a stationary m-dependent sequence is the time series model called the *moving average of order m* which is defined as follows. Let $\{Z_n\}$ be iid and define for given constants $c_1, \dots, c_m$ the process

$$X_t = \sum_{i=1}^{m} c_j Z_{t-j}, \quad t = 0, 1, \dots.$$

Theorem 8.6.3 (Hoeffding and Robbins) *Suppose $\{X_n, n \geq 1\}$ is a strictly stationary and m-dependent sequence with $E(X_1) = 0$ and*

$$Cov(X_t, X_{t+h}) = E(X_t X_{t+h}) =: \gamma(h).$$

Suppose

$$v_m := \gamma(0) + 2\sum_{j=1}^{m} \gamma(j) \neq 0.$$

Then

$$\frac{1}{\sqrt{n}} \sum_{i=1}^{n} X_i \Rightarrow N(0, v_m), \tag{8.15}$$

and

$$n Var(\bar{X}_n) \to v_m, \tag{8.16}$$

where $\bar{X}_n = \sum_{i=1}^{n} X_i / n$.

Proof. *Part 1: Variance calculation.* We have

$$\begin{aligned} n\mathrm{Var}(\bar{X}_n) &= \frac{1}{n} E\Big(\sum_{i=1}^{n} X_i\Big)^2 = \frac{1}{n} E\Big(\sum_{i=1}^{n}\sum_{j=1}^{n} X_i X_j\Big) \\ &= \frac{1}{n} \sum_{i=1}^{n}\sum_{j=1}^{n} \gamma(j-i) \\ &= \frac{1}{n} \sum_{|k|<n} (\#(i,j) : j - i = k)\gamma(k) \\ &= \sum_{|k|<n} \left(\frac{n-|k|}{n}\right) \gamma(k). \end{aligned}$$

This last step is justified by noting that, for example when $k > 0$, i could be $1, 2, \ldots, n-k$ and $j = k + i$. Thus we conclude that

$$n\mathrm{Var}(\bar{X}_n) = \sum_{|k|<n} \left(1 - \frac{|k|}{n}\right) \gamma(k).$$

Recall that $\gamma(l) = 0$ if $|l| > m$ and as $n \to \infty$

$$n\mathrm{Var}(\bar{X}_n) \to \sum_{|k|<\infty} \gamma(k) = v_m. \tag{8.17}$$

Part 2: The big block–little block method. Pick $u > 2m$ and consider the following diagram.

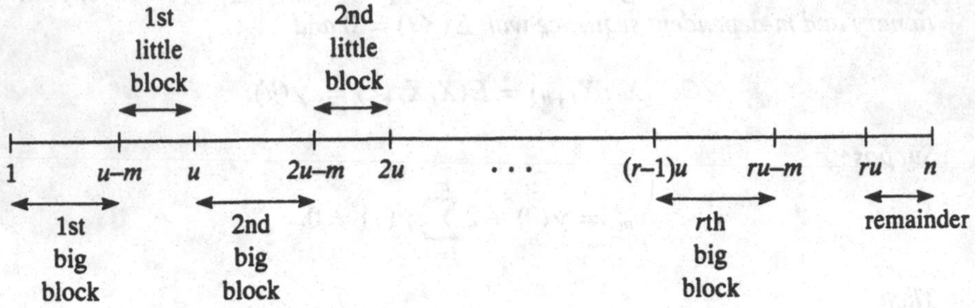

Let

$$r = \left[\frac{n}{u}\right]$$

so that $r/n \to 1/u$ and define

$$\begin{aligned} \xi_1 &= X_1 + \cdots + X_{u-m}, \\ \xi_2 &= X_{u+1} + \cdots + X_{2u-m}, \\ &\vdots \quad \vdots \\ \xi_r &= X_{(r-1)u+1} + \cdots + X_{ru-m} \end{aligned}$$

which are the "big block" sums. Note by stationarity and m-dependence that $\xi_1, \ldots, \xi_r$ are iid because the little blocks have been removed.

Define

$$\begin{aligned} X_{un} :&= \frac{\xi_1 + \cdots + \xi_r}{\sqrt{n}} \\ &= \frac{\xi_1 + \cdots + \xi_r}{\sqrt{r}} \sqrt{\frac{r}{n}}. \end{aligned}$$

Note

$$\sqrt{\frac{r}{n}} \to \sqrt{\frac{1}{u}}$$

as $n \to \infty$. From the CLT for iid summands, as $n \to \infty$

$$X_{un} \Rightarrow N(0, \frac{\text{Var}(\xi_1)}{u}) =: X_u.$$

Now observe, that as $u \to \infty$

$$\begin{aligned} \frac{\text{Var}(\xi_1)}{u} = \frac{\text{Var}(\sum_{i=1}^{u-m} X_i)}{u} &= \frac{(u-m)^2}{u} \text{Var}\left(\frac{\sum_{i=1}^{u-m} X_i}{u-m}\right) \\ &= (u-m)\text{Var}(\bar{X}_{u-m}) \cdot \frac{u-m}{u} \\ &\to v_m \cdot 1 = v_m, \end{aligned}$$

from the variance calculation in Part 1. Thus, as $u \to \infty$,

$$X_u = N(0, \frac{\mathrm{Var}(\xi_1)}{u}) \Rightarrow N(0, v_m) =: X,$$

since a sequence of normal distributions converges weakly if their means (in this case, all zero) and variances converge.

By the second converging together theorem, it remains to show that

$$\lim_{u\to\infty} \limsup_{n\to\infty} P[\left|X_{un} - \frac{\sum_{i=1}^n X_i}{\sqrt{n}}\right| > \epsilon] = 0. \tag{8.18}$$

For $i = 1, \ldots, r-1$, let

$$B_i = \{iu - m + 1, \ldots, iu\}$$

be the m integers in the ith little block, and let

$$B_r = \{ru - m + 1, \ldots, n\}$$

be the integers in the last little block coupled with the remainder due to u not dividing n exactly. Then we have

$$\left|\frac{\sum_{i=1}^n X_i}{\sqrt{n}} - X_{un}\right| = \frac{1}{\sqrt{n}}\left|\sum_{i\in B_1} X_i + \cdots + \sum_{i\in B_{r-1}} X_i + \sum_{i\in B_r} X_i\right|$$

and all sums on the right side are independent by m-dependence. So

$$\mathrm{Var}\left(\frac{\sum_{i=1}^n X_i}{\sqrt{n}} - X_{un}\right) = \frac{1}{n}\left((r-1)\mathrm{Var}(\sum_{i=1}^m X_i) + \mathrm{Var}(\sum_{i=1}^{n-ru+m+1} X_i)\right).$$

Note that

$$\begin{aligned} h(n) :&= n - ru + m + 1 = n - \left[\frac{n}{u}\right]u + m + 1 \\ &\le n - (\frac{n}{u} - 1)u + m + 1 \\ &= n - n + u + m + 1 = u + m + 1. \end{aligned}$$

Thus for fixed u, as $n \to \infty$,

$$\frac{1}{n}\mathrm{Var}(\sum_{i=1}^{h(n)} X_i) \le \frac{\sup_{1\le j\le u+m+1} \mathrm{Var}(\sum_{i=1}^j X_i)}{n} \to 0.$$

Also, since $r/n \to 1/u$ as $n \to \infty$

$$\frac{1}{n}\left((r-1)\mathrm{Var}(\sum_{i=1}^m X_i)\right) \sim \frac{1}{u}\mathrm{Var}(\sum_{i=1}^m X_i) \to 0$$

as $u \to \infty$ and we have by Chebychev's inequality

$$\begin{aligned}\lim_{u\to\infty}\limsup_{n\to\infty} P\left[\left|\frac{\sum_{i=1}^n X_i}{\sqrt{n}} - X_{un}\right| > \epsilon\right] &\le \lim_{u\to\infty}\limsup_{n\to\infty}\frac{1}{\epsilon^2}\mathrm{Var}\left(\frac{\sum_{i=1}^n X_i}{\sqrt{n}} - X_{un}\right)\\ &= \lim_{u\to\infty}\left(\frac{1}{u}\mathrm{Var}\left(\sum_{i=1}^m X_i\right) + 0\right)\\ &= 0.\end{aligned}$$

This completes the proof. □

8.7 The Convergence to Types Theorem

Many convergence in distribution results in probability and statistics are of the following form: Given a sequence of random variables $\{\xi_n, n \ge 1\}$ and $a_n > 0$ and $b_n \in \mathbb{R}$, we prove that

$$\frac{\xi_n - b_n}{a_n} \Rightarrow Y,$$

where Y is a non-degenerate random variable; that is, Y is not a constant a.s. This allows us to write

$$P[\frac{\xi_n - b_n}{a_n} \le x] \approx P[Y \le x] =: G(x),$$

or by setting $y = a_n x + b_n$

$$P[\xi_n \le y] \approx G(\frac{y - b_n}{a_n}).$$

This allows us to approximate the distribution of ξ_n with a location–scale family.

The question arises: In what sense, if any are the normalizing constants a_n and b_n unique? If we changed normalizations, could we get something significantly different?

The answer is contained in the *convergence to types theorem*. The normalizations are determined up to an asymptotic equivalence and the limit distribution is determined up to location and scale.

Example. As a standard example, supppse $\{X_n, n \ge 1\}$ are iid with $E(X_n) = \mu$ and $\mathrm{Var}(X_n) = \sigma^2$. The Central Limit Theorem states that for each $x \in \mathbb{R}$

$$P[\frac{S_n - n\mu}{\sigma\sqrt{n}} \le x] \to P[Y \le x] = \int_{-\infty}^{x} \frac{e^{-u^2/2}}{\sqrt{2\pi\sigma}} du$$

so that

$$P[S_n \le y] \approx N(\frac{y - n\mu}{\sigma\sqrt{n}}),$$

where $N(x)$ is the standard normal distribution function.

Definition. Two distribution functions $U(x)$ and $V(x)$ are of the same *type* if there exist constants $A > 0$ and $B \in \mathbb{R}$ such that

$$V(x) = U(Ax + B).$$

In terms of random variables, if X has distribution U and Y has distribution V, then

$$Y \stackrel{d}{=} \frac{X - B}{A}.$$

For example, we may speak of the normal type. If $X_{0,1}$ has $N(0, 1, x)$ as its distribution and $X_{\mu,\sigma}$ has $N(\mu, \sigma^2)$ as its distribution, then $X_{\mu,\sigma} \stackrel{d}{=} \sigma X_{0,1} + \mu$.

Now we state the theorem developed by Gnedenko and Khintchin.

Theorem 8.7.1 (Convergence to Types Theorem) *We suppose $U(x)$ and $V(x)$ are two proper distributions, neither of which is concentrated at a point. Suppose for $n \ge 0$ that X_n are random variables with distribution function F_n and the U, V are random variables with distribution functions $U(x)$, $V(x)$. We have constants $a_n > 0$, $\alpha_n > 0$, $b_n \in \mathbb{R}$, $\beta_n \in \mathbb{R}$.*

(a) If

$$F_n(a_n x + b_n) \stackrel{w}{\to} U(x), \quad F_n(\alpha_n x + \beta_n) \stackrel{w}{\to} V(x) \tag{8.19}$$

or equivalently

$$\frac{X_n - b_n}{a_n} \Rightarrow U, \quad \frac{X_n - \beta_n}{\alpha_n} \Rightarrow V, \tag{8.20}$$

then there exist constants $A > 0$, and $B \in \mathbb{R}$ such that as $n \to \infty$

$$\frac{\alpha_n}{a_n} \to A > 0, \quad \frac{\beta_n - b_n}{a_n} \to B, \tag{8.21}$$

and

$$V(x) = U(Ax + B), \quad V \stackrel{d}{=} \frac{U - B}{A}. \tag{8.22}$$

(b) Conversely, if (8.21) *holds, then either of the relations in* (8.19) *implies the other and* (8.22) *holds.*

Proof. (b) Suppose

$$G_n(x) := F_n(a_n x + b_n) \stackrel{w}{\to} U(x)$$

and

$$\alpha_n/a_n \to A > 0, \quad \frac{\beta_n - b_n}{a_n} \to B.$$

Then

$$F_n(\alpha_n x + \beta_n) = G_n\big(\frac{\alpha_n}{a_n}x + (\frac{\beta_n - b_n}{a_n})\big).$$

Pick x such that $x \in \mathcal{C}(U(A \cdot +B))$.

Suppose $x > 0$. A similar argument works if $x \leq 0$. Given $\epsilon > 0$ for large n, we have

$$(A-\epsilon)x + B - \epsilon \leq \frac{\alpha_n}{a_n}x + (\frac{\beta_n - b_n}{a_n}) \leq (A+\epsilon)x + (B+\epsilon),$$

so

$$\limsup_{n\to\infty} F_n(\alpha_n x + \beta_n) \leq \limsup_{n\to\infty} G_n((A+\epsilon)x + (B+\epsilon)).$$

Therefore, for any $z \in \mathcal{C}(U(\cdot))$ with $z > (A+\epsilon)x + (B+\epsilon)$. we have

$$\limsup_{n\to\infty} F_n(\alpha_n x + \beta_n) \leq \limsup_{n\to\infty} G_n(z) = U(z).$$

Thus

$$\limsup_{n\to\infty} F_n(\alpha_n x + \beta_n) \leq \inf_{\substack{z>(A+\epsilon)x+(B+\epsilon) \\ z\in\mathcal{C}(U(\cdot))}} U(z).$$

Since $\epsilon > 0$ is arbitrary,

$$\limsup_{n\to\infty} F_n(\alpha_n x + \beta_n) \leq \inf_{z>Ax+B} U(z) = U(Ax+B)$$

by right continuity of $U(\cdot)$. Likewise,

$$\begin{aligned}\liminf_{n\to\infty} F_n(\alpha_n x + \beta_n) &\geq \liminf_{n\to\infty} G_n((A-\epsilon)x + B - \epsilon) \\ &\geq \liminf_{n\to\infty} G_n(z) = U(z)\end{aligned}$$

for any $z < (A-\epsilon)x + B - \epsilon$ and $z \in \mathcal{C}(U(\cdot))$. Since this is true for all $\epsilon > 0$,

$$\liminf_{n\to\infty} F_n(\alpha_n x + \beta_n) \geq \sup_{\substack{z<Ax+B \\ z\in\mathcal{C}(U(\cdot))}} U(z) = U(Ax+B),$$

since $Ax + B \in \mathcal{C}(U(\cdot))$.

We now focus on the proof of part (a). Suppose

$$F_n(a_n x + b_n) \to U(x), \quad F_n(\alpha_n x + \beta_n) \to V(x).$$

Recall from Lemma 8.3.1 that if $G_n \xrightarrow{w} G$, then also $G_n^{\leftarrow} \xrightarrow{w} G^{\leftarrow}$. Thus we have

$$\frac{F_n^{\leftarrow}(y) - b_n}{a_n} \to U^{\leftarrow}(y), \quad y \in C(U^{\leftarrow})$$
$$\frac{F_n^{\leftarrow}(y) - \beta_n}{\alpha_n} \to V^{\leftarrow}(y), \quad y \in C(V^{\leftarrow}).$$

Since $U(x)$ and $V(x)$ do not concentrate at one point, we can find $y_1 < y_2$ with $y_i \in C(U^{\leftarrow}) \cap C(V^{\leftarrow})$, for $i = 1, 2$, such that

$$-\infty < U^{\leftarrow}(y_1) < U^{\leftarrow}(y_2) < \infty,$$

and

$$-\infty < V^{\leftarrow}(y_1) < V^{\leftarrow}(y_2) < \infty.$$

Therefore, for $i = 1, 2$ we have

$$\frac{F_n^{\leftarrow}(y_i) - b_n}{a_n} \to U^{\leftarrow}(y_i), \quad \frac{F_n^{\leftarrow}(y_i) - \beta_n}{\alpha_n} \to V^{\leftarrow}(y_i). \tag{8.23}$$

In (8.23) subtract the expressions with $i = 1$ from the ones with $i = 2$ to get

$$\frac{F_n^{\leftarrow}(y_2) - F_n^{\leftarrow}(y_1)}{a_n} \to U^{\leftarrow}(y_2) - U^{\leftarrow}(y_1),$$
$$\frac{F_n^{\leftarrow}(y_2) - F_n^{\leftarrow}(y_1)}{\alpha_n} \to V^{\leftarrow}(y_2) - V^{\leftarrow}(y_1).$$

Now divide the second convergence in the previous line into the first convergence. The result is

$$\frac{\alpha_n}{a_n} \to \frac{U^{\leftarrow}(y_2) - U^{\leftarrow}(y_1)}{V^{\leftarrow}(y_2) - V^{\leftarrow}(y_1)} =: A > 0.$$

Also from (8.23)

$$\frac{F_n^{\leftarrow}(y_1) - b_n}{a_n} \to U^{\leftarrow}(y_1),$$
$$\frac{F_n^{\leftarrow}(y_1) - \beta_n}{a_n} = \frac{F_n^{\leftarrow}(y_1) - \beta_n}{\alpha_n} \cdot \frac{\alpha_n}{a_n} \to V^{\leftarrow}(y_1)A,$$

so subtracting yields

$$\frac{\beta_n - b_n}{a_n} \to V^{\leftarrow}(y_1)A - U^{\leftarrow}(y_1) =: B,$$

as desired. So (8.21) holds. By part (b) we get (8.22). □

Remarks.

(1) The theorem shows that when

$$\frac{X_n - b_n}{a_n} \Rightarrow U$$

and U is non-constant, we can always center by choosing $b_n = F_n^{\leftarrow}(y_1)$ and we can always scale by choosing $a_n = F_n^{\leftarrow}(y_2) - F_n^{\leftarrow}(y_1)$. Thus quantiles can always be used to construct the centering and scaling necessary to produce convergence in distribution.

(2) Consider the following example which shows the importance of assuming limits are non-degenerate in the convergence to types theorem. Let

$$U(x) = \begin{cases} 0, & \text{if } t < c, \\ 1, & \text{if } t \geq c. \end{cases}$$

Then

$$U^{\leftarrow}(t) = \inf\{y : U(y) \geq t\} = \begin{cases} -\infty, & \text{if } t = 0, \\ c, & \text{if } 0 < t \leq 1, \\ \infty, & \text{if } t > 1. \end{cases}$$

8.7.1 *Application of Convergence to Types: Limit Distributions for Extremes*

A beautiful example of the use of the convergence to types theorem is the derivation of the *extreme value distributions.* These are the possible limit distributions of centered and scaled maxima of iid random variables.

Here is the problem: Suppose $\{X_n, n \geq 1\}$ is an iid sequence of random variables with common distribution F. The extreme observation among the first n is

$$M_n := \bigvee_{i=1}^{n} X_i.$$

Theorem 8.7.2 *Suppose there exist normalizing constants $a_n > 0$ and $b_n \in \mathbb{R}$ such that*

$$F^n(a_n x + b_n) = P[\frac{M_n - b_n}{a_n} \leq x] \stackrel{w}{\to} G(x), \tag{8.24}$$

where the limit distribution G is proper and non-degenerate. Then G is the type of one of the following extreme value distributions:

(i) $\Phi_\alpha(x) = \exp\{-x^{-\alpha}\}, \quad x > 0, \quad \alpha > 0,$

$$(ii)\ \Psi_\alpha(x) = \begin{cases} \exp\{-(x)^\alpha\}, & x < 0,\quad \alpha > 0 \\ 1 & x > 0, \end{cases}$$

$$(iii)\ \Lambda(x) = \exp\{-e^{-x}\}, \quad x \in \mathbb{R}.$$

The statistical significance is the following. The types of the three extreme value distributions can be united as a one parameter family indexed by a shape parameter $\gamma \in \mathbb{R}$:

$$G_\gamma(x) = \exp\{-(1+\gamma x)^{-1/\gamma}\}, \quad 1+\gamma x > 0$$

where we interpret the case of $\gamma = 0$ as

$$G_0 = \exp\{-e^{-x}\}, \quad x \in \mathbb{R}.$$

Often in practical contexts the distribution F is unknown and we must estimate the distribution of M_n or a quantile of M_n. For instance, we may wish to design a dam so that in 10,000 years, the probability that water level will exceed the dam height is 0.001. If we assume F is unknown but satisfies (8.24) with some G_γ as limit, then we may write

$$P[M_n \le x] \approx G_\gamma(a_n^{-1}(x-b_n)),$$

and now we have a three parameter estimation problem since we must estimate γ, a_n, b_n.

Proof. We proceed in a sequence of steps.

Step (i). We claim that there exist two functions $\alpha(t) > 0$ and $\beta(t)$, $t > 0$ such that for all $t > 0$,

$$\frac{a_n}{a_{[nt]}} \to \alpha(t), \quad \frac{b_n - b_{[nt]}}{a_{[nt]}} \to \beta(t), \tag{8.25}$$

and also

$$G^t(x) = G(\alpha(t)x + \beta(t)). \tag{8.26}$$

To see this, note that from (8.24), for every $t > 0$, we have on the one hand

$$F^{[nt]}(a_{[nt]}x + b_{[nt]}) \xrightarrow{w} G(x)$$

and on the other

$$F^{[nt]}(a_n x + b_n) = (F^n(a_n x + b_n))^{[nt]/n} \to G^t(x).$$

Thus G^t and G are of the same type and the convergence to types theorem is applicable. Applying it to (8.25) and (8.26) yields the claim.

Step (ii). We observe that the function $\alpha(t)$ and $\beta(t)$ are Lebesgue measurable. For instance, to prove $\alpha(\cdot)$ is measurable, it suffices (since limits of measurable functions are measurable) to show that the function

$$t \mapsto \frac{a_n}{a_{[nt]}}$$

is measurable for each n. Since a_n does not depend on t, the previous statement is true if the function

$$t \mapsto a_{[nt]}$$

is measurable. Since this function has a countable range $\{a_j, j \geq 1\}$ it suffices to show

$$\{t > 0 : a_{[nt]} = a_j\}$$

is measurable. But this set equals

$$\bigcup_{k: a_k = a_j} [\frac{k}{n}, \frac{k+1}{n}),$$

which, being a union of intervals, is certainly a measurable set.

Step (iii). **Facts about the Hamel Equation.** We need to use facts about possible solutions of functional equations called Hamel's equation and Cauchy's equation. If $f(x), x > 0$ is finite, measurable and real valued and satisfies the *Cauchy equation*

$$f(x + y) = f(x) + f(y), \quad x > 0, y > 0,$$

then f is necessarily of the form

$$f(x) = cx, \quad x > 0,$$

for some $c \in \mathbb{R}$. A variant of this is *Hamel's equation.* If $\phi(x), x > 0$ is finite, measurable, real valued and satisfies Hamel's equation

$$\phi(xy) = \phi(x)\phi(y), \quad x > 0, y > 0,$$

then ϕ is of the form

$$\phi(x) = x^\rho,$$

for some $\rho \in \mathbb{R}$.

Step (iv). Another useful fact. If F is a non-degenerate distribution function and

$$F(ax + b) = F(cx + d) \quad \forall x \in \mathbb{R},$$

for some $a > 0$, and $c > 0$, then $a = c$, and $b = d$. A proof of this is waiting for you in the exercises (Exercise 6).

Step (v). Now we claim that the functions $\alpha(\cdot)$ and $\beta(\cdot)$ satisfy ($t > 0,\ s > 0$)

$$\alpha(ts) = \alpha(t)\alpha(s), \tag{8.27}$$
$$\beta(ts) = \alpha(t)\beta(s) + \beta(t) \tag{8.28}$$
$$= \alpha(s)\beta(t) + \beta(s), \tag{8.29}$$

the last line following by symmetry.

To verify these assertions we use

$$G^t(x) = G(\alpha(t)x + \beta(t))$$

to conclude that

$$\begin{aligned} G(\alpha(ts)x + \beta(ts)) &= G^{ts}(x) = (G^s(x))^t \\ &= (G(\alpha(s)x + \beta(s)))^t \\ &= G(\alpha(t)[\alpha(s)x + \beta(s)] + \beta(t)) \\ &= G(\alpha(t)\alpha(s)x + \alpha(t)\beta(s) + \beta(t)). \end{aligned}$$

Now apply Step (iv).

Step (vi). Now we prove that there exists $\theta \in \mathbb{R}$ such that $\alpha(t) = t^\theta$. If $\theta = 0$, then $\beta(t) = c \log t$, for some $c \in \mathbb{R}$. If $\theta \neq 0$, then $\beta(t) = c(1 - t^\theta)$, for some $c \in \mathbb{R}$.

Proof of (vi): Since $\alpha(\cdot)$ satisfies the Hamel equation, $\alpha(t) = t^\theta$ for some $\theta \in \mathbb{R}$. If $\theta = 0$, then $\alpha(t) \equiv 1$ and $\beta(t)$ satisfies

$$\beta(ts) = \beta(s) + \beta(t).$$

So $\exp\{\beta(\cdot)\}$ satisfies the Hamel equation which implies that

$$\exp\{\beta(t)\} = t^c,$$

for some $c \in \mathbb{R}$ and thus $\beta(t) = c \log t$.

If $\theta \neq 0$, then

$$\beta(ts) = \alpha(t)\beta(s) + \beta(t) = \alpha(s)\beta(t) + \beta(s).$$

Fix $s_0 \neq 1$ and we get

$$\alpha(t)\beta(s_0) + \beta(t) = \alpha(s_0)\beta(t) + \beta(s_0),$$

and solving for $\beta(t)$ we get

$$\beta(t)(1 - \alpha(s_0)) = \beta(s_0)(1 - \alpha(t)).$$

Note that $1 - \alpha(s_0) \neq 0$. Thus we conclude

$$\beta(t) = \left(\frac{\beta(s_0)}{1 - \alpha(s_0)}\right)(1 - \alpha(t)) =: c(1 - t^\theta).$$

Step (vii). We conclude that either

$$\text{(a)}\quad G^t(x) = G(x + c\log t), \quad (\theta = 0),$$

or

$$\text{(b)}\quad G^t(x) = G(t^\theta x + c(1 - t^\theta)), \quad (\theta \neq 0).$$

Now we show that $\theta = 0$ corresponds to a limit distribution of type $\Lambda(x)$, that the case $\theta > 0$ corresponds to a limit distribution of type Φ_α and that $\theta < 0$ corresponds to Ψ_α.

Consider the case $\theta = 0$. Examine the equation in (a): For fixed x, the function $G^t(x)$ is non-increasing in t. So $c < 0$, since otherwise the right side of (a) would not be decreasing. If $x_0 \in \mathbb{R}$ such that $G(x_0) = 1$, then

$$1 = G^t(x_0) = G(x_0 + c\log t), \quad \forall t > 0,$$

which implies

$$G(y) = 1, \quad \forall y \in \mathbb{R},$$

and this contradicts G non-degenerate. If $x_0 \in \mathbb{R}$ such that $G(x_0) = 0$, then

$$0 = G^t(x_0) = G(x_0 + c\log t), \quad \forall t > 0,$$

which implies

$$G(x) = 0, \quad \forall x \in \mathbb{R},$$

again giving a contradiction. We conclude $0 < G(y) < 1$, for all $y \in \mathbb{R}$.

In (a), set $x = 0$ and set $G(0) = e^{-\kappa}$. Then

$$e^{-t\kappa} = G(c\log t).$$

Set $y = c\log t$, and we get

$$G(y) = \exp\{-\kappa e^{y/c}\} = \exp\{-e^{-(\frac{y}{|c|} - \log\kappa)}\}$$

which is the type of $\Lambda(x)$. The other cases $\theta > 0$ and $\theta < 0$ are handled similarly. □

8.8 Exercises

1. Let S_n have a binomial distribution with parameters n and $\theta \in [0, 1]$. What CLT does S_n satisfy? In statistical terms, $\hat{\theta} := S_n/n$ is an estimator of θ and

$$\frac{S_n - E(S_n)}{\sqrt{\mathrm{Var}(S_n)}}$$

is an approximate pivot for the parameter θ. If

$$g(\theta) = \log\left(\frac{\theta}{1-\theta}\right)$$

is the log–odds ratio, we would use $g(\hat{\theta})$ to estimate $g(\theta)$. What CLT does $g(\hat{\theta})$ satisfy? Use the delta method.

2. Suppose $\{X_n, n \geq 1\}$ is a sequence of random variables satisfying

$$P[X_n = n] = \frac{1}{n},$$

$$P[X_n = 0] = 1 - \frac{1}{n}.$$

(a) Does $\{X_n\}$ converge in probability? If so, to what? Why?

(b) Does $\{X_n\}$ converge in distribution? If so, to what? Why?

(c) Suppose in addition that $\{X_n\}$ is an independent sequence. Does $\{X_n\}$ converge almost surely? What is

$$\limsup_{n\to\infty} X_n \text{ and } \liminf_{n\to\infty} X_n$$

almost surely? Explain your answer.

3. Suppose $\{U_n, n \geq 1\}$ are iid $U(0, 1)$ random variables so that

$$P[U_j \leq x] = x, \quad 0 \leq x \leq 1.$$

(a) Show $\prod_{j=1}^n U_j^{1/n}$ converges almost surely. What is the limit?

(b) Center and scale the sequence $\{\prod_{j=1}^n U_j^{1/n}, n \geq 1\}$ and show the resulting sequence converges in distribution to a non-degenerate limit. To which one?

4. (a) Let $\{X_n, n \geq 0\}$ be positive integer valued random variables. Prove

$$X_n \Rightarrow X_0$$

iff for every $k \geq 0$

$$P[X_n = k] \to P[X_0 = k].$$

(b) Let $\{\mathbf{X}_n\}$ be a sequence of random vectors on $\mathbb{R}^d$ such that $\mathbf{X}_n$ has a discrete distribution having only points with integer components as possible values. Let $\mathbf{X}$ be another such random vector. Show

$$\mathbf{X}_n \Rightarrow \mathbf{X}$$

iff

$$\sum_{\mathbf{x}} |P[\mathbf{X}_n = \mathbf{x}] - P[\mathbf{X} = \mathbf{x}]| \to 0$$

as $n \to \infty$. (Use Scheffé's lemma.)

(c) For events $\{A_n, n \geq 0\}$, prove

$$1_{A_n} \Rightarrow 1_{A_0} \text{ iff } P(A_n) \to P(A_0).$$

(d) Let F_n concentrate all mass at x_n for $n \geq 0$. Prove

$$F_n \Rightarrow F_0 \text{ iff } x_n \to x_0.$$

(e) Let $X_n = 1 - 1/n$ or $1 + 1/n$ each with probability 1/2 and suppose $P[X = 1] = 1$. Show $X_n \Rightarrow X$ but that the mass function $f_n(x)$ of X_n does not converge for any x.

5. (a) If $u_n(x), x \in \mathbb{R}$ are non-decreasing functions for each n and $u_n(x) \to u_0(x)$ and $u_0(\cdot)$ is continuous, then for any $-\infty < a < b < \infty$

$$\sup_{x \in [a,b]} |u_n(x) - u_0(x)| \to 0.$$

Thus, convergence of monotone functions to a continuous limit implies local uniform convergence.

(b) Suppose $F_n, n \geq 0$ are proper df's and $F_n \Rightarrow F_0$. If F_0 is continuous, show

$$\sup_{x \in \mathbb{R}} |F_n(x) - F_0(x)| \to 0.$$

For instance, in the central limit theorem, where F_0 is the normal distribution, convergence is always uniform.

(c) Give a simple proof of the Glivenko–Cantelli lemma under the additional hypothesis that the underlying distribution is continuous.

6. Let F be a non-degenerate df and suppose for $a > 0, c > 0$ and $b \in \mathbb{R}$, $d \in \mathbb{R}$, that for all x

$$F(ax + b) = F(cx + d).$$

Prove that $a = c$ and $b = d$. Do this 2 ways:

(i) Considering inverse functions.

(ii) Showing it is enough to prove $F(Ax + B) = F(x)$ for all x implies $A = 1$, and $B = 17$ (just kidding, $B = 0$). If $Tx = Ax + B$ then iterate the relation F(Tx)=F(x) again and again.

7. Let $\{X_n, n \geq 1\}$ be iid with $E(X_n) = \mu$, $\text{Var}(X_n) = \sigma^2$ and suppose N is a $N(0, 1)$ random variable. Show

$$\sqrt{n}(\bar{X}_n^2 - \mu^2) \Rightarrow 2\mu\sigma N \tag{a}$$

$$\sqrt{n}(e^{\bar{X}_n} - e^{\mu}) \Rightarrow \sigma e^{\mu} N. \tag{b}$$

8. Suppose $X_1, \dots, X_n$ are iid exponentially distributed with mean 1. Let

$$X_{1,n} < \cdots < X_{n,n}$$

be the order statistics. Fix an integer l and show

$$nX_{l,n} \Rightarrow Y_l$$

where Y_l has a gamma $(l, 1)$ distribution.

Try doing this (a) in a straightforward way by brute force and then (b) try using the Renyi representation for the spacings of order statistics from the exponential density. See Exercise 32 on page 116.

9. Let $\{X_n, n \geq 0\}$ be random variables. Show $X_n \Rightarrow X_0$ iff $E(g(X_n)) \to E(g(X_0))$ for all continuous functions g with compact support.

10. Let X and Y be independent Bernoulli random variables on a probability space $(\Omega, \mathcal{B}, P)$ with $X \stackrel{d}{=} Y$ and

$$P[X = 0] = \frac{1}{2} = P[X = 1].$$

Let $X_n = Y$ for $n \geq 1$. Show that

$$X_n \Rightarrow X$$

but that X_n does NOT converge in probability to X.

11. **Lévy metric.** For two probability distributions F, G, define

$$d(F, G) := \inf\{\delta > 0 : \forall x \in \mathbb{R},\ F(x - \delta) - \delta \leq G(x) \leq F(x + \delta) + \delta\}.$$

Show this is a metric on the space of probability distribution functions which metrizes weak convergence; that is, $F_n \Rightarrow F_0$ iff $d(F_n, F_0) \to 0$.

12. Suppose F_n has density

$$f_n(x) = \begin{cases} 1 - \cos 2n\pi x, & \text{if } 0 \leq x \leq 1, \\ 0, & \text{otherwise.} \end{cases}$$

Show that F_n converges weakly to the uniform distribution on $[0, 1]$ but that the densities f_n do not converge.

13. Suppose $\{N_n, n \geq 0\}$ is a sequence of normal random variables. Show $N_n \Rightarrow N_0$ iff

$$E(N_n) \to E(N_0) \text{ and } \mathrm{Var}(N_n) \to \mathrm{Var}(N_0).$$

Derive a comparable result for (a) Poisson random variables; (b) exponential random variables.

14. Consider the sphere in $\mathbb{R}^n$ of radius $\sqrt{n}$ and suppose $\mathbf{X}_n$ is uniformly distributed on the surface of this sphere. Show that the first component of $\mathbf{X}_n$ converges in distribution to a standard normal. Hint: If $N_i, i \geq 1$ are iid $N(0, 1)$ random variables, show

$$\mathbf{X}_n \stackrel{d}{=} (N_1, \dots, N_n)\sqrt{\frac{n}{\sum_{i=1}^n N_i^2}}.$$

15. (Weissman) Suppose $Y_n, n \geq 1$ are random variables such that there exist $a_n > 0, b_n \in \mathbb{R}$ and

$$P[Y_n \leq a_n x + b_n] \to G(x),$$

non-degenerate, and for each $t > 0$

$$P[Y_{[nt]} \leq a_n x + b_n] \to G_t(x),$$

non-degenerate. Then there exists $\alpha(t) > 0$, $\beta(t) \in \mathbb{R}$ such that

$$G(x) = G_t(\alpha(t)x + \alpha(t))$$

and $\alpha(t) = t^\theta$. If $\theta = 0$, then $\beta(t) = c \log t$, and if $\theta \neq 0$, then $\beta(t) = c(1 - t^\theta)$.

16. Suppose $\{X_n, n \geq 1\}$ are iid non-negative random variables and set $M_n = \vee_{i=1}^n X_i, n \geq 1$. Show that there exists a sequence $a_n > 0$ such that $(x > 0)$

$$\lim_{n\to\infty} P[M_n/a_n \leq x] = \exp\{-x^{-\alpha}\}, \quad x > 0,\ \alpha > 0,$$

iff the regular variation condition holds:

$$\lim_{t\to\infty} \frac{P[X_1 > tx]}{P[X_1 > t]} = x^{-\alpha}, \quad x > 0. \tag{8.30}$$

In this case, what limit distribution exists for $\log M_n$? For M_n^2?

Verify (8.30) for the Cauchy density and the Pareto distribution?

17. If $\{X_n\}$ are iid $U(0, 1)$ random variables, find a non-degenerate limit distribution for $M_n = \vee_{i=1}^n X_i$ under suitable normalization.

18. Give an example of a sequence of discrete distributions that converge weakly to a limit distribution which possesses a density.

19. **(Second continuous mapping theorem)** Suppose that $X_n \Rightarrow X_0$ and that for $n \geq 0$, $\chi_n : \mathbb{R} \mapsto \mathbb{R}$ are measurable. Define

$$E := \{x : \exists x_n \to x \text{ but } \chi_n(x_n) \not\to \chi_0(x)\}.$$

Suppose E is measurable and $P[X_0 \in E] = 0$. Show $\chi_n(X_n) \Rightarrow \chi_0(X_0)$.

20. Suppose we have independent Bernoulli trials where the probability of success in a trial is p. Let ν_p be the geometrically distributed number of trials needed to get the first success so that

$$P[\nu_p \geq n] = (1-p)^{n-1}, \quad n \geq 1.$$

Show as $p \to 0$

$$p\nu_p \Rightarrow E,$$

where E is a unit exponential random variable.

21. **Sampling with replacement.** Let $\{X_n, n \geq 1\}$ be iid and uniformly distributed on the set $\{1, \ldots, m\}$. In repeated sampling, let ν_m be the time of the first coincidence; that is, the time when we first get a repeated outcome

$$\nu_m := \inf\{n \geq 2 : X_n \in \{X_1, \ldots, X_{n-1}\}\}.$$

Verify that

$$P[\nu_m > n] = \prod_{i=2}^{n}\left(1 - \frac{i-1}{m}\right).$$

Show as $m \to \infty$ that

$$\nu_m/\sqrt{m} \Rightarrow \nu$$

where $P[\nu > x] = \exp\{-x^2/2\}$, $x > 0$.

22. **Sample median; more order statistics.** Let $U_1, \ldots, U_n$ be iid $U(0,1)$ random variables and consider the order statistics $U_{1,n} \leq U_{2,n} \leq \cdots \leq U_{n,n}$. When n is odd, the middle order statistic is the sample median. Show that

$$2(U_{n+1,2n+1} - \frac{1}{2})\sqrt{2n}$$

has a limit distribution. What is it? (Hint: Use Scheffé's lemma 8.2.1 page 253.)

23. Suppose $\{X_n, n \geq 1\}$ are iid random variables satisfying

$$E(X_n) = \mu, \quad \text{Var}(X_n) = \sigma^2.$$

The central limit theorem is assumed known. Set $\bar{X}_n = \sum_{i=1}^{n} X_i/n$. Let $N(0,1)$ be a standard normal random variable. Prove

(i) $\sqrt{n}(\bar{X}_n^2 - \mu^2) \Rightarrow 2\mu\sigma N(0,1)$.

(ii) $\sqrt{n}(e^{\bar{X}_n} - e^{\mu}) \Rightarrow e^{\mu} N(0,1)$.

(iii) $\sqrt{n}(\log \bar{X}_n - \log \mu) \Rightarrow \frac{1}{\mu} N(0,1)$, assuming $\mu \neq 0$.

Now assume additionally that $E(X_1^4) < \infty$ and prove

(iv) $\sqrt{n}\left(\log(\frac{1}{n}\sum_{i=1}^n (X_i - \bar{X}_n)^2) - \log \sigma^2\right) \Rightarrow \frac{1}{\sigma^2}\sqrt{E(X_1^4)} N(0,1)$.

(v) Define the sample variance

$$S_n^2 = \frac{1}{n}\sum_{i=1}^n (X_i - \bar{X}_n)^2.$$

Show

$$\sqrt{n}(\sqrt{S_n^2} - \sigma) \Rightarrow \frac{1}{2\sigma}\sqrt{E(X_1^4)} N(0,1).$$

What is a limit law for S_n^2?

24. Show that the normal type is closed under convolution. In other words, if N_1, N_2 are two independent normally distributed random variables, show that $N_1 + N_2$ is also normally distributed. Prove a similar statement for the Poisson and Cauchy types.

Is a similar statement true for the exponential type?

25. (i) Suppose F is a distribution function and u is a bounded continuous function on $\mathbb{R}$. Define the convolution transform as

$$F * u(t) = \int_{\mathbb{R}} u(t-y) F(dy).$$

Let $\{F_n, n \geq 0\}$ be a sequence of probability distribution functions. Let $C[-\infty, \infty]$ be the class of bounded, continuous functions on $\mathbb{R}$ with finite limits existing at $\pm\infty$. Prove that $F_n \Rightarrow F_0$ iff for each $u \in C[-\infty, \infty]$, $U_n := F_n * u$ converges uniformly to a limit U. In this case, $U = F_0 * u$.

(ii) Suppose X is a random variable and set $F_n(x) = P[X/n \leq x]$. Prove $F_n * u \to u$ uniformly.

(iii) Specialize (ii) to the case where F is the standard normal distribution and verify the *approximation lemma*: Given any $\epsilon > 0$ and any $u \in C[-\infty, \infty]$, there exists an infinitely differentiable $v \in C[-\infty, \infty]$ such that

$$\sup_{x \in \mathbb{R}} |v(x) - u(x)| < \epsilon.$$

(iv) Suppose that $u(x, y)$ is a function on $\mathbb{R}^2$ vanishing at the infinities. Then u can be approximated uniformly by finite linear combinations

$\sum_k c_k g_k(x)h_k(y)$ with infinitely differentiable g_k, h_k. (Hint: Use normal distributions.)

(v) Suppose F_n is discrete with equal atoms at $-n, 0, n$. What is the vague limit of F_n as $n \to \infty$? What is the vague limit of $F_n * F_n$?

(vi) Suppose F_n concentrates all mass at $1/n$ and $u(x) = \sin(x^2)$. Then $F_n * u$ converges pointwise but not uniformly. (Is $u \in C[-\infty, \infty]$?)

26. Suppose $\{X_n, n \geq 1\}$ are iid random variables with common distribution F and set $S_n = \sum_{i=1}^n X_i$. Assume that there exist $a_n > 0,\ b_n \in \mathbb{R}$ such that

$$a_n^{-1} S_n - b_n \Rightarrow Y$$

where Y has a non-degenerate proper distribution. Use the convergence to types theorem to show that

$$a_n \to \infty, \quad a_n/a_{n+1} \to 1.$$

(Symmetrize to remove b_n. You may want to first consider a_{2n}/a_n.)

27. Suppose $\{X_n, n \geq 1\}$ are iid and non-negative random variables with common density $f(x)$ satisfying

$$\lambda := \lim_{t \downarrow 0} f(t) > 0.$$

Show $n \bigwedge_{i=1}^n X_i$ has a limit distribution. (This is extreme value theory, but for minima not maxima.)

28. Let $x \in (0, 1)$ have binary expansion

$$x = \sum_{n=1}^{\infty} \frac{d_n}{2^n}.$$

Set

$$f_n(x) = \begin{cases} 2, & \text{if } d_n = 0, \\ 1, & \text{if } d_n = 1. \end{cases}$$

Then show $\int_0^1 f_n(x)dx = 1$ so that f_n is a density. The sequence f_n only converges on a set of Lebesgue measure 0. If X_n is a random variable with density f_n then $X_n \Rightarrow U$, where U is $U(0, 1)$.

29. Suppose $\{X_t, t \geq 0\}$ is a family of random variables parameterized by a continuous variable t and assume there exist normalizing constants $a(t) > 0,\ b(t) \in \mathbb{R}$ such that as $t \to \infty$

$$\frac{X_t - b(t)}{a(t)} \Rightarrow Y,$$

where Y is non-degenerate. Show the normalizing functions can always be assumed continuous; that is, there exist continuous functions $\alpha(t) > 0$, $\beta(t) \in \mathbb{R}$ such that

$$\frac{X_t - \beta(t)}{\alpha(t)} \Rightarrow Y',$$

where Y' has a non-degenerate distribution. (Hint: The convergence to types theorem provides normalizing constants which can be smoothed by integration.)

30. Suppose $\{X_n, n \geq 1\}$ are random variables and there exist normalizing constants $a_n > 0$, $b_n \in \mathbb{R}$ such that

$$\frac{X_n - b_n}{a_n} \Rightarrow Y,$$

where Y is non-degenerate. Assume further that for $\delta > 0$

$$\sup_{n\geq 1} E\left(\left|\frac{X_n - b_n}{a_n}\right|^{2+\delta}\right) < \infty.$$

Show the mean and standard deviation can be used for centering and scaling:

$$\frac{X_n - E(X_n)}{\sqrt{\text{Var}(X_n)}} \Rightarrow Y',$$

where Y is non-degenerate.

It is enough for moment generating functions to converge:

$$E(e^{\gamma a_n^{-1}(X_n - b_n)}) \Rightarrow E(e^{\gamma Y}),$$

for $\gamma \in I$, an open interval containing 0.

31. If $X_n \Rightarrow X_0$ and

$$\sup_n E(|X_n|^{2+\delta}) < \infty,$$

show that

$$E(X_n) \to E(X_0), \quad \text{Var}(X_n) \to \text{Var}(X_0).$$

(Use Baby Skorohod and uniform integrability.)

32. Given random variables $\{X_n\}$ such that $0 \leq X_n \leq 1$. Suppose for all $x \in (0, 1)$

$$P[X_n \leq x] \to 1 - p.$$

Show $X_n \Rightarrow B$ where B is a Bernoulli random variable with success probability p.

33. Verify that a continuous distribution function is always uniformly continuous.

34. Suppose $\{E_n, n \geq 1\}$ are iid unit exponential random variables. Recall from Example 8.1.2 that

$$\bigvee_{i=1}^{n} E_i - \log n \Rightarrow Y,$$

where Y has a Gumbel distribution.

Let $\{W_n, n \geq 1\}$ be iid Weibull random variables satisfying

$$P[W_n > x] = e^{-x^{\alpha}}, \quad \alpha > 0,\ x > 0.$$

Use the delta method to derive a weak limit theorem for $\vee_{i=1}^{n} W_i$. (Hint: Express W_i as a function of E_i which then requires that $\vee_{i=1}^{n} W_i$ is a function of $\vee_{i=1}^{n} E_i$.)

35. Suppose $\{F_n, n \geq 0\}$ are probability distributions such that $F_n \Rightarrow F_0$. For $t > 0$, let $u_t(\cdot) : \mathbb{R} \mapsto \mathbb{R}$ be an equicontinuous family of functions that are uniformly bounded; that is

$$\sup_{t>0, x\in\mathbb{R}} u_t(x) \leq M,$$

for some constant M. Then show

$$\lim_{n\to\infty} \int_{\mathbb{R}} u_t(x) F_n(dx) = \int_{\mathbb{R}} u_t(x) F_0(dx),$$

uniformly in t.

36. Suppose $F_n \Rightarrow F_0$ and $g \geq 0$ is continuous and satisfies $\int_{\mathbb{R}} g dF_n = 1$ for all $n \geq 0$. Define the new measure $G_n(A) = \int_A g dF_n$. Show $G_n \Rightarrow G_0$. (You can either do this directly or use Scheffé's lemma.)

9

Characteristic Functions and the Central Limit Theorem

This chapter develops a transform method called *characteristic functions* for dealing with sums of independent random variables. The basic problem is that the distribution of a sum of independent random variables is rather complex and hard to deal with. If X_1, X_2 are independent random variables with distributions F_1, F_2, set

$$g(u, v) = 1_{(-\infty,t]}(u + v).$$

Then using the transformation theorem and Fubini's theorem (Theorems 5.5.1 and 5.9.2), we get for $t \in \mathbb{R}$

$$\begin{aligned}
P[X_1 + X_2 \le t] &= E(g(X_1, X_2)) \\
&= \iint g F_1 \times F_2 \qquad \text{(transformation theorem)} \\
&= \iint_{\{(x,y)\in\mathbb{R}^2 : x+y\le t\}} F_1 \times F_2 \\
&= \int_{\mathbb{R}} \Big[\int_{\{x\in\mathbb{R}: x\le t-y\}} F_1(dx)\Big] F_2(dy) \qquad \text{(Fubini's theorem)} \\
&= \int_{\mathbb{R}} F_1(t-y) F_2(dy) \\
&=: F_1 * F_2(t),
\end{aligned}$$

where the last line defines the *convolution* between two distributions on the real line. Convolution is a fairly complex operation and somewhat difficult to deal with. Transform methods convert convolution of distributions into products of transforms and products are easier to handle than convolutions.

The next section reviews the relation between transforms and the central limit theorem.

9.1 Review of Moment Generating Functions and the Central Limit Theorem

The *moment generating function* (mgf) $\widehat{F}(t)$ of a random variable X with distribution F exists if

$$\widehat{F}(t) := Ee^{tX} = \int_{\mathbb{R}} e^{tx} F(dx) < \infty, \quad \forall t \in I,$$

where I is an interval containing 0 in its interior. The problem with using the mgf of a distribution is that it does not always exist. For instance, the mgf of the Cauchy distribution does not exist and in fact existence of the mgf is equivalent to the tail of the distribution of X being exponentially bounded:

$$P[|X| > x] \leq Ke^{-cx}, \quad \text{for some } K > 0 \text{ and } c > 0.$$

(So what do we do if the mgf does not exist?)

The mgf, if it exists, uniquely determines the distribution of X, and we hope that we can relate convergence in distribution to convergence of the transforms; that is, we hope $X_n \Rightarrow X$ if

$$Ee^{tX_n} \to Ee^{tX}, \quad \forall t \in I,$$

where I is a neighborhood of 0. This allows us to think about the central limit theorem in terms of transforms as follows.

Suppose $\{X_n, n \geq 1\}$ is an iid sequence of random variables satisfying

$$E(X_n) = 0, \quad \text{Var}(X_n) = E(X_n^2) = \sigma^2.$$

Suppose the mgf of X_i exists. Then

$$\begin{aligned} Ee^{tS_n/\sqrt{n}} &= Ee^{t\sum_{i=1}^n X_i/\sqrt{n}} \\ &= E\prod_{i=1}^n e^{tX_i/\sqrt{n}} = \left(Ee^{tX_1/\sqrt{n}}\right)^n \\ &= \left(\widehat{F}(t/\sqrt{n})\right)^n \end{aligned}$$

and expanding in a Taylor series about 0, we get

$$= \left(1 + \frac{tE(X_1)}{\sqrt{n}} + \frac{t^2\sigma^2}{2n} + \text{junk}\right)^n$$

where "junk" represents the remainder in the expansion which we will not worry about now. Hence, as $n \to \infty$, if we can neglect "junk" we get

$$\to e^{t^2\sigma^2/2},$$

which is the mgf of a $N(0, \sigma^2)$ random variable. Thus we hope that

$$\frac{S_n}{\sqrt{n}} \Rightarrow N(0, \sigma^2).$$

How do we justify all this rigorously? Here is the program.

1. We first need to replace the mgf by the *characteristic function* (chf) which is a more robust transform. It always exists and shares many of the algebraic advantages of the mgf.
2. We need to examine the properties of chf and feel comfortable with this transform.
3. We need to understand the connection between moments of the distribution and expansions of the chf.
4. We need to prove uniqueness; that is that the chf uniquely determines the distribution.
5. We need a test for weak convergence using chf's. This is called the *continuity theorem.*
6. We need to prove the CLT for the iid case.
7. We need to prove the CLT for independent, non-identically distributed random variables.

This is the program. Now for the details.

9.2 Characteristic Functions: Definition and First Properties

We begin with the definition.

Definition 9.2.1 The characteristic function (chf) of a random variable X with distribution F is the complex valued function of a real variable t defined by

$$\begin{aligned}\phi(t) &:= Ee^{itX}, \quad t \in \mathbb{R} \\ &= E(\cos(tX)) + iE(\sin(tX)) \\ &= \int_{\mathbb{R}} \cos(tx)F(dx) + i\int_{\mathbb{R}} \sin(tx)F(dx).\end{aligned}$$

A big advantage of the chf as a transform is that it always exists:

$$|Ee^{itX}| \le E|e^{itX}| = 1.$$

Note that

$$|E(U+iV)|^2 = |E(U)+iE(V)|^2 = (EU)^2 + (EV)^2$$

and applying the Schwartz Inequality we get that this is bounded above by

$$\begin{aligned} &\le E(U^2) + E(V^2) = E(U^2+V^2) \\ &= E|U+iV|^2. \end{aligned}$$

We now list some **elementary properties** of chf's.

1. The chf $\phi(t)$ is uniformly continuous on $\mathbb{R}$. For any $t \in \mathbb{R}$, we have

$$\begin{aligned} |\phi(t+h) - \phi(t)| &= |Ee^{i(t+h)X} - Ee^{itX}| \\ &= |Ee^{itX}(e^{ihX} - 1)| \\ &\le E|e^{ihX} - 1| \to 0 \end{aligned}$$

as $h \downarrow 0$ by the dominated convergence theorem. Note that the upper bound is independent of t which accounts for the uniform continuity.

2. The chf satisfies $|\phi(t)| \le 1$ and $\phi(0) = 1$.

3. The effect on the chf of scaling and centering of the random variable is given by

$$Ee^{it(aX+b)} = \phi(at)e^{ibt}.$$

4. Let $\bar{\phi}(t)$ be the complex conjugate of $\phi(t)$. Then

$$\begin{aligned} \phi(-t) = \bar{\phi}(t) &= Re(\phi(t)) - iIm(\phi(t)) \\ &= \text{chf of } -X. \end{aligned}$$

5. We have that

$$Re(\phi(t)) = \int \cos(tx)F(dx)$$

is an even function, while

$$Im(\phi(t)) = \int \sin(tx)F(dx)$$

is an odd function.

6. The chf ϕ is real iff
$$X \stackrel{d}{=} -X$$
iff F is a symmetric function. This follows since ϕ is real iff $\phi = \bar{\phi}$ iff X and $-X$ have the same chf. As we will see (anticipating the uniqueness theorem), this implies $X \stackrel{d}{=} -X$.

7. If X_i has chf ϕ_i, $i = 1, 2$ and X_1 and X_2 are independent, then the chf of $X_1 + X_2$ is the product $\phi_1(t)\phi_2(t)$ since
$$\begin{aligned} Ee^{it(X_1+X_2)} &= Ee^{itX_1}e^{itX_2} \\ &= Ee^{itX_1}Ee^{itX_2}. \end{aligned}$$
Compare this with the definition of convolution at the beginning of the chapter.

8. We generalize the previous property in a direction useful for the central limit theorem by noting that if $X_1, \dots, X_n$ are iid with common chf ϕ, then
$$Ee^{it(a_n^{-1}S_n - nb_n)} = e^{-itnb_n}(\phi(t/a_n))^n.$$

9.3 Expansions

The CLT is dependent on good expansions of the chf ϕ. These in turn depend on good expansions of e^{ix} so we first take this up.

9.3.1 *Expansion of e^{ix}*

Start with an integration by parts. For $n \geq 0$ we have the following identity:
$$\int_0^x e^{is}(x-s)^n ds = \frac{x^{n+1}}{n+1} + \frac{i}{n+1}\int_0^x (x-s)^{n+1}e^{is}ds. \tag{9.1}$$
For $n = 0$, (9.1) gives
$$\int_0^x e^{is}ds = \frac{e^{ix}-1}{i} = x + i\int_0^x (x-s)e^{is}ds.$$
So we have
$$\begin{aligned} e^{ix} &= 1 + ix + i^2\int_0^x (x-s)e^{is}ds \\ &= 1 + ix + i^2[\frac{x^2}{2} + \frac{i}{2}\int_0^x (x-s)^2 e^{is}ds] \end{aligned}$$

(from (9.1) with $n = 1$)

$$= 1 + ix + \frac{(ix)^2}{2} + \frac{i^3}{2}\Big[\frac{x^3}{3} + \frac{i}{3}\int_0^x (x-s)^3 e^{is} ds\Big]$$

$$\vdots$$

where the last expression before the vertical dots comes from applying (9.1) with $n = 2$. In general, we get for $n \geq 0$ and $x \in \mathbb{R}$,

$$e^{ix} = \sum_{k=0}^{n} \frac{(ix)^k}{k!} + \frac{i^{n+1}}{n!}\int_0^x (x-s)^n e^{is} ds. \tag{9.2}$$

Thus

$$\Big|e^{ix} - \sum_{k=0}^{n} \frac{(ix)^k}{k!}\Big| \leq \frac{|x|^{n+1}}{(n+1)!} \tag{9.3}$$

where we have used the fact that $|e^{ix}| = 1$. Therefore we conclude that chopping the expansion of e^{ix} after a finite number of terms gives an error bounded by the modulus of the first neglected term.

Now write (9.1) with $n-1$ in place of n and transpose to get

$$\int_0^x (x-s)^{n-1} e^{is} ds - \frac{x^n}{n} = \frac{i}{n}\int_0^x (x-s)^n e^{is} ds.$$

If we multiply through by $\frac{i^n}{(n-1)!}$ and interchange left and right sides of the equation, we obtain

$$\frac{i^{n+1}}{n!}\int_0^x (x-s)^n e^{is} ds = \frac{i^n}{(n-1)!}\int_0^x (x-s)^{n-1} e^{is} ds - \frac{(ix)^n}{n!}.$$

Substitute this in the right side of (9.2) and we get

$$e^{ix} - \sum_{k=0}^{n} \frac{(ix)^k}{k!} = \frac{i^n}{(n-1)!}\int_0^x (x-s)^{n-1} e^{is} ds - \frac{(ix)^n}{n!},$$

and thus

$$\Big|e^{ix} - \sum_{k=0}^{n} \frac{(ix)^k}{k!}\Big| \leq \frac{|x|^n}{n!} + \frac{|x|^n}{n!} = \frac{2|x|^n}{n!}. \tag{9.4}$$

Combining (9.3) and (9.4) gives

$$\Big|e^{ix} - \sum_{k=0}^{n} \frac{(ix)^k}{k!}\Big| \leq \frac{|x|^{n+1}}{(n+1)!} \wedge \frac{2|x|^n}{n!}. \tag{9.5}$$

Note that the first term in the minimum gives a better estimate for small x, while the second term gives a better estimate for large x.

Now suppose that X is a random variable whose first n absolute moments are finite:

$$E(|X|) < \infty, \ldots, E(|X|^n) < \infty.$$

Then

$$\begin{aligned}|\phi(t) - \sum_{k=0}^{n} \frac{(it)^k E(X^k)}{k!}| &= |E(e^{itX}) - E\left(\sum_{k=0}^{n} \frac{(it)^k}{k!} X^k\right)| \\ &\leq E|e^{itX} - \sum_{k=0}^{n} \frac{(itX)^k}{k!}|\end{aligned}$$

and applying (9.5) with x replaced by tX, we get

$$|\phi(t) - \sum_{k=0}^{n} \frac{(it)^k}{k!} E(X^k)| \leq E\left(\frac{|tX|^{n+1}}{(n+1)!} \wedge \frac{2|tX|^n}{n!}\right). \tag{9.6}$$

Next, suppose all moments exist and that for all $t \in \mathbb{R}$

$$\lim_{n\to\infty} \frac{|t|^n E(|X|^n)}{n!} = 0. \tag{9.7}$$

In (9.6), let $n \to \infty$ to get

$$\phi(t) = \sum_{k=0}^{\infty} \frac{(it)^k}{k!} E(X^k).$$

A sufficient condition for (9.7) is

$$\sum_{k=0}^{\infty} \frac{|t|^k E|X|^k}{k!} = Ee^{|t||X|} < \infty, \tag{9.8}$$

which holds if

$$\Psi(t) = Ee^{tX} < \infty, \quad \forall t \in \mathbb{R};$$

that is, if the mgf exists on all of $\mathbb{R}$. (To check this last statement, note that if $Ee^{tX} < \infty$ for all t, we have

$$\begin{aligned}Ee^{|t||X|} &= E(e^{|t|X} 1_{[X>0]}) + E(e^{-|t|X} 1_{[X<0]}) \\ &\leq \Psi(|t|) + \Psi(-|t|) < \infty.\end{aligned}$$

This verifies the assertion.)

Example 9.3.1 Let X be a random variable with $N(0,1)$ distribution. For any $t \in \mathbb{R}$,

$$Ee^{tX} = \int_{-\infty}^{\infty} e^{tu} \frac{e^{-u^2/2}}{\sqrt{2\pi}} du$$
$$= \int_{-\infty}^{\infty} \frac{1}{\sqrt{2\pi}} \exp\{-\frac{1}{2}(u^2 - 2tu + t^2)\} du\, e^{t^2/2}$$

(from completing the square)

$$= e^{t^2/2} \int_{-\infty}^{\infty} \frac{e^{-\frac{1}{2}(u-t)^2}}{\sqrt{2\pi}} du$$
$$= e^{t^2/2} \int_{-\infty}^{\infty} n(t,1,u) du = e^{t^2/2}.$$

Here, $n(t,1,u)$ represents the normal density with mean t and variance 1 which integrates to 1. Thus we conclude that for all $t \in \mathbb{R}$

$$Ee^{tX} < \infty.$$

We may therefore expand the mgf as well as $e^{t^2/2}$ to get

$$Ee^{tX} = \sum_{k=0}^{\infty} \frac{t^k E(X^k)}{k!} = e^{t^2/2} = \sum_{k=0}^{\infty} \left(\frac{t^2}{2}\right)^k / k!;$$

that is,

$$\sum_{k=0}^{\infty} \frac{E(X^k)}{k!} t^k = \sum_{l=0}^{\infty} \frac{(\frac{1}{2})^l}{l!} t^{2l}.$$

Equating coefficients yields that the

$$\text{coefficient of } t^{2n} = \frac{E(X^{2n})}{(2n)!} = \frac{(\frac{1}{2})^n}{n!}.$$

So we conclude that

$$E(X^{2n}) = \frac{(2n)!}{n!} (\frac{1}{2})^n, \quad E(X^{2n+1}) = 0.$$

Thus, since

$$\phi(t) = Ee^{itX} = \sum_{k=0}^{\infty} \frac{(it)^k}{k!} E(X^k),$$

we get

$$\begin{aligned}\phi(t) &= \sum_{l=0}^{\infty} \frac{(it)^{2l}}{(2l)!} E(X^{2l}) = \sum_{l=0}^{\infty} \frac{(it)^{2l}}{(2l)!} \cdot \frac{(2l)!}{l!} (\frac{1}{2})^l \\ &= \sum_{l=0}^{\infty} \frac{(i^2 t^2 \frac{1}{2})^l}{l!} = \sum_{l=0}^{\infty} \frac{(-t^2/2)^l}{l!} \\ &= e^{-t^2/2}. \end{aligned} \tag{9.9}$$

This shows one way of computing the chf of the N(0,1) distribution.

Note that the chf of the normal density has the property that the chf and the density are the same, apart from multiplicative constants. This is a useful and unusual feature.

9.4 Moments and Derivatives

When the kth absolute moment of a random variable exists, it can be computed by taking k-fold derivatives of the chf.

Suppose X is a random variable with finite first absolute moment; that is, $E(|X|) < \infty$. Then

$$\begin{aligned}\frac{\phi(t+h) - \phi(t)}{h} - E(iXe^{itX}) &= E\left(\frac{e^{i(t+h)X} - e^{itX} - ihXe^{itX}}{h}\right) \\ &= E\left(e^{itX} \frac{\{e^{ihX} - 1 - ihX\}}{h}\right).\end{aligned}$$

Apply (9.4) with $n = 1$ to get

$$|e^{itX}| \quad |\frac{e^{ihX} - 1 - ihX}{h}| \leq 2|X| \in L_1.$$

Since by (9.3) or (9.5) we have

$$|\frac{e^{ihX} - 1 - ihX}{h}| \leq \frac{h^2 X^2}{2h} = h\frac{X^2}{2} \to 0$$

as $h \downarrow 0$, we get by dominated convergence that

$$\begin{aligned}&\lim_{h \downarrow 0} \left(\frac{\phi(t+h) - \phi(t)}{h} - E(iXe^{itX})\right) \\ &= E\left(\lim_{h \downarrow 0} e^{itX} (\frac{e^{ihX} - 1 - ihX}{h})\right) \\ &= 0.\end{aligned}$$

Thus

$$\phi'(t) = E(iXe^{itX}). \tag{9.10}$$

In general, we have that if $E(|X|^k) < \infty$,

$$\phi^{(k)}(t) = E\left((iX)^k e^{itX}\right), \quad \forall t \in \mathbb{R} \tag{9.11}$$

and hence

$$\phi^{(k)}(0) = i^k E(X^k).$$

9.5 Two Big Theorems: Uniqueness and Continuity

We seek to prove the central limit theorem. Our program for doing this is to show that the chf of centered and scaled sums of independent random variables converges as the sample size increases to the chf of the $N(0, 1)$ distribution which we know from Example 9.3.1 is $\exp\{-t^2/2\}$. To make this method work, we need to know that the chf uniquely determines the distribution and that when chf's of distributions converge, their distributions converge weakly. The goal of this section is to prove these two facts.

Theorem 9.5.1 (Uniqueness Theorem) *The chf of a probability distribution uniquely determines the probability distribution.*

Proof. We use the fact that the chf and density of the normal distribution are the same apart from multiplicative constants.

Let X be a random variable with distribution F and chf ϕ. We show that ϕ determines F. For any distribution G with chf γ and any real $\theta \in \mathbb{R}$, we have by applying Fubini's theorem the *Parseval relation*

$$\begin{aligned}
\int_{\mathbb{R}} e^{-i\theta y}\phi(y)G(dy) &= \int_{y\in\mathbb{R}} e^{-i\theta y}\left[\int_{x\in\mathbb{R}} e^{iyx}F(dx)\right]G(dy) \\
&= \iint_{\mathbb{R}^2} e^{iy(x-\theta)}F(dx)G(dy) \\
&= \int_{x\in\mathbb{R}}\left[\int_{y\in\mathbb{R}} e^{i(x-\theta)y}G(dy)\right]F(dx) \qquad (9.12)\\
&= \int_{x\in\mathbb{R}} \gamma(x-\theta)F(dx). \qquad (9.13)
\end{aligned}$$

Now let N have a $N(0, 1)$ distribution with density $n(x)$ so that σN has a normal density with variance σ^2. Replace $G(dy)$ by this normal density $\sigma^{-1}n(\sigma^{-1}y)$. After changing variables on the left side and taking account of the form of the normal chf γ given by (9.9) on the right side, we get

$$\int_{\mathbb{R}} e^{-i\theta\sigma y}\phi(\sigma y)n(y)dy = \int_{z\in\mathbb{R}} e^{-\sigma^2(z-\theta)^2/2}F(dz). \tag{9.14}$$

Now, integrate both sides of (9.14) over θ from $-\infty$ to x to get

$$\int_{\theta=-\infty}^{x}\int_{\mathbb{R}} e^{-i\theta\sigma y}\phi(\sigma y)n(y)dyd\theta = \int_{\theta=-\infty}^{x}\int_{z\in\mathbb{R}} e^{-\sigma^2(z-\theta)^2/2}F(dz)d\theta,$$

and using Fubini's theorem to reverse the order of integration on the right side yields

$$= \int_{z\in\mathbb{R}} \sqrt{2\pi}\Big[\int_{\theta=-\infty}^{x} \frac{e^{-\sigma^2(z-\theta)^2/2}}{\sqrt{2\pi}}d\theta\Big]F(dz).$$

In the inner integral on the right side, make the change of variable $s = \theta - z$ to get

$$\begin{aligned}\int_{\theta=-\infty}^{x}&\int_{\mathbb{R}} e^{-i\theta\sigma y}\phi(\sigma y)n(y)dyd\theta\\ &= \frac{1}{\sigma}\int_{z\in\mathbb{R}}\sqrt{2\pi}\Big[\int_{s=-\infty}^{x-z}\frac{e^{-\sigma^2 s^2/2}}{\sqrt{2\pi}\sigma^{-1}}ds\Big]F(dz)\\ &= \sqrt{2\pi}\sigma^{-1}\int_{z\in\mathbb{R}}\Big[\int_{-\infty}^{x-z} n(0,\sigma^{-2},z)dz\Big]F(dz)\\ &= \sqrt{2\pi}\sigma^{-1}P[\sigma^{-1}N + X \le x].\end{aligned}$$

Divide through by $\sqrt{2\pi}\sigma^{-1}$. Let $\sigma \to \infty$. Given the chf ϕ, we find

$$\begin{aligned}\lim_{\sigma\to\infty}&\frac{\sigma}{\sqrt{2\pi}}\int_{\theta=-\infty}^{x}\int_{\mathbb{R}} e^{-i\theta\sigma y}\phi(\sigma y)n(y)dyd\theta\\ &= \lim_{\sigma\to\infty} P[\sigma^{-1}N + X \le x] = F(x), \quad \forall x \in C(F), \qquad (9.15)\end{aligned}$$

by Slutsky's theorem 8.6.1 of Chapter 8. So for any $x \in C(F)$, ϕ determines $F(x)$ which is sufficient for proving the result. □

A quick corollary gives **Fourier inversion.**

Corollary 9.5.1 *Suppose F is a probability distribution with an integrable chf ϕ; that is, $|\phi| \in L_1$ so that*

$$\int_{\mathbb{R}} |\phi(t)|dt < \infty.$$

Then F has a bounded continuous density f given by

$$f(x) = \frac{1}{2\pi}\int_{\mathbb{R}} e^{-iyx}\phi(y)dy.$$

Proof. Examine (9.15). Note

$$P[\sigma^{-1}N + X \le x] =: F_\sigma(x)$$

has a density f_σ since $\sigma^{-1}N$ has a density. From the left side of (9.15),

$$f_\sigma(\theta) = \frac{1}{\sqrt{2\pi}} \int_{\mathbb{R}} e^{-i\theta y} \phi(y) n(\sigma^{-1} y) dy$$
$$= \frac{1}{2\pi} \int_{\mathbb{R}} e^{-i\theta y} \phi(y) e^{-\sigma^{-2} y^2/2} dy.$$

Note that

$$\left| e^{-i\theta y} \phi(y) e^{-\sigma^{-2} y^2/2} \right| \le |\phi(y)| \in L_1$$

and as $\sigma \to \infty$

$$e^{-i\theta y} \phi(y) e^{-\sigma^{-2} y^2/2} \to e^{-i\theta y} \phi(y).$$

So by dominated convergence, $f_\sigma(\theta) \to f(\theta)$. Furthermore, for any finite interval I

$$\sup_{\theta \in I} f_\sigma(\theta) \le \frac{1}{2\pi} \int_{\mathbb{R}} |\phi(y)| e^{-\sigma^{-2} y^2/2} dy$$
$$\le \frac{1}{2\pi} \int_{\mathbb{R}} |\phi(y)| dy < \infty.$$

So as $\sigma \to \infty$, we have by Slutsky's theorem 8.6.1 and bounded convergence

$$F(I) = \lim_{\sigma \to \infty} P[\sigma^{-1} N + X \in I] = \lim_{\sigma \to \infty} \int_I f_\sigma(\theta) d\theta = \int_I f(\theta) d\theta.$$

Thus f is the density of F. □

We now state and prove the continuity theorem which allows the conclusion of weak convergence of probability distributions from pointwise convergence of their chf's.

Theorem 9.5.2 (Continuity Theorem) *(i) Easy part: We let $\{X_n, n \ge 1\}$ be a sequence of random variables with X_n having distribution F_n and chf ϕ_n. If as $n \to \infty$ we have*

$$X_n \Rightarrow X_0,$$

then

$$\phi_n(t) \to \phi_0(t), \quad \forall t \in \mathbb{R}.$$

(ii) Deeper part: Suppose

(a) $\lim_{n\to\infty} \phi_n(t)$ exists for all t. Call the limit $\phi_\infty(t)$.

(b) $\phi_\infty(t)$ is continuous at 0.

Then for some distribution function F_∞,

$$F_n \Rightarrow F_\infty$$

and ϕ_∞ is the chf of F_∞. If $\phi_\infty(0) = 1$, then F_∞ is proper.

Proof of the easy part (i). If $X_n \Rightarrow X_0$, then by the continuous mapping theorem, we have $e^{itX_n} \Rightarrow e^{itX_0}$ and since $|e^{itX_n}| \leq 1$, we have by dominated convergence that

$$\phi_n(t) = Ee^{itX_n} \to Ee^{itX_0} = \phi_0(t).$$

□

The proof of the harder part (ii) is given in the next section.

We close this section with some simple but illustrative examples of the use of the continuity theorem where we make use of the harder half to prove convergence.

Example 9.5.1 (WLLN) *Suppose $\{X_n, n \geq 1\}$ is iid with common chf $\phi(t)$ and assume $E(|X_1|) < \infty$ and $E(X_1) = \mu$. Then*

$$S_n/n \xrightarrow{P} \mu.$$

Since convergence in probability to a constant is equivalent to weak convergence to the constant, it suffices to show the chf of S_n/n converges to the chf of μ, namely $e^{i\mu t}$. We have

$$Ee^{itS_n/n} = \phi^n(t/n) = \left(1 + \frac{it\mu}{n} + o(\frac{t}{n})\right)^n. \tag{9.16}$$

The last equality needs justification, but suppose for the moment it is true; this would lead to

$$\phi^n(t/n) = \left(1 + \frac{it\mu + o(1)}{n}\right)^n \to e^{it\mu},$$

as desired.

To justify the representation in (9.16), note from (9.6) with $n = 1$ that

$$|\phi(\frac{t}{n}) - 1 - i\frac{t}{n}\mu| \leq E\left(\frac{t^2|X_1|^2}{2n^2} \wedge 2\frac{t}{n}|X_1|\right),$$

so it suffices to show that

$$nE\left(\frac{t^2|X_1|^2}{2n^2} \wedge 2\frac{t}{n}|X_1|\right) \to 0. \tag{9.17}$$

Bring the factor n inside the expectation. On the one hand

$$n\left(\frac{t^2|X_1|^2}{2n^2} \wedge 2\frac{t}{n}|X_1|\right) \leq 2t|X_1| \in L_1,$$

and on the other

$$n\left(\frac{t^2|X_1|^2}{2n^2}\wedge 2\frac{t}{n}|X_1|\right)\le \frac{t^2}{2n}|X_1|^2 \stackrel{a.s.}{\to} 0,$$

as $n\to\infty$. So by dominated convergence, (9.17) follows as desired. □

Example 9.5.2 (Poisson approximation to the binomial) *Suppose the random variable S_n has binomial mass function so that*

$$P[S_n=k]=\binom{n}{k}p^k(1-p)^{n-k},\quad k=0,\dots,n.$$

If $p=p(n)\to 0$ as $n\to\infty$ in such a way that $np\to\lambda>0$, then

$$S_n\Rightarrow PO(\lambda)$$

where the limit is a Poisson random variable with parameter λ.

To verify this, we first calculate the chf of $PO(\lambda)$. We have

$$\begin{aligned}Ee^{it(PO(\lambda))}&=\sum_{k=0}^{\infty}e^{itk}\frac{e^{-\lambda}\lambda^k}{k!}\\ &=e^{-\lambda}\sum_{k=0}^{\infty}(\lambda e^{it})^k/k!=e^{-\lambda}e^{\lambda e^{it}}\\ &=e^{\lambda(e^{it}-1)}.\end{aligned}$$

Recall we can represent a binomial random variable as a sum of iid Bernoulli random variables $\xi_1,\dots,\xi_n$ where $P[\xi_1=1]=p=1-P[\xi_1=0]$. So

$$\begin{aligned}Ee^{itS_n}=&(Ee^{it\xi_1})^n=(1-p+e^{it}p)^n\\ =&(1+p(e^{it}-1))^n=\left(1+\frac{np(e^{it}-1)}{n}\right)^n\\ \to& e^{\lambda(e^{it}-1)}.\end{aligned}$$

The limit is the chf of $PO(\lambda)$ just computed. □

The final example is a more sophisticated version of Example 9.5.2. In queueing theory, this example is often used to justify an assumption about traffic inputs being a Poisson process.

Example 9.5.3 *Suppose we have a doubly indexed array of random variables such that for each $n=1,2,\dots$, $\{\xi_{n,k},k\ge 1\}$ is a sequence of independent (but not necessarily identically distributed) Bernoulli random variables satisfying*

$$P[\xi_{n,k}=1]=p_k(n)=1-P[\xi_{n,k}=0],\tag{9.18}$$

$$\bigvee_{1\le k\le n} p_k(n) =: \delta(n) \to 0, \quad n \to \infty, \tag{9.19}$$

$$\sum_{k=1}^{n} p_k(n) = E(\sum_{k=1}^{n} \xi_{n,k}) \to \lambda \in (0,\infty), \quad n \to \infty. \tag{9.20}$$

Then

$$\sum_{k=1}^{n} \xi_{n,k} \Rightarrow PO(\lambda).$$

The proof is left to Exercise 13.

9.6 The Selection Theorem, Tightness, and Prohorov's theorem

This section collects several important results on subsequential convergence of probability distributions and culminates with the rest of the proof of the continuity theorem.

9.6.1 *The Selection Theorem*

We seek to show that every infinite family of distributions contains a weakly convergent subseqence. We begin with a lemma.

Lemma 9.6.1 (Diagonalization) *Given a sequence $\{a_j, j \ge 1\}$ of distinct real numbers and a family $\{u_n(\cdot), n \ge 1\}$ of real functions from $\mathbb{R} \mapsto \mathbb{R}$, there exists a subsequence $\{u_{n_k}(\cdot)\}$ converging at each a_j for every j. (Note that $\pm\infty$ is an acceptable limit.)*

Proof. The proof uses a diagonalization argument.

There exists a subsequence $\{n_k\}$ such that $\{u_{n_k}(a_1)\}$ converges. We call this $\{u_k^{(1)}(\cdot), k \ge 1\}$ so that $\{u_k^{(1)}(a_1), k \ge 1\}$ converges.

Now there exists a subsequence k_j such that $\{u_{k_j}^{(1)}(a_2), j \ge 1\}$ converges. Call this subfamily of functions $\{u_j^{(2)}(\cdot), j \ge 1\}$ so that

$$\{u_j^{(2)}(a_1), j \ge 1\} \text{ and } \{u_j^{(2)}(a_2), j \ge 1\}$$

are both convergent.

Now continue by induction: Construct a subsequence $\{u_j^{(n)}(\cdot), j \ge 1\}$ for each n, which converges at a_n and is a subsequence of previous sequences so that

$$\{u_j^{(n)}(a_i), j \ge 1\}$$

converges for $i = 1, \ldots, n$. Now consider the diagonal sequence of functions

$$\{u_j^{(j)}, j \geq 1\}.$$

For any a_i

$$\{u_n^{(n)}(a_i), n \geq i\} \subset \{u_j^{(i)}(a_i), j \geq i\}$$

where the sequence on the right is convergent so

$$\lim_{n\to\infty} u_n^{(n)}(a_i) \text{ exists}$$

for $i = 1, 2, \ldots$. □

Remark. If $|u_n(\cdot)| \leq M$ for all n, then

$$\lim_{n\to\infty} |u_n^{(n)}(a_i)| \leq M.$$

Lemma 9.6.2 *If $D = \{a_i\}$ is a countable dense subset of $\mathbb{R}$ and if $\{F_n\}$ are df's such that*

$$\lim_{n\to\infty} F_n(a_i) \text{ exists}$$

for all i, then define

$$F_\infty(a_i) = \lim_{n\to\infty} F_n(a_i).$$

This determines a df F_∞ on $\mathbb{R}$ and

$$F_n \stackrel{w}{\to} F_\infty.$$

Proof. Extend F_∞ to $\mathbb{R}$ by right continuity: Define for any x,

$$F_\infty(x) = \lim_{a_i \downarrow x} \downarrow F_\infty(a_i).$$

This makes F_∞ right continuous and monotone. Let $x \in C(F_\infty)$. Since D is dense, there exist $a_i, a_i' \in D$ such that

$$a_i \uparrow x, \quad a_i' \downarrow x,$$

and for every k and i

$$F_k(a_i) \leq F_k(x) \leq F_k(a_i').$$

Take the limit on k:

$$F_\infty(a_i) \leq \liminf_{k\to\infty} F_k(x) \leq \limsup_{k\to\infty} F_k(x) \leq F(a_i').$$

Let $a_i \uparrow x$ and $a_i' \downarrow x$ and use the fact that $x \in C(F_\infty)$ to get

$$F_\infty(x) \leq \liminf_{k\to\infty} F_k(x) \leq \limsup_{k\to\infty} F_k(x) \leq F(x).$$

□

We are now in a position to state and prove the Selection Theorem.

Theorem 9.6.1 (Selection Theorem) *Any sequence of probability distributions* $\{F_n\}$ *contains a weakly convergent subsequence (but the limit may be defective).*

Proof. Let $D = \{a_i\}$ be countable and dense in $\mathbb{R}$. There exists a subsequence $\{F_{n_k}\}$ such that

$$\lim_{k\to\infty} F_{n_k}(a_j)$$

exists for all j. Hence $\{F_{n_k}\}$ converges weakly from Lemma 9.6.2. □

9.6.2 Tightness, Relative Compactness, and Prohorov's theorem

How can we guarantee that subsequential limits of a sequence of distributions will be non-defective?

Example. Let $X_n \equiv n$. If F_n is the df of X_n, then $X_n \to \infty$ and so $F_n(x) \to 0$ for all x.

Probability mass escapes to infinity. *Tightness* is a concept designed to prevent this. Let Π be a family of non-defective probability df's.

Definition. Π is a *relatively compact* family of distributions if every sequence of df's in Π has a subsequence weakly converging to a proper limit; that is, if $\{F_n\} \subset \Pi$, there exists $\{n_k\}$ and a proper df F_0 such that $F_{n_k} \Rightarrow F_0$.

Definition. Π is *tight*, if for all $\epsilon > 0$, there exists a compact set $K \subset \mathbb{R}$ such that

$$F(K) > 1 - \epsilon, \quad \forall F \in \Pi;$$

or equivalently, if for all $\epsilon > 0$, there exists a finite interval I such that

$$F(I^c) \le \epsilon, \quad \forall F \in \Pi;$$

or equivalently, if for all $\epsilon > 0$, there exists M_ϵ such that

$$F(M_\epsilon) - F(-M_\epsilon) \le \epsilon, \quad \forall F \in \Pi.$$

Most of the mass is contained in a big interval for all the distributions. This prevents mass from slip sliding away to one of the infinities.

Random variables $\{X_n\}$ whose df's $\{F_n\}$ are tight are called *stochastically bounded*. This means, for every $\epsilon > 0$, there exists M_ϵ such that

$$\sup_n P[|X_n| > M_\epsilon] < \epsilon.$$

Prohorov's theorem shows the equivalence of relative compactness and tightness.

Theorem 9.6.2 (Prohorov's theorem) *The family* Π *of probability distributions is relatively compact iff* Π *is tight.*

Proof. Suppose Π is tight. Let $\{F_n\} \subset \Pi$. By the selection theorem 9.6.1, for some n_k, we have $F_{n_k} \xrightarrow{w} F_\infty$. We must show F_∞ is proper. Given $\epsilon > 0$, there exists M such that

$$\sup_n F_n([-M, M]^c) < \epsilon.$$

Pick $M' > M$ such that $M' \in C(F_\infty)$. Then

$$\epsilon > 1 - F_n(M') + F_n(-M') \to 1 - F_\infty(M') + F_\infty(-M').$$

So $F_\infty([-M', M']^c) < \epsilon$, and therefore $F_\infty([-M', M']) > 1 - \epsilon$. Since this is true for all ϵ, we have $F_\infty(\mathbb{R}) = 1$.

Conversely, if Π is not tight, then there exists $\epsilon > 0$, such that for all M, there exists $F \in \Pi$ such that $F([-M, M]) \leq 1 - \epsilon$. So there exist $\{F_n\} \subset \Pi$ with the property that $F_n([-n, n]) \leq 1 - \epsilon$. There exists a convergent subsequence n_k such that $F_{n_k} \xrightarrow{w} F_\infty$. For any $a, b \in C(F_\infty)$,

$$[a, b] \subset [-n, n]$$

for large n, and so

$$F_\infty([a, b]) = \lim_{n_k \to \infty} F_{n_k}([a, b]) \leq \lim_{n_k \to \infty} F_{n_k}([-n_k, n_k]) \leq 1 - \epsilon.$$

So $F_\infty(\mathbb{R}) \leq 1 - \epsilon$ and F_∞ is not proper so Π is not relatively compact. □

Here are common *criteria for tightness*: Let $\{X_n\}$ be random variables with df's $\{F_n\}$.

1. If there exists $r > 0$ such that

$$\limsup_{n \to \infty} E(|X_n|^r) < \infty$$

then $\{F_n\}$ is tight by Chebychev's inequality.

2. If $\{X_n\}$ and $\{Y_n\}$ are two stochastically bounded sequences, then $\{X_n + Y_n\}$ is stochastically bounded. This follows from the inequality

$$P[|X_n + Y_n| > M] \leq P[|X_n| > M/2] + P[|Y_n| > M/2].$$

3. If F_n concentrates on $[a, b]$ for all n, then $\{F_n\}$ is tight. So for example, if U_n are identically distributed uniform random variables on $(0, 1)$, then $\{c_n U_n\}$ is stochastically bounded if $\{c_n\}$ is bounded.

4. If $X_n \stackrel{d}{=} \sigma_n N_n + \mu_n$, where N_n are identically distributed $N(0, 1)$ random variables, then $\{X_n\}$ is stochastically bounded if $\{\sigma_n\}$ and $\{\mu_n\}$ are bounded.

9.6.3 Proof of the Continuity Theorem

Before proving the rest of the continuity theorem, we discuss a method which relates distribution tails and chf's.

Lemma 9.6.3 *If F is a distribution with chf ϕ, then there exists $\alpha \in (0, \infty)$ such that for all $x > 0$*

$$F([-x, x]^c) \leq \alpha x \int_0^{x^{-1}} (1 - Re\, \phi(t))dt.$$

Proof. Since

$$\text{Re}\, \phi(t) = \int_{-\infty}^{\infty} \cos ty F(dy),$$

we have

$$x \int_0^{x^{-1}} (1 - \text{Re}\, \phi(t))dt = x \int_0^{x^{-1}} \int_{-\infty}^{\infty} (1 - \cos ty) F(dy)dt$$

which by Fubini is

$$= x \int_{-\infty}^{\infty} \left[\int_{t=0}^{x^{-1}} (1 - \cos ty)dt \right] F(dy)$$

$$= x \int_{-\infty}^{\infty} \left(x^{-1} - \frac{\sin x^{-1} y}{y} \right) F(dy)$$

$$= \int_{-\infty}^{\infty} \left(1 - \frac{\sin x^{-1} y}{x^{-1} y} \right) F(dy).$$

Since the integrand is non-negative, this is greater than

$$\int_{|y|>x} \left(1 - \frac{\sin x^{-1} y}{x^{-1} y} \right) F(dy)$$

$$\geq \alpha^{-1} F([-x, x]^c),$$

where

$$\alpha^{-1} = \inf_{|x^{-1} y| \geq 1} \left(1 - \frac{\sin x^{-1} y}{x^{-1} y} \right).$$

□

This is what is needed to proceed with the deferred proof of the continuity theorem.

Proof of the Continuity Theorem. Suppose for all $t \in \mathbb{R}$, we have $\phi_n(t) \to \phi_\infty(t)$ where ϕ_∞ is continuous at 0. Then we assert $\{F_n\}$ is tight. To understand why tightness is present, assume $M > 0$ and apply Lemma 9.6.3:

$$\limsup_{n\to\infty} F_n([-M, M]^c) \le \limsup_{n\to\infty} \alpha M \int_0^{M^{-1}} (1 - \text{Re}\,\phi_n(t))dt.$$

Now $\phi_n(t) \to \phi_\infty(t)$ implies that

$$\text{Re}\,\phi_n(t) \to \text{Re}\,\phi_\infty(t), \quad 1 - \text{Re}\,\phi_n(t) \to 1 - \text{Re}\,\phi_\infty(t),$$

and since $1 - \phi_n$ is bounded, so is $\text{Re}\,(1 - \phi_n) = (1 - \text{Re}\,\phi_n)$. By dominated convergence,

$$\limsup_{n\to\infty} F_n([-M, M]^c) \le \alpha M \int_0^{M^{-1}} (1 - \text{Re}\,\phi_\infty(t))dt.$$

Since ϕ_∞ is continuous at 0, $\lim_{t\to 0} \phi_\infty(t) = \phi_\infty(0) = \lim_{n\to\infty} \phi_n(0) = 1$ as $t \to 0$. So $1 - \text{Re}\,\phi_\infty(t) \to 0$ as $t \to 0$, and thus for given $\epsilon > 0$ and M sufficiently large, we have

$$\alpha M \int_0^{M^{-1}} (1 - \text{Re}\,\phi_\infty(t))dt \le \alpha M \int_0^{M^{-1}} \epsilon dt = \alpha\epsilon.$$

Hence $\{F_n\}$ is tight. Since $\{F_n\}$ is tight, any two convergent subsequences of $\{F_n\}$ must converge to the same limit, because if

$$F_{n'} \Rightarrow F, \text{ and } F_{n''} \Rightarrow G,$$

then F and G are proper. So by part (i) of the continuity theorem already proved,

$$\phi_{n'} \to \phi_F = \phi_\infty$$

and

$$\phi_{n''} \to \phi_G = \phi_\infty,$$

and hence $\phi_F = \phi_G$. By the Uniqueness Theorem 9.5.1, $F = G$. Thus any two convergent subsequences converge to the same limit and hence $\{F_n\}$ converges to a limit whose chf is ϕ_∞. □

9.7 The Classical CLT for iid Random Variables

We now turn to the proof of the CLT for sums of iid random variables. If $\{X_n\}$ are iid random variables with finite mean $E(X_n) = \mu$ and variance $\text{Var}(X_n) = \sigma^2$, we will show that as $n \to \infty$

$$\frac{\sum_{i=1}^n X_i - n\mu}{\sigma\sqrt{n}} \Rightarrow N(0, 1).$$

The method will be to show that the chf of the left side converges to the standard normal chf $e^{-t^2/2}$.

We begin with a lemma which allows us to compare products.

Lemma 9.7.1 (Product Comparison) *For $i = 1, \dots, n$, suppose that $a_i \in \mathbb{C}$, $b_i \in \mathbb{C}$, with $|a_i| \le 1$ and $|b_i| \le 1$. Then*

$$\Big|\prod_{i=1}^n a_i - \prod_{i=1}^n b_i\Big| \le \sum_{i=1}^n |a_i - b_i|.$$

Proof. For $n = 2$, we merely have to write

$$a_1a_2 - b_1b_2 = a_1(a_2 - b_2) + (a_1 - b_1)b_2.$$

Finish by taking absolute values and using the fact that the moduli of a_i and b_i are bounded by 1. For general n, use induction. □

Theorem 9.7.1 (CLT for iid random variables) *Let $\{X_n, n \ge 1\}$ be iid random variables with $E(X_n) = \mu$ and $Var(X_n) = \sigma^2$. Suppose N is a random variable with $N(0, 1)$ distribution. If $S_n = X_1 + \cdots + X_n$, then*

$$\frac{S_n - n\mu}{\sigma\sqrt{n}} \Rightarrow N.$$

Proof. Without loss of generality let $E(X_n) = 0$, $E(X_n^2) = 1$, (otherwise prove the result for

$$X_i^* = \frac{X_i - \mu}{\sigma}$$

and

$$E(X_i^*) = 0, \quad E(X_i^*)^2 = 1.)$$

Let

$$\phi_n(t) = Ee^{itS_n/\sqrt{n}}, \quad \phi(t) = Ee^{itX_1}.$$

Then

$$\phi_n(t) = (Ee^{itX_1/\sqrt{n}})^n = \phi^n(t/\sqrt{n}).$$

Since the first two moments exist, we use (9.6) and expand ϕ:

$$\begin{aligned}\phi\left(\frac{t}{\sqrt{n}}\right) &= 1 + \frac{itE(X_1)}{\sqrt{n}} + \frac{i^2t^2E(X_1)^2}{2n} + o(\frac{t^2}{n}) \\ &= 1 + 0 - \frac{t^2}{2n} + o(\frac{t^2}{n}) \end{aligned} \tag{9.21}$$

where

$$|o(t^2/n)| \le E\left(\frac{|tX_1|^3}{n^{3/2}3!} \wedge \frac{2|tX_1|^2}{2n}\right).$$

We claim that

$$no(t^2/n) \le E\left(\frac{|tX_1|^3}{\sqrt{n}3!} \wedge |tX_1|^2\right) \to 0, \quad n \to \infty. \tag{9.22}$$

To see this, observe that on the one hand

$$\frac{|tX_1|^3}{\sqrt{n}3!} \wedge |tX_1|^2 \le |tX_1|^2 \in L_1$$

and

$$\frac{|tX_1|^3}{\sqrt{n}3!} \wedge |tX_1|^2 x \le \frac{|tX_1|^3}{\sqrt{n}3!} \to 0,$$

as $n \to \infty$. So by dominated convergence

$$E\left(\frac{|tX_1|^3}{\sqrt{n}3!} \wedge |tX_1|^2\right) \to 0.$$

Now

$$\left|(\phi(t/\sqrt{n}))^n - \left(1 - \frac{t^2/2}{n}\right)^n\right| \le n|\phi(\frac{t}{\sqrt{n}}) - (1 - \frac{t^2/2}{n})|$$

(where we have applied the product comparison Lemma 9.7.1)

$$= no(t^2/n) \to 0.$$

Since

$$\left(1 - \frac{t^2/2}{n}\right)^n \to e^{-t^2/2},$$

the chf of the $N(0, 1)$ distribution, the result follows. □

9.8 The Lindeberg–Feller CLT

We now generalize the CLT for iid summands given in Section 9.7 to the case where the summands are independent but not identically distributed.

Let $\{X_n, n \ge 1\}$ be independent (but not necessarily identically distributed) and suppose X_k has distribution F_k and chf ϕ_k, and that $E(X_k) = 0$, $\text{Var}(X_k) = \sigma_k^2$. Define

$$s_n^2 = \sigma_1^2 + \cdots + \sigma_n^2 = \text{Var}(\sum_{i=1}^n X_i).$$

We say that $\{X_k\}$ satisfies the *Lindeberg condition* if for all $t > 0$ as $n \to \infty$ we have

$$\frac{1}{s_n^2}\sum_{k=1}^n E\left(X_k^2 1_{[|X_k/s_n|>t]}\right) = \frac{1}{s_n^2}\sum_{k=1}^n \int_{|x|>ts_n} x^2 F_k(dx) \to 0. \tag{9.23}$$

Remarks.

- The Lindeberg condition (9.23) says for each k, most of the mass of X_k is centered in an interval about the mean (= 0) and this interval is small relative to s_n.

- The Lindeberg condition (9.23) implies

$$\max_{k\le n}\frac{\sigma_k^2}{s_n^2}\to 0,\quad (n\to\infty). \tag{9.24}$$

To see this, note that

$$\begin{aligned}\frac{\sigma_k^2}{s_n^2}&=\frac{1}{s_n^2}E(X_k^2)\\&=\frac{1}{s_n^2}E\big(X_k^2 1_{[|X_k/s_n|\le t]}\big)+\frac{1}{s_n^2}E\big(X_k^2 1_{[|X_k/s_n|>t]}\big)\\&\le t^2+\frac{1}{s_n^2}\sum_{k=1}^n E(X_k^2 1_{[|X_k/s_n|>t]})\\&\to t^2.\end{aligned}$$

So for any $t>0$,

$$\lim_{n\to\infty}\left(\max_{k\le n}\frac{\sigma_k^2}{s_n^2}\right)\le t^2.$$

To finish the proof, let $t\downarrow 0$. □

- Condition (9.24) implies

$$\max_{k\le n}P[|X_k|/s_n>\epsilon]\to 0 \tag{9.25}$$

by Chebychev's inequality. This condition (9.25) is called uniform asymptotic negligibility (UAN). It is typical in central limit theorems that the UAN condition holds so that no one summand dominates but each sumand contributes a small amount to the total.

We now state and prove the sufficiency part of the Lindeberg–Feller central limit theorem.

Theorem 9.8.1 (Lindeberg–Feller CLT) *With the notation given at the beginning of this section, The Lindeberg condition* (9.23) *implies*

$$\frac{S_n}{s_n}\Rightarrow N(0,1),$$

where $N(0,1)$ is a normal random variable with mean 0 and variance 1.

Although we shall not prove it, the converse is true in the following sense. If

(i) $\vee_{k=1}^n \sigma_k^2/s_n^2 \to 0$ and

(ii) $S_n/\sqrt{n} \Rightarrow N(0,1)$,

then the Lindeberg condition (9.23) holds.

Proof. We proceed in a series of steps.

(1) We begin with a preliminary result. Suppose $\{Y_n, n \geq 1\}$ is an iid sequence of random variables with common distribution F and chf ϕ. Let N be independent of $\{Y_k\}$ and assume N is Poisson distributed with parameter c. Define $\chi_n = \sum_{i=1}^n Y_i$. We compute the chf of χ_N as follows: For $t \in \mathbb{R}$,

$$\begin{aligned} E(e^{it\chi_N}) &= \sum_{k=0}^{\infty} E(e^{it\chi_N} 1_{[N=k]}) \\ &= \sum_{k=0}^{\infty} E(e^{it\chi_k} 1_{[N=k]}), \end{aligned}$$

and since N is independent of $\{Y_k\}$, this equals

$$\begin{aligned} &= \sum_{k=0}^{\infty} E(e^{it\chi_k}) P[N=k] \\ &= \sum_{k=0}^{\infty} \phi^k(t) \frac{e^{-c}c^k}{k!} \\ &= e^{-c} e^{c\phi(t)} = e^{c(\phi(t)-1)}. \end{aligned}$$

We may also conclude that

$$e^{c(\phi(t)-1)}$$

must be a chf.

(2) To show $S_n/s_n \Rightarrow N(0,1)$, we need to show that the chf of S_n/s_n satisfies

$$\phi_{S_n/s_n}(t) = \prod_{k=1}^{n} \phi_k(t/s_n) \to e^{-t^2/2} = \text{chf of } N(0,1). \tag{9.26}$$

This follows from the continuity theorem 9.5.2. We claim that (9.26) holds if

$$\sum_{k=1}^{n} (\phi_k(t/s_n) - 1) + t^2/2 \to 0 \tag{9.27}$$

because

$$|\exp\{\sum_{k=1}^{n} (\phi_k(t/s_n) - 1)\} - \prod_{k=1}^{n} \phi_k(t/s_n)| \to 0. \tag{9.28}$$

Thus, assuming (9.28) is true, it will suffice to prove (9.27).

Here is the verification of (9.28). Recall that $\exp\{\phi_k(t/s_n) - 1\}$ is a chf from step (1) and hence bounded in modulus by 1. Then

$$\left|\exp\{\sum_{k=1}^n(\phi_k(t/s_n) - 1)\} - \prod_{k=1}^n \phi_k(t/s_n)\right|$$
$$= \left|\prod_{k=1}^n e^{\phi_k(t/s_n)-1} - \prod_{k=1}^n \phi_k(t/s_n)\right|,$$

and applying the product comparison lemma 9.7.1, we have the bound

$$\le \sum_{k=1}^n |e^{\phi_k(t/s_n)-1} - \phi_k(t/s_n)|$$
$$= \sum_{k=1}^n |e^{\phi_k(t/s_n)-1} - 1 - (\phi_k(t/s_n) - 1)|.$$

Note that for $z \in \mathbb{C}$,

$$|e^z - 1 - z| = |\sum_{k=2}^\infty \frac{z^k}{k!}| \le \sum_{k=2}^\infty |z|^k = \frac{|z|^2}{1-|z|}$$
$$\le 2|z|^2, \text{ if } |z| \le \frac{1}{2},$$
$$\le \delta|z|, \text{ if } |z| \le \frac{\delta}{2} < \frac{1}{2}. \tag{9.29}$$

Now for fixed $t \in \mathbb{R}$, we apply (9.6) with $n = 1$ to get the inequalities

$$|\phi_k(t/s_n) - 1| \le \frac{t^2}{2s_n^2}\sigma_k^2 \tag{9.30}$$
$$\le \frac{t^2}{2}\bigvee_{k=1}^n \frac{\sigma_k^2}{s_n^2}. \tag{9.31}$$

Recall (9.24). We conclude from (9.31) that given $\delta > 0$, if n is sufficiently large, then for $k = 1, \dots, n$

$$|\phi_k(t/s_n) - 1| \le \frac{\delta}{2}. \tag{9.32}$$

Let $z_k = \phi_k(t/s_n) - 1$ and

$$|\exp\{\sum_{k=1}^n(\phi_k(t/s_n) - 1)\} - \prod_{k=1}^n \phi_k(t/s_n)|$$
$$\le \sum_{k=1}^n |e^{z_k} - 1 - z_k| \le \sum_{k=1}^n \delta|z_k|$$

for n large, because $\vee_{k=1}^{n}|z_k| \le \delta/2$ for n large, and applying (9.30), we get the bound

$$\le \delta \frac{t^2}{2s_n^2}\sum_{k=1}^{n}\sigma_k^2 = \delta\frac{t^2}{2}$$

and since δ is arbitrary, we have (9.28) as desired. Thus it suffices to prove (9.27).

(3) We now concentrate on proving (9.27). Write

$$\begin{aligned}&\sum_{k=1}^{n}(\phi_k(t/s_n)-1)+t^2/2\\ &\qquad=\sum_{k=1}^{n}E\left(e^{itX_k/s_n}-1-i\frac{t}{s_n}X_k-\frac{1}{2}(\frac{it}{s_n})^2X_k^2\right).\end{aligned}$$

Let $(\cdot)$ represent what is inside the expectation on the previous line. We get the decomposition

$$\begin{aligned}&=\sum_{k=1}^{n}\{E(\cdot)1_{[|X_k|/s_n\le\epsilon]}+E(\cdot)1_{[|X_k|/s_n>\epsilon]}\}\\ &=I+II.\end{aligned}$$

We now show I and II are small. For I we have using (9.6) with $n=2$:

$$\begin{aligned}|I| &\le \sum_{k=1}^{n}E|(\cdot)|1_{[|X_k/s_n|\le\epsilon]}\\ &\le \sum_{k=1}^{n}E\left(\frac{1}{3!}\left|\frac{t}{s_n}X_k\right|^3 1_{[|X_k/s_n|\le\epsilon]}\right)\\ &=\frac{|t|^3}{6}\sum_{k=1}^{n}E\left(\left|\frac{X_k}{s_n}\right|^3 1_{[|X_k/s_n|\le\epsilon]}\right)\\ &\le \frac{|t|^3}{6}\epsilon\sum_{k=1}^{n}E\left(\left|\frac{X_k}{s_n}\right|^2 1_{[|X_k/s_n|\le\epsilon}\right)\\ &\le \frac{|t|^3}{6}\epsilon\sum_{k=1}^{n}\frac{\sigma_k^2}{s_n^2}\\ &=\epsilon\frac{|t|^3}{6}\end{aligned}$$

where we used the fact that

$$\sum_{k=1}^{n}\frac{\sigma_k^2}{s_n^2}=1.$$

Now we show why II is small. We have

$$
\begin{aligned}
|II| &\le \sum_{k=1}^{n} E\left(|\cdot|1_{[|X_k/s_n|>\epsilon]}\right)\\
&\le 2\sum_{k=1}^{n}\frac{1}{2}E\left(\left|\frac{tX_k}{s_n}\right|^2 1_{[|X_k/s_n|>\epsilon]}\right)
\end{aligned}
$$

(from (9.6) with $n = 2$)

$$
= \frac{t^2}{s_n^2}\sum_{k=1}^{n} E\left(X_k^2 1_{[|X_k/s_n|>\epsilon]}\right) \to 0
$$

by the Lindeberg condition (9.23). This completes the proof of (9.27) and hence the theorem is proved. □

We next present a sufficient condition for the Lindeberg condition (9.23) called the Liapunov condition which is relatively easy to verify.

Corollary 9.8.1 (Liapunov Condition) *Let $\{X_k, k \ge 1\}$ be an independent sequence of random variables satisfying $E(X_k) = 0$, $Var(X_k) = \sigma_k^2 < \infty$, $s_n^2 = \sum_{k=1}^n \sigma_k^2$. If for some $\delta > 0$*

$$
\frac{\sum_{k=1}^n E|X_k|^{2+\delta}}{s_n^{2+\delta}} \to 0,
$$

then the Lindeberg condition (9.23) holds and hence the CLT.

Remark. A useful special case of the Liapunov condition is when $\delta = 1$:

$$
\frac{\sum_{k=1}^n E|X_k|^3}{s_n^3} \to 0.
$$

Proof. We have

$$
\begin{aligned}
\frac{1}{s_n^2}\sum_{k=1}^{n} E\left(X_k^2 1_{[|X_k/s_n|>t]}\right) &= \sum_{k=1}^{n} E\left(\left|\frac{X_k}{s_n}\right|^2 \cdot 1 \cdot 1_{[|X_k/ts_n|>1]}\right)\\
&\le \sum_{k=1}^{n} E\left(\left|\frac{X_k}{s_n}\right|^2 \left|\frac{X_k}{ts_n}\right|^{\delta} 1_{[X_k/ts_n>1]}\right)\\
&\le \frac{1}{t^\delta}\frac{\sum_{k=1}^n E|X_k|^{2+\delta}}{s_n^{2+\delta}}\\
&\to 0.
\end{aligned}
$$

□

Example: Record counts are asymptotically normal. We now to examine the weak limit behavior of the record count process. Suppose $\{X_n, n \geq 1\}$ is an iid sequence of random variables with common continuous distribution F, and define

$$1_k = 1_{[X_k \text{ is a record }]}, \quad \mu_n = \sum_{i=1}^{n} 1_k.$$

So μ_n is the number of records among $X_1, \ldots, X_n$. We know from Chapter 8 that as $n \to \infty$

$$\frac{\mu_n}{\log n} \stackrel{a.s.}{\to} 1.$$

Here we will prove

$$\frac{\mu_n - \log n}{\sqrt{\log n}} \Rightarrow N(0, 1).$$

To check this, recall

$$E(1_k) = \frac{1}{k}, \quad \text{Var}(1_k) = \frac{1}{k} - \frac{1}{k^2}.$$

Thus

$$\begin{aligned} s_n^2 = \text{Var}(\mu_n) &= \sum_{k=1}^{n} \left(\frac{1}{k} - \frac{1}{k^2}\right) \\ &= \sum_{k=1}^{n} \frac{1}{k} - \sum_{k=1}^{n} \frac{1}{k^2} \\ &\sim \log n. \end{aligned}$$

So

$$s_n^3 \sim (\log n)^{3/2}.$$

Now

$$\begin{aligned} E|1_k - E(1_k)|^3 &= E|1_k - \frac{1}{k}|^3 \\ &= |1 - \frac{1}{k}|^3 \frac{1}{k} + \left(\frac{1}{k}\right)^3 \left(1 - \frac{1}{k}\right) \\ &\leq \frac{1}{k} + \frac{1}{k^3}, \end{aligned}$$

and therefore

$$\begin{aligned} \frac{\sum_{k=1}^{n} E|1_k - E(1_k)|^3}{s_n^3} &\leq \frac{\sum_{k=1}^{n} \left(\frac{1}{k} + \frac{1}{k^3}\right)}{(\log n)^{3/2}} \\ &\sim \frac{\log n}{(\log n)^{3/2}} \to 0. \end{aligned}$$

So the Liapunov condition is valid and thus

$$\frac{\mu_n - E(\mu_n)}{\sqrt{\mathrm{Var}(\mu_n)}} \Rightarrow N(0,1).$$

Note

$$\sqrt{\mathrm{Var}(\mu_n)} \sim s_n \sim \sqrt{\log n}$$

and

$$\frac{E(\mu_n) - \log n}{\sqrt{\log n}} = \frac{\sum_{k=1}^{n} \frac{1}{k} - \log n}{\sqrt{\log n}} \sim \frac{\gamma}{\sqrt{\log n}} \to 0,$$

where γ is Euler's constant. So by the convergence to types theorem

$$\frac{\mu_n - \log n}{\sqrt{\log n}} \Rightarrow N(0,1).$$

□

9.9 Exercises

1. **Triangular arrays.** Suppose for each n, that $\{X_{k,n}, 1 \le k \le n\}$ are independent and define $S_n = \sum_{k=1}^{n} X_{k,n}$. Assume $E(X_{k,n}) = 0$ and $\mathrm{Var}(S_n) = 1$, and

$$\sum_{k=1}^{n} E\left(|X_{k,n}|^2 1_{[|X_{k,n}|>t]}\right) \to 0$$

as $n \to \infty$ for every $t > 0$. Adapt the proof of the sufficiency part of the Lindeberg–Feller CLT to show $S_n \Rightarrow N(0,1)$.

2. Let $\{X_n, n \ge 0\}$ be a sequence of random variables.

(a) Suppose $\{X_n, n \ge 0\}$ are Poisson distributed random variables so that for $n \ge 0$ there exist constants λ_n and

$$P[X_n = k] = \frac{e^{-\lambda_n}\lambda_n^k}{k!}, \quad k \ge 0.$$

Compute the characteristic function and give necessary and sufficient conditions for

$$X_n \Rightarrow X_0.$$

(b) Suppose the $\{X_n\}$ are each normally distributed and

$$E(X_n) = \mu_n \in R, \quad \mathrm{Var}(X_n) = \sigma_n^2.$$

Give necessary and sufficient conditions for

$$X_n \Rightarrow X_0.$$

3. Let $\{X_k, k \geq 1\}$ be independent with the range of X_k equal to $\{\pm 1, \pm k\}$ and
$$P[X_k = \pm 1] = \frac{1}{2}(1 - \frac{1}{k^2}), \quad P[X_k = \pm k] = \frac{1}{2k^2}.$$
By simple truncation, prove that $S_n/\sqrt{n}$ behaves asymptotically in the same way as if $X_k = \pm 1$ with probability 1/2. Thus the distribution of $S_n/\sqrt{n}$ tends to $N(0, 1)$ but
$$\text{Var}(S_n/\sqrt{n}) \to 2.$$

4. Let $\{U_k\}$ be an independent sequence of random variables with U_k uniformly distributed on $[-a_k, a_k]$.
(a) Show that if there exists $M > 0$ such that $|a_k| \leq M$ but $\sum_k a_k^2 = \infty$, then the Lindeberg condition and thus the CLT holds.
(b) If $\sum_k a_k^2 < \infty$, then the Lindeberg condition does not hold.

5. Suppose X_n and Y_n are independent for each n and
$$X_n \Rightarrow X_0, \quad Y_n \Rightarrow Y_0.$$
Prove using characteristic functions that
$$X_n + Y_n \Rightarrow X_0 + Y_0.$$

6. (a) Suppose X_n has a uniform distribution on $(-n, n)$. Find the chf of X_n.
(b) Show $\lim_{n\to\infty} \phi_n(t)$ exists.
(c) Is there a proper, non-degenerate random variable X_0 such that
$$X_n \Rightarrow X_0?$$
Why or why not? What does this say about the continuity theorem for characteristic functions?

7. Suppose $\{X_n, n \geq 1\}$ are iid with common density
$$f(x) = |x|^{-3}, \quad |x| > 1.$$
(a) Check that $E(X_1) = 0$ but $E(X_n^2) = \infty$.
(b) Despite the alarming news in (a), we still have
$$\frac{S_n}{\sqrt{n \log n}} \Rightarrow N(0, 1).$$
Hint: Define
$$Y_n = X_n 1_{[|X_n| \leq \sqrt{n}]}$$

and check Liapunov's condition for $\{Y_n\}$ for $\delta = 1$. Then show

$$\sum_n P[X_n \neq Y_n] < \infty.$$

(c) It turns out that for iid random variables $\{X_n\}$ with $E(X_n) = 0$, the necessary and sufficient condition for the CLT is that

$$\lim_{t\to\infty} \frac{U(tx)}{U(t)} = 1,$$

where

$$U(t) := E\big(X_1^2 1_{[|X_1|\le t]}\big).$$

Check this condition for the example in part (a).

8. *Probabilistic proof of Stirling's formula.* Suppose $\{X_n\}$ are iid, Poisson distributed random variables with parameter 1 and as usual suppose $S_n = \sum_{i=1}^n X_i$. Prove the following:

(a) $$E\left[\left(\frac{S_n - n}{\sqrt{n}}\right)^-\right] = e^{-n}\sum_{k=0}^{n}\left(\frac{n-k}{\sqrt{n}}\right)\frac{n^k}{k!} = \frac{n^{n+1/2}e^{-n}}{n!}.$$

(b) $$\left(\frac{S_n - n}{\sqrt{n}}\right)^- \Rightarrow N^-,$$

where N is a N(0,1) random variable.

(c) Show

$$\left(\frac{S_n - n}{\sqrt{n}}\right)^- \text{ is u.i.}$$

(d) $$E\left[\left(\frac{S_n - n}{\sqrt{n}}\right)^-\right] \to E(N^-) = \frac{1}{\sqrt{2\pi}}.$$

(e) $n! \sim \sqrt{2\pi}\, n^{n+1/2}e^{-n}, \quad n \to \infty.$

9. (a) Suppose X is exponentially distributed with density

$$f(x) = e^{-x}, \quad x > 0.$$

What is the chf of X? Is the function

$$\frac{1}{1+it}$$

a chf? If so, of what random variable?

(b) Let X_1 be Bernoulli random variable with possible values ± 1 with probability $1/2$ each. What is the chf of X_1?

(c) Is $(\cos t)^{17}$ a chf? Of what random variable?

(d) Is $|\cos t|$ a chf? (Try differentiating twice.)

(e) Is $|\cos t|^2$ a chf?

The modulus of a chf need not be a chf but the modulus square is a chf.

(f) Prove that if X is a random variable with $E(|X|) < \infty$ and with chf ϕ, then

$$\int_{\mathbb{R}} |x| F(dx) = \frac{2}{\pi} \int_0^\infty \frac{1 - \operatorname{Re} \phi(t)}{t^2} dt.$$

10. Suppose Y_s is Poisson distributed with parameter s so that

$$P[Y_s = k] = e^{-s} \frac{s^k}{k!}.$$

Compute the chf of Y_s. Prove

$$\frac{Y_s - s}{\sqrt{s}} \Rightarrow N,$$

where N is a $N(0, 1)$ random variable.

11. If $\{X_n, n \geq 1\}$ is iid with $E(X_n) = 0$ and $\operatorname{Var}(X_n) = 1$, then $S_n/\sqrt{n} \Rightarrow N(0, 1)$. Show this cannot be strengthened to $S_n/\sqrt{n}$ converges in probability. (If $S_n/\sqrt{n} \stackrel{P}{\to} X$, then $S_{2n}/\sqrt{2n} \stackrel{P}{\to} X$. Subtract.)

12. Suppose $\{X_n, n \geq 1\}$ are independent and symmetric random variables so that

$$X_k \stackrel{d}{=} -X_k.$$

If for every $t > 0$, as $n \to \infty$

$$\sum_{k=1}^n P[|X_k| > t a_n] \to 0$$

$$a_n^{-2} \sum_{k=1}^n E\left(X_k^2 1_{[|X_k| \leq t a_n]}\right) \to 1,$$

where $a_n > 0$, then show

$$S_n / a_n \Rightarrow N(0, 1).$$

Here $S_n = \sum_{i=1}^n X_i$.

Hint: Try truncating at level $a_n t$: Set

$$X_j' = X_j 1_{[|X_j| \leq t a_n]}.$$

Consider S_n' and show it is enough for $S_n'/a_n \Rightarrow N(0, 1)$.

13. Prove the law of rare events stated in Example 9.5.3.

14. Assume $\phi(t)$ is a chf and G is the distribution of a positive random variable Y. Show all of the following are chf's and interpret probabilistically:

(a) $\int_0^1 \phi(ut)du$,
(b) $\int_0^\infty \phi(ut)e^{-u}du$,
(c) $\int_0^\infty e^{-|t|u}G(du)$,
(d) $\int_0^\infty \phi(ut)G(du)$.

(For example, if X has chf ϕ and U is uniform on $(0, 1)$ and independent of X, what is the chf of XU?)

15. (i) Suppose $\{E_n, n \geq 1\}$ are iid unit exponential random variables so that $P[E_1 > x] = e^{-x}$, $x > 0$. Show $(\sum_{i=1}^n E_i - n)/\sqrt{n}$ is asymptotically normal.

(ii) Now suppose X_t is a random variable with gamma density

$$F_t(x) = e^{-x}x^{t-1}/\Gamma(t), \quad t > 0,\ x > 0.$$

Use characteristic functions to show that

$$(X_t - t)/\sqrt{t} \Rightarrow N$$

where N is a $N(0, 1)$ random variable.

(iii) Show total variation convergence of the distribution of $(X_t - t)/\sqrt{t}$ to the distribution of N:

$$\sup_{B \in \mathcal{B}(\mathbb{R})} |P[\frac{X_t - t}{\sqrt{t}} \in B] - P[N \in B]| \to 0$$

as $\alpha \to \infty$. (Hint: Use Scheffe; approximate $\Gamma(t)$ via Stirling's formula.

16. (a) Suppose X and Y are iid $N(0, 1)$ random variables. Show

$$\frac{X + Y}{\sqrt{2}} \stackrel{d}{=} X \stackrel{d}{=} Y.$$

(b) Conversely: Suppose X and Y are independent with common distribution function $F(x)$ having mean zero and variance 1, and suppose further that

$$\frac{X + Y}{\sqrt{2}} \stackrel{d}{=} X \stackrel{d}{=} Y.$$

Show that both X and Y have a $N(0, 1)$ distribution. (Use the central limit theorem.)

17. (a) Give an example of a random variable Y such that $E(Y) = 0$ and

$$EY^2 < \infty, \quad E|Y^{2+\delta}| = \infty,$$

for all $\delta > 0$. (This means finding a probability density.)

(b) Suppose $\{Y_n, n \geq 1\}$ are iid with $EY_1 = 0$, and $EY_1^2 = \sigma^2 < \infty$. Suppose the common distribution is the distribution found in (a). Show that Lindeberg's condition holds but Liapunov's condition fails.

18. Use the central limit theorem to evaluate

$$\frac{1}{(n-1)!}\int_0^n e^{-x}x^{n-1}dx.$$

Hint: consider $P[S_n < x]$ where S_n is a sum of n iid unit exponentially distributed random variables.

19. Suppose $\{e_n, n \geq 1\}$ are independent exponentially distributed random variables with $E(e_n) = \mu_n$. If

$$\lim_{n\to\infty} \bigvee_{i=1}^{n} \frac{\mu_i}{\sum_{j=1}^{n}\mu_j} = 0,$$

then

$$\sum_{i=1}^{n}(e_i - \mu_i) / \sqrt{\sum_{j=1}^{n}\mu_j^2} \Rightarrow N(0,1).$$

20. Use the method of the selection theorem to prove the Arzela–Ascoli theorem: Let $\{u_n(x), n \geq 1\}$ be an equicontinuous sequence of real valued functions defined on $\mathbb{R}$, which is uniformly bounded; that is, $\sup_{n,x}|u_n(x)| \leq 1$. Then there exists a subsequence $\{u_{n'}\}$ which is converging locally uniformly to continuous limit u.

21. (a) Suppose $\{F_\lambda, \lambda \in \Lambda\}$ is a family of probability distributions and suppose the chf of F_λ is ϕ_λ. If $\{\phi_\lambda, \lambda \in \Lambda\}$ is equicontinuous, then $\{F_\lambda, \lambda \in \Lambda\}$ is tight.

(b) If $\{F_n, n \geq 0\}$ is a sequence of probability distributions such that $F_n \Rightarrow F_0$, then the corresponding chf's are equicontinuous. By the Arzela–Ascoli theorem, uniformly bounded equicontinuous functions converge locally uniformly. Thus weak convergence of $\{F_n\}$ means the chf's converge locally uniformly.

22. A continuous function which is a pointwise limit of chf's is a chf. (Use the continuity theorem.)

23. A complex valued function $\phi(\cdot)$ of a real variable is called *non-negative definite* if for every choice of integer n and reals $t_1, \dots, t_n$ and complex numbers $c_1, \dots, c_n$, we have

$$\sum_{r,s=1}^{n} \phi(t_r - t_s) c_r \bar{c}_s \geq 0.$$

Show that every chf is non-negative definite.

24. (a) Suppose $K(\cdot)$ is a complex valued function on the integers such that $\sum_{n=-\infty}^{\infty} |K(n)| < \infty$. Define

$$f(\lambda) = \frac{1}{2\pi} \sum_{n=-\infty}^{\infty} e^{-in\lambda} K(n) \tag{9.33}$$

and show that

$$K(h) = \int_{-\pi}^{\pi} e^{ihx} f(x) dx, \quad h = 0, \pm 1, \pm 2, \dots. \tag{9.34}$$

(b) Let $\{\{X_n, n = 0, \pm 1, \pm 2, \dots\}$ be a zero mean weakly stationary process. This means $E(X_m) = 0$ for all m and

$$\gamma(h) = E(X_m X_{m+h})$$

is independent of m. The function γ is called the autocovariance (acf) function of the process $\{X_n\}$.

Prove the following: An absolutely summable complex valued function $\gamma(\cdot)$ defined on the integers is the autocovariance function of a weakly stationary process iff

$$f(\lambda) = \frac{1}{2\pi} \sum_{n=-\infty}^{\infty} e^{-in\lambda} \gamma(n) \geq 0, \text{ for all } \lambda \in [-\pi, \pi],$$

in which case

$$\gamma(h) = \int_{-\pi}^{\pi} e^{ihx} f(x) dx.$$

(So $\gamma(\cdot)$ is a chf.)

Hint: If $\gamma(\cdot)$ is an acf, check that

$$f_N(\lambda) = \frac{1}{2\pi N} \sum_{r,s=1}^{N} e^{-ir\lambda} \gamma(r - s) e^{is\lambda} \geq 0$$

and $f_N(\lambda) \to f(\lambda)$ as $N \to \infty$. Use (9.34). Conversely, if $\gamma(\cdot)$ is absolutely summable, use (9.34) to write γ as a Fourier transform or chf of f.

Check that this makes γ non-negative definite and thus there is a Gaussian process with this γ as its acf.

(c) Suppose

$$X_n = \sum_{i=0}^{q} \theta_i Z_{n-i},$$

where $\{Z_n\}$ are iid $N(0, 1)$ random variables. Compute $\gamma(h)$ and $f(\lambda)$.

(d) Suppose $\{X_n\}$ and $\{Y_n\}$ are two uncorrelated processes (which means $E(X_m Y_n) = 0$ for all m, n), and that each has absolutely summable acfs. Compute $\gamma(h)$ and $f(\lambda)$ for $\{X_n + Y_n\}$.

25. Show the chf of the uniform density on (a, b) is

$$\frac{e^{itb} - e^{ita}}{it(b-a)}.$$

If $\phi(t)$ is the chf of the distribution F and $\phi(t)(1-e^{ith})/(ith)$ is integrable in t, show the inversion formula

$$h^{-1}F(x, x+h] = \frac{1}{2\pi}\int_{-\infty}^{\infty} \phi(t)\left(\frac{1-e^{-ith}}{ith}\right)e^{-itx}dt.$$

Hint: Let $U_{-h,0}$ be the uniform distribution on $(-h, 0)$. What is the chf of $F * U_{-h,0}$? The convolution has a density; what is it? Express this density using Fourier inversion (Corollary 9.5.1).

26. Why does the Fourier inversion formula for densities (Corollary 9.5.1) not apply to the uniform density?

27. Suppose for each $n \geq 0$ that $\phi_n(t)$ is an integrable chf corresponding to a distribution F_n, which by Fourier inversion (Corollary 9.5.1) has a density f_n. If as $n \to \infty$

$$\int_{-\infty}^{\infty} |\phi_n(t) - \phi_0(t)|dt \to 0,$$

then show $f_n \to f_0$ uniformly.

28. Show the chf of $F(x) = 1 - e^{-x},\ x > 0$ is $1/(1 - it)$. If E_1, E_2 are iid with this distribution, then the symmetrized variable $E_1 - E_2$ has a bilateral exponential density. Show that the chf of $E_1 - E_2$ is $1/(1+t^2)$.

Consider the Cauchy density

$$f(x) = \frac{1}{\pi}\left(\frac{1}{1+x^2}\right), \qquad x \in \mathbb{R}.$$

Note that apart from a constant, $f(x)$ is the same as the chf of the bilateral exponential density. Use this fact to show the chf of the Cauchy density is $\phi(t) = e^{-|t|}$. Verify that the convolution of two Cauchy densities, is a density of the same type.

29. **Triangle density.** (a) Suppose $U_{a,b}$ is the uniform distribution on (a, b). The distribution $U_{(-1,0)} * U_{(0,1)}$ has a density called the triangle density. Show the chf of the triangle density is $2(1 - \cos t)/t^2$. Verify that this chf is integrable.

Check that

$$f(x) = (1 - \cos x)/(\pi x^2), \quad x \in \mathbb{R}$$

is a probability density. Hint: Use (a) and Fourier inversion to show $1 - |x|$ is a chf. Set $x = 0$.

30. Suppose $U_1, \ldots, U_n$ are iid $U(0, 1)$ random variables. Use the uniqueness theorem to show that $\sum_{i=1}^n U_i$ has density

$$f(x) = \frac{1}{(n-1)!} \sum_{j=0}^{n} (-1)^j \binom{n}{j} (x - j)_+^n, \quad x > 0.$$

31. Suppose F is a probability distribution with chf $\phi(t)$. Prove for all $\alpha > 0$

$$\int_0^\alpha (F(x + u) - F(x - u))du = \frac{1}{\pi} \int_{-\infty}^{\infty} \frac{1 - \cos \alpha t}{t^2} e^{-itx} \phi(t)dt,$$

and

$$\int_0^\alpha \int_{-u}^{u} \phi(t)dt = 2 \int_{-\infty}^{\infty} \frac{1 - \cos \alpha x}{x^2} F(dx).$$

32. Suppose X has chf

$$\phi(t) = \frac{3 \sin t}{t^3} - \frac{3 \cos t}{t^2}, \quad t \neq 0.$$

(a) Why is X symmetric?

(b) Why is the distribution of X absolutely continuous?

(c) Why is $P[|X| > 1] = 0$?

(d) Show $E(X^{2n}) = 3/(2n + 1)(2n + 3)$. (Try expanding $\phi(t)$.)

33. The convergence to types theorem could be used to prove that if $X_n \Rightarrow X$ and $a_n \to a$ and $b_n \to b$, then $a_n X_n + b_n \Rightarrow aX + b$. Prove this directly using chf's.

34. Suppose $\{X_n, n \geq 1\}$ are independent random variables and suppose X_n has a $N(0, \sigma_n^2)$ distribution. Choose σ_n^2 so that $\vee_{i=1}^n \sigma_i^2 / s_n^2 \not\to 0$. (Give an example of this.) Then

$$S_n / s_n \stackrel{d}{=} N(0, 1)$$

and hence $S_n / s_n \Rightarrow N(0, 1)$. Conclusion: sums of independent random variables can be asymptotically normal even if the Lindeberg condition fails.

35. Let $\{X_n, n \geq 1\}$ be independent random variables satisfying the Lindeberg condition so that $\sum_{i=1}^n X_i$ is asymptotically normal. As usual, set $s_n^2 = \text{Var}(\sum_{i=1}^n X_i$. Now define random variables $\{\xi_n, n \geq 1\}$ to be independent and independent of $\{X_n\}$ so that the distribution of ξ_n is symmetric about 0 with

$$P[\xi_n = 0] = 1 - \frac{1}{n^2},$$

$$P[|\xi_n| > x] = \frac{1}{n^2} x^{-1}, \quad x > 1.$$

Does the mean or variance of ξ_n exist?

Prove

$$\sum_{i=1}^n \frac{(X_i + \xi_i)}{s_n} \Rightarrow N(0, 1).$$

Thus asymptotic normality is possible even when neither a mean nor a second moment exist.

36. Suppose X is a random variable with the property that X is irrational with probability 1. (For instance, this holds if X has a continuous distribution function.) Let F_n be the distribution of $nX - [nX]$, the fractional part of nX. Prove $n^{-1} \sum_{i=1}^n F_i \Rightarrow U$, the uniform distribution on $[0, 1]$. Hint: You will need the continuity theorem and the following fact: If θ is irrational, then the sequence $\{n\theta - [n\theta], n \geq 1\}$ is uniformly distributed modulo 1. A sequence $\{x_n\}$ is uniformly distributed if the sequence contains elements of $[0, 1]$ such that

$$\frac{1}{n} \sum_{i=1}^n \epsilon_{x_i}(\cdot) \xrightarrow{v} \lambda(\cdot),$$

where $\lambda(\cdot)$ is Lebesgue measure on $[0, 1]$, and for $B \in \mathcal{B}([0, 1])$,

$$\epsilon_x(B) = \begin{cases} 1, & \text{if } x \in B, \\ 0, & \text{if } x \notin B. \end{cases}$$

37. Between 1871 and 1900, 1,359,670 boys and 1,285,086 girls were born. Is this data consistent with the hypothesis that boys and girls are equally likely to be born?

38. (a) If $\{X_n, n \geq 1\}$ are independent and X_n has chf ϕ_n, then if $\sum_{i=1}^\infty X_i$ is convergent, $\prod_{n=1}^\infty \phi_n(t)$ is also convergent in the sense of infinite products.

(b) Interpret and prove probabilistically the trigonometric identity

$$\frac{\sin t}{t} = \prod_{n=1}^\infty \cos(t/2^n).$$

(Think of picking a number at random in $(0, 1)$.)

39. **Renewal theory.** Suppose $\{X_n, n \geq 1\}$ are iid non-negative random variables with common mean μ and variance σ^2. Use the central limit theorem to derive an asymptotic normality result for

$$N(t) = \sup\{n : S_n \leq t\},$$

namely,

$$\frac{N(t) - \frac{t}{\mu}}{\sigma t^{1/2} \mu^{-3/2}} \Rightarrow N(0, 1).$$

40. Suppose X and Y are iid with mean 0 and variance 1. If

$$X + Y \perp\!\!\!\perp X - Y,$$

then both X and Y are $N(0, 1)$.

41. If $\phi_k, k \geq 0$ are chf's, then so is $\sum_{k=0}^{\infty} p_k \phi_k$ for any probability mass function $\{p_k, k \geq 0\}$.

42. (a) For $n \in \mathbb{Z}$ define

$$e_n(t) = \frac{1}{\sqrt{2\pi}} e^{int}, \quad t \in (-\pi, \pi].$$

Show that $\{e_n, n = 0, \pm 1, \pm 2, \ldots\}$ are orthonormal; that is, show

$$\frac{1}{2\pi} \int_{-\pi}^{\pi} e^{ikt} dt = \begin{cases} 1, & \text{if } k = 0, \\ 0, & \text{if } k \neq 0. \end{cases}$$

(b) Suppose X is integer valued with chf ϕ. Show

$$P[X = k] = \frac{1}{2\pi} \int_{-\pi}^{\pi} e^{-ikt} \phi(t) dt.$$

(c) If $X_1, \ldots, X_n$ are iid, integer valued, with common chf $\phi(t)$, show

$$P[S_n = k] = \frac{1}{2\pi} \int_{-\pi}^{\pi} e^{-ikt} (\phi(t))^n dt.$$

43. Suppose $\{X_n, n \geq 1\}$ are independent random variables, satisfying

$$E(X_n) = 0, \quad \mathrm{Var}(X_n) = \sigma_n^2 < \infty.$$

Set $s_n^2 = \sum_{i=1}^n \sigma_i^2$. Assume

(a) $S_n / s_n \Rightarrow N(0, 1)$,

(b) $\sigma_n / s_n \to p$.

Prove $X_n/s_n \Rightarrow N(0, p^2)$. (Hint: Assume as known a theorem of Cramér and Lévy which says that if X, Y are independent and the sum $X + Y$ is normally distribute, then each of X and Y is normally distributed.)

44. **Approximating roulette probabilities.** The probability of winning \$1 in roulette is 18/38 and the probability of losing \$1 is thus 20/38. Let $\{X_n, n \geq 1\}$ be the outcomes of successive plays; so each random variable has range ± 1 with probabilities 18/38, 20/38. Find an approximation by the central limit theorem for $P[S_n \geq 0]$, the probability that after n plays, the gambler is not worse off than when he/she started.

45. Suppose $f(x)$ is an even probability density so that $f(x) = f(-x)$. Define

$$g(x) = \begin{cases} \int_x^\infty \frac{f(s)}{s} ds, & \text{if } x > 0, \\ g(-x), & \text{if } x < 0. \end{cases}$$

Why is g a probability density?

If f has chf $\phi(t)$, how can you express the chf of g in terms of ϕ?

46. Suppose $\{X_n, n \geq 1\}$ are independent gamma distributed random variables and that the shape parameter of X_n is α_n. Give conditions on $\{\alpha_n\}$ which guarantee that the Lindeberg condition satisfied.

10 Martingales

Martingales are a class of stochastic processes which has had profound influence on the development of probability and stochastic processes. There are few areas of the subject untouched by martingales. We will survey the theory and applications of discrete time martingales and end with some recent developments in mathematical finance. Here is what to expect in this chapter:

- Absolute continuity and the Radon–Nikodym Theorem.
- Conditional expectation.
- Martingale definitions and elementary properties and examples.
- Martingale stopping theorems and applications.
- Martingale convergence theorems and applications.
- The fundamental theorems of mathematical finance.

10.1 Prelude to Conditional Expectation: The Radon–Nikodym Theorem

We begin with *absolute continuity* and relate this to differentiation of measures. These concepts are necessary for a full appreciation of the mathematics of conditional expectations.

Let $(\Omega, \mathcal{B})$ be a measurable space. Let μ and λ be positive bounded measures on $(\Omega, \mathcal{B})$. We say that λ is *absolutely continuous* (AC) with respect to μ, written

$\lambda << \mu$, if $\mu(A) = 0$ implies $\lambda(A) = 0$. We say that λ *concentrates* on $A \in \mathcal{B}$ if $\lambda(A^c) = 0$. We say that λ and μ are *mutually singular*, written $\lambda \perp \mu$, if there exist events $A, B \in \mathcal{B}$, such that $A \cap B = \emptyset$ and λ concentrates on A, μ concentrates on B.

Example. If $U_{[0,1]}$, $U_{[2,3]}$ are uniform distributions on [0, 1], and [2, 3] respectively, then $U_{[0,1]} \perp U_{[2,3]}$. It is also true that $U_{[0,1]} \perp U_{[1,2]}$.

Theorem 10.1.1 (Lebesgue Decomposition) *Suppose that μ and λ are positive bounded measures on $(\Omega, \mathcal{B})$.*

(a) There exists a unique pair of positive, bounded measures λ_a, λ_s on $\mathcal{B}$ such that

$$\lambda = \lambda_a + \lambda_s$$

where

$$\lambda_a << \mu, \quad \lambda_s \perp \mu, \quad \lambda_a \perp \lambda_s.$$

(b) There exists a non-negative $\mathcal{B}$-measurable function X with

$$\int X d\mu < \infty$$

such that

$$\lambda_a(E) = \int_E X d\mu, \quad E \in \mathcal{B}.$$

X is unique up to sets of μ measure 0.

We will not prove Theorem 10.1.1 but rather focus on the specialization known as the Radon–Nikodym theorem.

Theorem 10.1.2 (Radon–Nikodym Theorem) *Let $(\Omega, \mathcal{B}, P)$ be the probability space. Suppose ν is a positive bounded measure and $\nu << P$. Then there exists an integrable random variable $X \in \mathcal{B}$, such that*

$$\nu(E) = \int_E X dP, \quad \forall E \in \mathcal{B}.$$

X is a.s. unique (P) and is written

$$X = \frac{d\nu}{dP}.$$

We also write $d\nu = X dP$.

A nice proof of the Radon–Nikodym theorem is based on the following Hilbert space result (see, for example, Rudin (1966)).

Proposition 10.1.3 *Let $\mathbb{H}$ be a Hilbert space. For $x, y \in \mathbb{H}$, denote the inner product by (x, y). If $L : \mathbb{H} \mapsto \mathbb{R}$ is a continuous linear functional on $\mathbb{H}$, then there exists a unique $y \in \mathbb{H}$ such that*

$$L(x) = (x, y), \quad \forall x \in \mathbb{H}.$$

Proof. Case 1: If $L(x) \equiv 0$, $\forall x$, then we may take $y = 0$ and we are done.

Case 2: Suppose $L(x) \not\equiv 0$ and assume, for simplicity that $\mathbb{H}$ is real and define

$$M = \{x \in \mathbb{H} : L(x) = 0\}.$$

L is linear so M is a subspace. L is continuous so M is closed. Since L is not identically 0, we have $M \neq \mathbb{H}$. Thus there exists some $z' \notin M$ and by the projection theorem (Rudin (1966); Brockwell and Davis (1991))

$$z' = z_1 + z_2,$$

where $z_1 \in M$ and $z_2 \in M^{\perp}$, the orthogonal complement of M, and $z_2 \neq 0$. Then there exists $z \in M^{\perp}$, $z \neq 0$. Thus, it is also true that $z \notin M$ and hence $L(z) \neq 0$. Define

$$y = \frac{L(z)}{(z, z)} \cdot z, \tag{10.1}$$

so that

$$L(y) = L(z)^2/(z, z). \tag{10.2}$$

So $y \neq 0$, $y \in M^{\perp}$ and

$$(y, y) = \left(\frac{L(z)}{(z, z)}\right)^2 (z, z) = \frac{(L(z))^2}{(z, z)} = L(y). \tag{10.3}$$

from (10.2). Now write

$$x = \left(x - \frac{L(x)}{(y, y)} y\right) + \frac{L(x)}{(y, y)} y =: x' + x''$$

and note

$$L(x') = L(x) - \frac{L(x)L(y)}{(y, y)} = L(x) - L(x) = 0,$$

from (10.3). So $x' \in M$. Since $y \in M^{\perp}$ we have $(x', y) = 0$, and therefore

$$(x, y) = (x'', y) = L(x)$$

from the definition of x''. Thus $L(x) = (x, y)$ as required.

To show uniqueness of y: Suppose there exists y' and for all x

$$L(x) = (x, y) = (x, y').$$

Then for all x

$$(x, y - y') = 0,$$

and so

$$(y - y', y - y') = 0.$$

Thus $y - y' = 0$ and $y = y'$. □

Before the proof of the Radon–Nikodym theorem, we recall the Integral Comparison Lemma which is given in slightly expanded form as Lemma 10.1.1. This is needed for conditional expectations and martingales.

Lemma 10.1.1 (Integral Comparison Lemma) *Suppose* $(\Omega, \mathcal{B}, P)$ *is a probability space with* $\mathcal{G} \subset \mathcal{B}$ *a sub* σ*-field of* $\mathcal{B}$*. Suppose* $X \in \mathcal{G}$, $Y \in \mathcal{G}$*, and that* X *and* Y *are integrable. Then*

$$\begin{array}{llll} X & = Y \quad \text{a.s. iff} \quad \forall A \in \mathcal{G}, & \int_A X dP & = \int_A Y dP. \\ & \geq & & \geq \\ & \leq & & \leq \end{array}$$

Proof of the Radon–Nikodym Theorem. Suppose $\nu << P$ and define

$$Q(A) = \frac{\nu(A)}{\nu(\Omega)}$$

so Q is a probability measure and $Q << P$. Set

$$P^* = \frac{P + Q}{2},$$

which is also a probability measure. Then

$$H := L_2(P^*) = L_2(\Omega, \mathcal{B}, P^*)$$

is a Hilbert space with inner product

$$(Y_1, Y_2) = \int Y_1 Y_2 dP^*.$$

Note that all elements of $L_2(P^*) = L_2(\Omega, \mathcal{B}, P^*)$ are $\mathcal{B}$-measurable. On H, define the functional

$$L(Y) = \int_\Omega Y dQ, \tag{10.4}$$

so that $L : L_2(P^*) \mapsto \mathbb{R}$ is

(a) linear,

(b) bounded (and hence continuous).

To check (b) we must show

$$|L(Y)| \leq (\text{const})\|Y\|_2$$

where

$$\|Y\|_2^2 = (Y, Y) = \int_\Omega Y^2 dP^*.$$

However,

$$\begin{aligned}|L(Y)| &\leq \int |Y| dQ \leq \int |Y| dQ + \int |Y| dP \\ &= 2\int |Y| dP^* \leq 2(\int |Y|^2 dP^*)^{1/2}\end{aligned}$$

(by, for instance, Example 6.5.2 on page 189)

$$= 2\|Y\|_2.$$

Now, L is continuous and linear on $H = L_2(P^*)$. Thus from Proposition 10.1.3, there exists $Z \in L_2(P^*)$ such that for all $Y \in L_2(P^*)$

$$\begin{aligned}L(Y) = (Y, Z) &= \int YZ dP^* \\ &= \int \frac{1}{2} YZ dP + \int \frac{1}{2} YZ dQ \\ &= \int Y dQ. \end{aligned} \tag{10.5}$$

Consequently, from the definition (10.4), for all $Y \in L_2(P^*)$

$$\int Y(1 - \frac{Z}{2}) dQ = \int \frac{YZ}{2} dP. \tag{10.6}$$

Pick any set $A \in \mathcal{B}$ and substituting $Y = 1_A$ in (10.5) gives

$$\int Y dQ = Q(A) = \int_A Z dP^*. \tag{10.7}$$

Then we get from (10.7) that

$$0 \leq \frac{Q(A)}{P^*(A)} = \frac{\int_A Z dP^*}{P^*(A)} \leq \frac{\int_A Z dP^*}{Q(A)/2} = 2,$$

where, to get the right inequality, we applied the fact that $2P^* = P + Q \geq Q$. Therefore, for all $A \in \mathcal{B}$

$$0 \leq \int_A Z dP^* \leq 2P^*(A)$$

that is,

$$0 \leq \int_A Z dP^* \leq \int_A 2dP^*.$$

From the Integral Comparison Lemma 10.1.1

$$0 \leq Z \leq 2, \quad a.s.(P^*).$$

In (10.6) set $Y = 1_{[Z=2]}$ to get

$$\int_{[Z=2]} (1 - Z/2) dQ = \int_{[Z=2]} \frac{Z}{2} dP,$$

that is,

$$0 = P[Z = 2].$$

Since $Q << P$, we have $0 = Q[Z = 2]$ by the definition of absolute continuity and hence $P^*[Z = 2] = 0$. So $0 \leq Z < 2$, a.s. (P^*).

In (10.6), set

$$Y = \left(\frac{Z}{2}\right)^n 1_A, \quad A \in \mathcal{B}.$$

Then $Y \in L_2(P^*)$ and, in fact, $0 \leq Y < 1$ a.s. (P or Q or P^*). From (10.6),

$$\int_A \left(\frac{Z}{2}\right)^n \left(1 - \frac{Z}{2}\right) dQ = \int_A \left(\frac{Z}{2}\right)^n \frac{Z}{2} dP.$$

Sum both sides over $n = 0$ to $n = N$ to get

$$\int_A \left(1 - \left(\frac{Z}{2}\right)^{N+1}\right) dQ = \int_A \frac{Z}{2} \sum_{j=0}^{N} \left(\frac{Z}{2}\right)^j dP. \tag{10.8}$$

Note, as $N \to \infty$,

$$1 - (Z/2)^{N+1} \nearrow 1, \quad \text{a.s. } P^*$$

and hence a.s. Q. If LHS refers to the left side of (10.8), then dominated convergence implies

$$\text{LHS} \to \int_A dQ = Q(A).$$

If RHS refers to the right side of (10.8), then monotone convergence implies

$$\text{RHS} \nearrow \int_A \frac{Z/2}{1 - Z/2} dP = \int_A \frac{Z}{2 - Z} dP.$$

Set $X = Z/(2-Z)$ and for all $A \in \mathcal{B}$

$$Q(A) = \int_A X dP.$$ □

In subsequent work we use the notation $\mu|_A$ for the restriction of the measure μ to the set A. Next we extend the Radon–Nikodym theorem to the case of σ-finite measures. (Recall that a measure μ is σ-finite if the space Ω can be decomposed $\Omega = \sum_{i=1}^{\infty} \Omega_i$ where on each piece Ω_i, μ is finite: $\mu(\Omega_i) < \infty$. Lebesgue measure λ on (R) is σ-finite since $\lambda((n, n+1]) = 1 < \infty$ and $\sum_{n=1}^{\infty}(n, n+1] = \mathbb{R}$.)

Corollary 10.1.1 *If μ, ν are σ-finite measures on $(\Omega, \mathcal{B})$, there exists a measurable $X \in \mathcal{B}$ such that*

$$\nu(A) = \int_A X d\mu, \quad \forall A \in \mathcal{B},$$

iff

$$\nu << \mu.$$

Proof. Write $\Omega = \sum_{i=1}^{\infty} \Omega_i$ where $\mu(\Omega_i) < \infty$, and $\nu(\Omega_i) < \infty$, for all i. On Ω_i, μ and ν are finite and if $\nu << \mu$, then

$$\nu|_{\Omega_i} << \mu|_{\Omega_i}$$

on $(\Omega_i, \Omega_i \cap \mathcal{B})$. Apply the Radon–Nikodym theorem to each Ω_i piece and put the pieces back together. □

The next corollary is important for the definition of conditional expectation.

Corollary 10.1.2 *Suppose Q and P are probability measures on $(\Omega, \mathcal{B})$ such that $Q << P$. Let $\mathcal{G} \subset \mathcal{B}$ be a sub-σ-algebra. Let $Q|_{\mathcal{G}}$, $P|_{\mathcal{G}}$ be the restrictions of Q and P to $\mathcal{G}$. Then in $(\Omega, \mathcal{G})$*

$$Q|_{\mathcal{G}} << P|_{\mathcal{G}}$$

and

$$\frac{dQ|_{\mathcal{G}}}{dP|_{\mathcal{G}}} \text{ is } \mathcal{G}\text{–measurable.}$$

Proof. Check the proof of the Radon–Nikodym theorem. In the construction, we deal with $L_2(\Omega, \mathcal{G}, P^*)$. □

10.2 Definition of Conditional Expectation

This section develops the definition of conditional expectation with respect to a σ-field and explains why the definition makes sense. The mathematics depends on Radon–Nikodym differentiation.

Suppose $X \in L_1(\Omega, \mathcal{B}, P)$ and let $\mathcal{G} \subset \mathcal{B}$ be a sub-σ-field. Then there exists a random variable $E(X|\mathcal{G})$, called the *conditional expectation* of X with respect to $\mathcal{G}$, such that

(i) $E(X|\mathcal{G})$ is $\mathcal{G}$-measurable and integrable.

(ii) For all $G \in \mathcal{G}$ we have

$$\int_G X dP = \int_G E(X|\mathcal{G}) dP.$$

To test that a random variable is the conditional expectation with respect to $\mathcal{G}$, one has to check two conditions: (i) the measurability condition and (ii) the integral condition.

There are (at least) two questions one can ask about this definition.

(a) Why does this definition of conditional expectation make mathematical sense?

(b) Why does this definition make intuitive sense?

It is relatively easy to answer (a) given the development of Radon–Nikodym differentiation. Suppose initially that $X \geq 0$. Define

$$\nu(A) = \int_A X dP, \quad A \in \mathcal{B}.$$

Then ν is finite and $\nu << P$. So

$$\nu|_{\mathcal{G}} << P|_{\mathcal{G}}.$$

From the Radon–Nikodym theorem, the derivative exists and we set

$$E(X|\mathcal{G}) = \frac{d\nu|_{\mathcal{G}}}{dP|_{\mathcal{G}}}$$

which by Corollary 10.1.2 is $\mathcal{G}$–measurable, and so for all $G \in \mathcal{G}$

$$\begin{aligned}
\nu|_{\mathcal{G}}(G) = \nu(G) &= \int_G \frac{d\nu|_{\mathcal{G}}}{dP|_{\mathcal{G}}} dP|_{\mathcal{G}} \\
&= \int_G \frac{d\nu|_{\mathcal{G}}}{dP|_{\mathcal{G}}} dP \qquad \text{since } P = P|_{\mathcal{G}} \text{ on } \mathcal{G} \\
&= \int_G E(X|\mathcal{G}) dP
\end{aligned}$$

which is (ii) of the definition of conditional expectation.

If $X \in L_1$ is not necessarily non-negative, then

$$E(X^+|\mathcal{G}) - E(X^-|\mathcal{G})$$

satisfies (i) and (ii) in the definition. □

Notes.

(1) *Definition of conditional probability*: Given $(\Omega, \mathcal{B}, P)$, a probability space, with $\mathcal{G}$ a sub-σ-field of $\mathcal{B}$, define

$$P(A|\mathcal{G}) = E(1_A|\mathcal{G}), \quad A \in \mathcal{B}.$$

Thus $P(A|\mathcal{G})$ is a random variable such that

(a) $P(A|\mathcal{G})$ is $\mathcal{G}$-measurable and integrable.

(b) $P(A|\mathcal{G})$ satisfies

$$\int_G P(A|\mathcal{G})dP = P(A \cap G), \quad \forall G \in \mathcal{G}.$$

(2) *Conditioning on random variables*: Suppose $\{X_t, t \in T\}$ is a family of random variables defined on $(\Omega, \mathcal{B})$ and indexed by some index set T. Define

$$\mathcal{G} := \sigma(X_t, t \in T)$$

to be the σ-field generated by the process $\{X_t, t \in T\}$. Then define

$$E(X|X_t, t \in T) = E(X|\mathcal{G}).$$

Note (1) continues the duality of probability and expectation but seems to place expectation in a somewhat more basic position, since conditional probability is defined in terms of conditional expectation. Note (2) saves us from having to make separate definitions for $E(X|X_1)$, $E(X|X_1, X_2)$, etc.

We now show the definition makes some intuitive sense and begin with an example.

Example 10.2.1 (Countable partitions) Let $\{\Lambda_n, n \geq 1\}$ be a partition of Ω so that $\Lambda_i \cap \Lambda_j = \emptyset, i \neq j$, and $\sum_n \Lambda_n = \Omega$. (See Exercise 26 of Chapter 1.) Define

$$\mathcal{G} = \sigma(\Lambda_n, n \geq 1)$$

so that

$$\mathcal{G} = \left\{ \sum_{i \in J} \Lambda_i : J \subset \{1, 2, \dots\} \right\}.$$

For $X \in L_1(P)$, define

$$E_{\Lambda_n}(X) = \int XP(d\omega|\Lambda_n) = \int_{\Lambda_n} XdP/P\Lambda_n,$$

if $P(\Lambda_n) > 0$ and $E_{\Lambda_n}(X) = 17$ if $P(\Lambda_n) = 0$. We claim

$$(a) \quad E(X|\mathcal{G}) \stackrel{a.s.}{=} \sum_{n=1}^{\infty} E_{\Lambda_n}(X) 1_{\Lambda_n}$$

and for any $A \in \mathcal{B}$

$$(b) \quad P(A|\mathcal{G}) \stackrel{a.s.}{=} \sum_{n=1}^{\infty} P(A|\Lambda_n) 1_{\Lambda_n}.$$

Proof of (a) and (b). We first check (a). Begin by observing

$$\sum_{n=1}^{\infty} E_{\Lambda_n}(X) 1_{\Lambda_n} \in \mathcal{G}.$$

Now pick $\Lambda \in \mathcal{G}$ and it suffices to show for our proposed form of $E(X|\mathcal{G})$ that

$$\int_{\Lambda} E(X|\mathcal{G}) dP = \int_{\Lambda} \left(\sum_{n=1}^{\infty} E_{\Lambda_n}(X) 1_{\Lambda_n} \right) dP = \int_{\Lambda} X dP. \tag{10.9}$$

Since $\Lambda \in \mathcal{G}$, Λ has the form $\Lambda = \sum_{i \in J} \Lambda_i$ for some $J \subset \{1, 2, \dots\}$. Now we see if our proposed form of $E(X|\mathcal{G})$ satisfies (10.9). We have

$$\begin{aligned}
&\int_{\Lambda} \sum_{n=1}^{\infty} E_{\Lambda_n}(X) 1_{\Lambda_n} dP \\
&\quad = \sum_{n \geq 1} \sum_{i \in J} \int_{\Lambda_i} E_{\Lambda_n}(X) 1_{\Lambda_n} dP \quad \text{(form of } \Lambda) \\
&\quad = \sum_{n \geq 1} \sum_{i \in J} E_{\Lambda_n}(X) P(\Lambda_i \Lambda_n) \\
&\quad = \sum_{i \in J} E_{\Lambda_i}(X) \cdot P(\Lambda_i) \quad (\{\Lambda_n\} \text{ are disjoint}) \\
&\quad = \sum_{i \in J} \frac{\int_{\Lambda_i} X dP}{P(\Lambda_i)} \cdot P(\Lambda_i) \quad (\text{definition of } E_{\Lambda}(X)) \\
&\quad = \sum_{i \in J} \int_{\Lambda_i} X dP = \int_{\sum_{i \in J} \Lambda_i} X dP \\
&\quad = \int_{\Lambda} X dP.
\end{aligned}$$

This proves (a). We get (b) from (a) by substituting $X = 1_A$. □

Interpretation: Consider an experiment with sample space Ω. Condition on the information that "some event in $\mathcal{G}$ occurs." Imagine that at a future time you will be told which set Λ_n the outcome ω falls in (but you will not be told ω). At time 0

$$\sum_{n=1}^{\infty} P(A|\Lambda_n) 1_{\Lambda_n}$$

is the best you can do to evaluate conditional probabilities.

Example 10.2.2 (Discrete case) Let X be a discrete random variable with possible values $x_1, x_2, \ldots$. Then for $A \in \mathcal{B}$

$$\begin{aligned} P(A|X) &= P(A|\sigma(X)) \\ &= P(A|\sigma([X = x_i], i = 1, 2, \ldots)) \\ &= \sum_{i=1}^{\infty} P(A|X = x_i) 1_{[X=x_i]} \end{aligned}$$

where we applied Example 10.2.1(b).

Note that if we attempted to develop conditioning by first starting with a definition for discrete random variables X, how would we extend the definition to continuous X's? What would be $P(A|X = x)$ if $P(X = x) = 0$ for all x? We could try to define

$$P(A|X = x) = \lim_{h \downarrow 0} P\left(A|X \in (x - h, x + h)\right)$$

but

(a) How do we know the limit exists for any x?

(b) How do we know the limit exists for all x?

The approach using Radon–Nikodym derivatives avoids many of these problems.

Example 10.2.3 (Absolutely continuous case) Let $\Omega = \mathbb{R}^2$ and suppose X and Y are random variables whose joint distribution is absolutely continuous with density $f(x, y)$ so that for $A \in \mathcal{B}(\mathbb{R}^2)$

$$P[(X, Y) \in A] = \iint_A f(x, y) dx dy.$$

What is $P[Y \in C|X]$ for $C \in \mathcal{B}(\mathbb{R})$? We use $\mathcal{G} = \sigma(X)$. Let

$$I(x) := \int_{\mathbb{R}} f(x, t) dt$$

be the marginal density of X and define and

$$\phi(X) = \begin{cases} \frac{\int_C f(X,t)dt}{I(x)}, & \text{if } I(x) > 0, \\ 17, & \text{if } I(x) = 0. \end{cases}$$

We claim that

$$P[Y \in C|X] = \phi(X).$$

First of all, note by Theorem 5.9.1 page 149 and composition (Proposition 3.2.2, page 77) that $\int_C f(X,t)dt$ is $\sigma(X)$-measurable and hence $\phi(X)$ is $\sigma(X)$-measurable. So it remains to show for any $\Lambda \in \sigma(X)$ that

$$\int_\Lambda \phi(X)dP = P([Y \in C] \cap \Lambda).$$

Since $\Lambda \in \sigma(X)$, the form of Λ is $\Lambda = [X \in A]$ for some $A \in \mathcal{B}(\mathbb{R})$. By the Transformation Theorem 5.5.1, page 135,

$$\int_\Lambda \phi(X)dP = \int_{X^{-1}(A)} \phi(X)dP = \int_A \phi(x)P[X \in dx]$$

and because a density exists for the joint distribution of (X, Y), we get this equal to

$$\begin{aligned}
&= \int_A \phi(x)\Big(\int_{\mathbb{R}} f(x,t)dt\Big)dx \\
&= \int_{A\cap\{x:I(x)>0\}} \phi(x)I(x)dx + \int_{A\cap\{x:I(x)=0\}} \phi(x)I(x)dx \\
&= \int_{A\cap\{x:I(x)>0\}} \phi(x)I(x)dx + 0 \\
&= \int_{A\cap\{x:I(x)>0\}} \frac{\int_C f(x,t)dt}{I(x)} I(x)dx \\
&= \int_{A\cap\{x:I(x)>0\}} \Big(\int_C f(x,t)dt\Big)dx \\
&= \int_A\Big(\int_C f(x,t)dt\Big)dx = P[X \in A, Y \in C] \\
&= P([Y \in C] \cap \Lambda)
\end{aligned}$$

as required. □

10.3 Properties of Conditional Expectation

This section itemizes the basic properties of conditional expectation. Many of these parallel those of ordinary expectation.

(1) *Linearity*. If $X, Y \in L_1$ and $\alpha, \beta \in \mathbb{R}$, we have

$$E((\alpha X + \beta Y)|\mathcal{G}) \overset{a.s.}{=} \alpha E(X|\mathcal{G}) + \beta E(Y|\mathcal{G}).$$

To verify this, observe that the right side is $\mathcal{G}$–measurable and for $\Lambda \in \mathcal{G}$

$$\int_\Lambda (\alpha E(X|\mathcal{G}) + \beta E(Y|\mathcal{G}))dP = \alpha \int_\Lambda E(X|\mathcal{G})dP + \beta \int_\Lambda E(Y|\mathcal{G})dP$$
$$= \alpha \int_\Lambda XdP + \beta \int_\Lambda YdP$$

(from the definition of conditional expectation)

$$= \int_\Lambda (\alpha X + \beta Y)dP.$$

(2) If $X \in \mathcal{G}$; $X \in L_1$, then

$$E(X|\mathcal{G}) \stackrel{a.s.}{=} X.$$

We prove this by merely noting that X is $\mathcal{G}$–measurable and

$$\int_\Lambda XdP = \int_\Lambda XdP, \quad \forall \Lambda \in \mathcal{G}.$$

In particular, for a constant c, c is $\mathcal{G}$-measurable so

$$E(c|\mathcal{G}) \stackrel{a.s.}{=} c.$$

(3) We have

$$E(X|\{\phi, \Omega\}) = E(X).$$

The reason is that $E(X)$ is measurable with respect to the σ-field $\{\emptyset, \Omega\}$ and for every $\Lambda \in \{\emptyset, \Omega\}$ (that is, $\Lambda = \emptyset$ or $\Lambda = \Omega$)

$$\int_\Lambda E(X)dP = \int_\Lambda XdP.$$

(4) *Monotonicity*. If $X \geq 0$, and $X \in L_1$, then $E(X|\mathcal{G}) \geq 0$ almost surely. The reason is that for all $\Lambda \in \mathcal{G}$

$$\int_\Lambda E(X|\mathcal{G})dP = \int_\Lambda XdP \geq 0 = \int_\Lambda 0dP.$$

The conclusion follows from Lemma 10.1.1. So if $X, Y \in L_1$, and $X \leq Y$, then

$$E(X|\mathcal{G}) \stackrel{a.s.}{\leq} E(Y|\mathcal{G}).$$

Thus conditional expectation is monotone.

(5) *Modulus inequality*. If $X \in L_1$

$$|E(X|\mathcal{G})| \leq E(|X||\mathcal{G})$$

since

$$\begin{aligned}|E(X)|\mathcal{G})| &= |E(X^+|\mathcal{G}) - E(X^-|\mathcal{G})| \\ &\le E(X^+|\mathcal{G}) + E(X^-|\mathcal{G})\end{aligned}$$

and using linearity, we get

$$= E(X^+ + X^-)|\mathcal{G}) = E(|X||\mathcal{G}).$$

(6) *Monotone convergence theorem.* If $X \in L_1$, $0 \le X_n \uparrow X$, then

$$E(X_n|\mathcal{G}) \uparrow E(X|\mathcal{G})$$

almost surely. Note that $\{E(X_n|\mathcal{G})\}$ is monotone by item (4). Set

$$Z := \lim_{n\to\infty} \uparrow E(X_n|\mathcal{G}).$$

Then $Z \in \mathcal{G}$ and for $\Lambda \in \mathcal{G}$

$$\begin{aligned}\int_\Lambda Z dP &= \int_\Lambda \lim_{n\to\infty} \uparrow E(X_n|\mathcal{G}) dP \\ &= \lim_{n\to\infty} \int_\Lambda E(X_n|\mathcal{G}) dP\end{aligned}$$

(by the monotone convergence theorem for integrals)

$$= \lim_{n\to\infty} \int_\Lambda X_n dP.$$

Again applying the monotone convergence theorem for integrals yields

$$= \int_\Lambda X dP.$$

Since $Z \in \mathcal{G}$ and

$$\int_\Lambda Z dP = \int_\Lambda X dP, \quad \forall \Lambda \in \mathcal{G},$$

we have by definition that $Z = E(X|\mathcal{G})$ which means

$$E(X|\mathcal{G}) = \lim_{n\to\infty} \uparrow E(X_n|\mathcal{G}).$$

(7) *Monotone convergence implies the Fatou lemma.* We have the conditional version of Fatou's lemma: If $0 \le X_n \in L_1$, then

$$E(\liminf_{n\to\infty} X_n|\mathcal{G}) \le \liminf_{n\to\infty} E(X_n|\mathcal{G}),$$

while if $X_n \le Z \in L_1$, then

$$E(\limsup_{n\to\infty} X_n|\mathcal{G}) \ge \limsup_{n\to\infty} E(X_n|\mathcal{G}).$$

For the proof of these conditional Fatou statements we note that

$$\begin{aligned} E\left(\liminf_{n\to\infty} X_n|\mathcal{G}\right) &= E\left(\lim_{n\to\infty} \bigwedge_{k\ge n} X_k|\mathcal{G}\right) \\ &= \lim_{n\to\infty} E\left(\bigwedge_{k\ge n} X_k|\mathcal{G}\right) \quad \text{(monotone convergence)} \\ &\le \liminf_{n\to\infty} E(X_n|\mathcal{G}). \end{aligned}$$

(8) *Fatou implies dominated convergence.* We have the conditional version of the dominated convergence theorem: If $X_n \in L_1$, $|X_n| \le Z \in L_1$ and $X_n \to X_\infty$, then

$$E\left(\lim_{n\to\infty} X_n|\mathcal{G}\right) \overset{a.s.}{=} \lim_{n\to\infty} E(X_n|\mathcal{G}).$$

(9) *Product rule.* Let X, Y be random variables satisfying $X, YX \in L_1$. If $Y \in \mathcal{G}$, then

$$E(XY|\mathcal{G}) \overset{a.s.}{=} YE(X|\mathcal{G}). \tag{10.10}$$

Note the right side of (10.10) is $\mathcal{G}$–measurable. Suppose we know that for all $\Lambda \in \mathcal{G}$ that

$$\int_\Lambda YE(X|\mathcal{G})dP = \int_\Lambda XYdP. \tag{10.11}$$

Then

$$\int_\Lambda YE(X|\mathcal{G})dP = \int_\Lambda XYdP = \int_\Lambda E(XY|\mathcal{G})dP,$$

and the result follows from Lemma 10.1.1.

Thus we have to only show (10.11). Start by assuming $Y = 1_\Delta$, $\Delta \in \mathcal{G}$. Then $\Lambda \cap \Delta \in \mathcal{G}$ and

$$\begin{aligned} \int_\Lambda YE(X|\mathcal{G})dP &= \int_{\Lambda\cap\Delta} E(X|\mathcal{G})dP \\ &= \int_{\Lambda\cap\Delta} XdP \\ &= \int_\Lambda XYdP. \end{aligned}$$

So (10.11) holds for $Y = 1_\Delta$ and hence (10.11) holds for

$$Y = \sum_{i=1}^{k} c_i 1_{\Delta i}$$

where $\Delta_i \in \mathcal{G}$.

Now suppose X, Y are non-negative. There exist $Y_n \uparrow Y$, and

$$Y_n = \sum_{i=1}^{k_n} c_i^{(n)} 1_{\Delta_i^{(n)}}$$

and

$$\int_\Lambda Y_n E(X|\mathcal{G})dP = \int_\Lambda XY_n dP. \tag{10.12}$$

By monotone convergence, $XY_n \nearrow XY$, and

$$Y_n E(X|\mathcal{G}) \nearrow YE(X|\mathcal{G}).$$

Letting $n \to \infty$ in (10.12) and using monotone convergence yields

$$\int_\Lambda YE(X|\mathcal{G})dP = \int_\Lambda XYdP.$$

If X, Y are not necessarily non-negative, write $X = X^+ - X^-$, $Y = Y^+ - Y^-$.

(10) *Smoothing*. If

$$\mathcal{G}_1 \subset \mathcal{G}_2 \subset \mathcal{B},$$

then for $X \in L_1$

$$E\big(E(X|\mathcal{G}_2)|\mathcal{G}_1\big) = E(X|\mathcal{G}_1) \tag{10.13}$$

$$E\big(E(X|\mathcal{G}_1)|\mathcal{G}_2\big) = E(X|\mathcal{G}_1). \tag{10.14}$$

Statement (10.14) follows from item (9) or item (2).

For the verification of (10.13), let $\Lambda \in \mathcal{G}_1$. Then $E(X|\mathcal{G}_1)$ is $\mathcal{G}_1$–measurable and

$$\begin{aligned}\int_\Lambda E(E(X|\mathcal{G}_2)|\mathcal{G}_1)dP &= \int_\Lambda E(X|\mathcal{G}_2)dP \quad \text{(definition)}\\ &= \int_\Lambda XdP \quad (\text{since } \Lambda \in \mathcal{G}_1 \subset \mathcal{G}_2)\\ &= \int_\Lambda E(X|\mathcal{G}_1)dP \quad \text{(by definition.)}\end{aligned}$$

A special case: $\mathcal{G}_1 = \{\emptyset, \Omega\}$. Then $E(X|\{\emptyset, \Omega\}) = E(X)$. So

$$E\big(E(X|\mathcal{G}_2)\big) = E\big(E(X|\mathcal{G}_2)|\{\emptyset, \Omega\}\big) = E(X|\{\emptyset, \Omega\}) = E(X). \tag{10.15}$$

To understand why (10.13) and (10.14) are called the smoothing equalities, recall Example 10.2.1 where $\mathcal{G} = \sigma(\Lambda_n, n \geq 1)$ and $\{\Lambda_n, n \geq 1\}$ is a countable partition. Then

$$E(X|\mathcal{G}) = \sum_{n=1}^{\infty} E_{\Lambda_n}(X)1_{\Lambda_n},$$

so that $E(X|\mathcal{G})(\cdot)$ is constant on each set Λ_n.

If $\mathcal{G}_1 \subset \mathcal{G}_2$ and both are generated by countable partitions $\{\Lambda_n^{(1)}, n \geq 1\}$ and $\{\Lambda_n^{(2)}, n \geq 1\}$, then for any $\Lambda_n^{(1)}$, $\Lambda_n^{(1)} \in \mathcal{G}_2$, so there exists an index set $J \subset \{1, 2, \dots\}$ and $\Lambda_n^{(1)} = \sum_{j \in J} \Lambda_j^{(2)}$. Thus, $E(X|\mathcal{G}_1)$ is constant on $\Lambda_n^{(1)}$ but $E(X|\mathcal{G}_2)$ may change values as ω moves from one element of $\{\Lambda_j^{(2)}, j \in J\}$ to another. Thus, as a function, $E(X|\mathcal{G}_1)$ is smoother than $E(X|\mathcal{G}_2)$.

(11) *Projections.* Suppose $\mathcal{G}$ is a sub σ-field of $\mathcal{B}$. Let $L_2(\mathcal{G})$ be the square integrable random variables which are $\mathcal{G}$-measurable. If $X \in L_2(\mathcal{B})$, then $E(X|\mathcal{G})$ is the projection of X onto $L_2(\mathcal{G})$, a subspace of $L_2(\mathcal{B})$. The projection of X onto $L_2(\mathcal{G})$ is the unique element of $L_2(\mathcal{G})$ achieving

$$\inf_{Z \in L_2(\mathcal{G})} \|X - Z\|_2.$$

It is computed by solving the prediction equations (Brockwell and Davis, 1991) for $Z \in L_2(\mathcal{G})$:

$$(Y, X - Z) = 0, \quad \forall Y \in L_2(\mathcal{G}).$$

This says that

$$\int Y(X - Z)dP = 0, \quad \forall Y \in L_2(\mathcal{G}).$$

But trying a solution of $Z = E(X|\mathcal{G})$, we get

$$\begin{aligned}
\int Y(X - Z)dP &= E\left(Y(X - E(X|\mathcal{G}))\right) \\
&= E(YX) - E(YE(X|\mathcal{G})) \\
&= E(YX) - E\left(E(YX|\mathcal{G})\right) \quad (\text{since } Y \in \mathcal{G}) \\
&= E(YX) - E(YX) = 0.
\end{aligned}$$

In time series analysis, $E(X|\mathcal{G})$ is the best predictor of X in $L_2(\mathcal{G})$. It is not often used when $\mathcal{G} = \sigma(X_1, \dots, X_n)$ and $X = X_{n+1}$ because of its lack of linearity and hence its computational difficulty.

(12) *Conditioning and independence.*

(a) If $X \in L_1$, then we claim

$$X \perp\!\!\!\perp \mathcal{G} \text{ implies } E(X|\mathcal{G}) = EX. \tag{10.16}$$

To check this note

(i) $E(X)$ is measurable $\mathcal{G}$.

(ii) For $\Lambda \in \mathcal{G}$,

$$\int_{\Lambda} E(X)dP = E(X)P(\Lambda)$$

and

$$\int_{\Lambda} XdP = E(X1_{\Lambda}) = E(X)P(\Lambda)$$

by independence.

(b) Let $\phi : \mathbb{R}^j \times \mathbb{R}^k \mapsto \mathbb{R}$ be a bounded Borel function. Suppose also that $X : \Omega \mapsto \mathbb{R}^j$, $Y : \Omega \mapsto \mathbb{R}^k$, $X \in \mathcal{G}$ and Y is independent of $\mathcal{G}$. Define

$$f_{\phi}(x) = E(\phi(x, Y)).$$

Then

$$E(\phi(X, Y)|\mathcal{G}) = f_{\phi}(X). \tag{10.17}$$

Proof of (10.17). Case 1. Suppose $\phi = 1_J$, where $J \in \mathcal{B}(\mathbb{R}^j \times \mathbb{R}^k)$.

Case 1a. Suppose $J = K \times L$, where $K \in \mathcal{B}(\mathbb{R}^j)$, and $L \in \mathcal{B}(\mathbb{R}^k)$. Then

$$E(\phi(X, Y)|\mathcal{G}) = P(X \in K, Y \in L|\mathcal{G}),$$

and because $[X \in K] \in \mathcal{G}$, this is

$$= 1_{[X \in K]} P(Y \in L|\mathcal{G}).$$

Since Y is independent of $\mathcal{G}$, this is

$$= 1_{[X \in K]} P[Y \in L] = f_{1_{K \times L}}(X).$$

Case 1b. Let

$$\mathcal{C} = \{J \in \mathcal{B}(\mathbb{R}^j \times \mathbb{R}^k) : \text{(10.17) holds for } \phi = 1_J\}.$$

Then $\mathcal{C} \supset$ RECTS, the measurable rectangles, by Case 1a. We now show $\mathcal{C}$ is a λ-system; that is,

(i) $\mathbb{R}^{j+k} \in \mathcal{C}$, which follows since $\mathbb{R}^{j+k} \in$ RECTS.

(ii) $J \in \mathcal{C}$ implies $J^c \in \mathcal{C}$, which follows since

$$\begin{aligned} P((X, Y) \in J^c|\mathcal{G}) &= 1 - P((X, Y) \in J|\mathcal{G}) \\ &= 1 - f_{1_J}(X) = f_{1_{J^c}}(X). \end{aligned}$$

(iii) If $A_n \in \mathcal{C}$ and A_n are disjoint, we may (but will not) show that $\sum_n A_n \in \mathcal{C}$.

Thus, $\mathcal{C}$ is a λ-system, and $\mathcal{C} \supset$ RECTS . Since RECTS is a π–class Dynkin's theorem implies that

$$\mathcal{C} \supset \sigma(\text{RECTS}) = \mathcal{B}(\mathbb{R}^{j+k}).$$

Case 2. We observe that (10.17) holds for for $\phi = \sum_{i=1}^{k} c_i 1_{J_i}$.
Case 3. We finish the argument with the usual induction. □

(13) *The conditional Jensen's inequality.* Let ϕ be a convex function, $X \in L_1$, and $\phi(X) \in L_1$. Then almost surely

$$\phi\big(E(X|\mathcal{G})\big) \leq E\big(\phi(X)|\mathcal{G}\big).$$

Proof of Jensen's inequality. Take the support line at x_0; it must lie under the graph of ϕ so that

$$\phi(x_0) + \lambda(x_0)(x - x_0) \leq \phi(x) \tag{10.18}$$

where $\lambda(x_0)$ is the slope of the support line through $(x_0, \phi(x_0))$. Replace x_0 by $E(X|\mathcal{G})$ and x by X so that

$$\phi(E(X|\mathcal{G})) + \lambda(E(X|\mathcal{G}))(X - E(X|\mathcal{G})) \leq \phi(X). \tag{10.19}$$

If there are no integrability problems (if!!!), we can take $E(\cdot|\mathcal{G})$ on both sides of (10.19). This yields for LHS, the left side of (10.19),

$$\begin{aligned} E(\text{LHS}|\mathcal{G}) &= \phi(E(X|\mathcal{G})) + E(\lambda(E(X|\mathcal{G}))(X - E(X|\mathcal{G}))|\mathcal{G}) \\ &= \phi(E(X|\mathcal{G})) + \lambda(E(X|\mathcal{G}))E(X - E(X|\mathcal{G}))|\mathcal{G}), \end{aligned}$$

and since $E\big((X - E(X|\mathcal{G}))|\mathcal{G}\big) = 0$, we have

$$= \phi(E(X|\mathcal{G})).$$

For RHS, the right side of (10.19) we have

$$E(\text{RHS}|\mathcal{G}) = E(\phi(X)|\mathcal{G})$$

and thus

$$\phi(E(X|\mathcal{G})) = E(\text{LHS}|\mathcal{G}) \leq E(\text{RHS}|\mathcal{G}) = E(\phi(X)|\mathcal{G}),$$

which is the conditional Jensen inequality.

Note that $\lambda(x)$ can be taken to be the right hand derivative

$$\lim_{h \downarrow 0} \frac{\phi(x+h) - \phi(x)}{h}$$

and so by convexity is non-decreasing in x. If $E(X|\mathcal{G})(\omega)$ were bounded as ω varies, then $\phi(E(X|\mathcal{G}))$ would be bounded and $\lambda(E(X|\mathcal{G}))$ would also be bounded and all terms in (10.19) would be integrable and the result would follow.

Now let

$$X' = X1_{[|E(X|\mathcal{G})|\leq n]}$$

and observe

$$\begin{aligned} E(X'|\mathcal{G}) &= E(X1_{[|E(X|\mathcal{G})|\leq n]}|\mathcal{G}) \\ &= 1_{[|E(X|\mathcal{G})|\leq n]}E(X|\mathcal{G}) \end{aligned}$$

is bounded, so the discussion of Jensen's inequality for the case of bounded conditional expectation applies, and

$$\phi(E(X'|\mathcal{G})) \leq E(\phi(X')|\mathcal{G}).$$

Thus, as $n \to \infty$

$$\begin{aligned} E\big(\phi(X')|\mathcal{G}\big) &= E\left(\phi(X1_{[|E(X|\mathcal{G})|\leq n]})|\mathcal{G}\right) \\ &= E\left(\phi(X)1_{[|E(X|\mathcal{G})|\leq n]} + \phi(0)1_{[|E(X|\mathcal{G})|>n]}|\mathcal{G}\right) \\ &= 1_{[|E(X|\mathcal{G})|\leq n]}E(\phi(X)|\mathcal{G}) + \phi(0)1_{[|E(X|\mathcal{G})|>n]} \\ &\to E(\phi(X)|\mathcal{G}) \end{aligned}$$

Also, as $n \to \infty$,

$$\begin{aligned} \phi(E(X'|\mathcal{G})) &= \phi(1_{[|E(X|\mathcal{G})|\leq n]}E(X|\mathcal{G})) \\ &\to \phi(E(X|\mathcal{G})) \end{aligned}$$

since ϕ is continuous. □

(14) *Conditional expectation is L_p norm reducing and hence continuous.* For $X \in L_p$, define $\|X\|_p = (E|X|^p)^{1/p}$ and suppose $p \geq 1$. Then

$$\|E(X|\mathcal{B})\|_p \leq \|X\|_p. \tag{10.20}$$

and conditional expectation is L_p continuous: If $X_n \stackrel{L_p}{\to} X_\infty$, then

$$E(X_n|\mathcal{B}) \stackrel{L_p}{\to} E(X_\infty|\mathcal{B}). \tag{10.21}$$

Proof of (10.20) **and** (10.21). The inequality (10.20) holds iff

$$(E|E(X|\mathcal{B})|^p)^{1/p} \leq \big(E(|X|^p)\big)^{1/p},$$

that is,

$$E(|E(X|\mathcal{B})|^p) \leq E(|X|^p).$$

From Jensen's inequality

$$\phi(E(X|\mathcal{B})) \leq E(\phi(X)|\mathcal{B})$$

if ϕ is convex. Since $\phi(x) = |x|^p$ is convex for $p \ge 1$, we get

$$\begin{aligned} E|E(X)|\mathcal{B})|^p &= E\phi(E(X|\mathcal{B})) \\ &\le E\,(E(\phi(X)|\mathcal{B})) = E(\phi(X)) \\ &= E(|X|^p). \end{aligned}$$

To prove (10.21), observe that

$$\begin{aligned} \|E(X_n|\mathcal{B}) - E(X_\infty|\mathcal{B})\|_p &= \|E((X_n - X_\infty)|\mathcal{B})\|_p \\ &\le \|X_n - X_\infty\|_p \to 0. \end{aligned}$$

where we have applied (10.20). □

10.4 Martingales

Suppose we are given integrable random variables $\{X_n, n \ge 0\}$ and σ-fields $\{\mathcal{B}_n, n \ge 0\}$ which are sub σ-fields of $\mathcal{B}$. Then $\{(X_n, \mathcal{B}_n), n \ge 0\}$ is a *martingale* (mg) if

(i) Information accumulates as time progresses in the sense that

$$\mathcal{B}_0 \subset \mathcal{B}_1 \subset \mathcal{B}_2 \subset \cdots \subset \mathcal{B}.$$

(ii) X_n is adapted in the sense that for each n, $X_n \in \mathcal{B}_n$; that is, X_n is $\mathcal{B}_n$-measurable.

(iii) For $0 \le m < n$,

$$E(X_n|\mathcal{B}_m) \overset{a.s.}{=} X_m.$$

If in (iii) equality is replaced by $\ge$; that is, things are getting better on the average:

$$E(X_n|\mathcal{B}_m) \overset{a.s.}{\ge} X_m,$$

then $\{X_n\}$ is called a *submartingale* (submg) while if things are getting worse on the average

$$E(X_n|\mathcal{B}_m) \overset{a.s.}{\le} X_m,$$

$\{X_n\}$ is called a *supermartingale* (supermg).

Here are some *elementary remarks*:

(i) $\{X_n\}$ is martingale if it is both a sub and supermartingale. $\{X_n\}$ is a supermartingale iff $\{-X_n\}$ is a submartingale.

(ii) By Lemma 10.1.1, postulate (iii) holds iff

$$\int_\Lambda X_n = \int_\Lambda X_m, \quad \forall \Lambda \in \mathcal{B}_m.$$

Similarly for the inequality versions of (iii).

(iii) Postulate (iii) could be replaced by

$$E(X_{n+1}|\mathcal{B}_n) = X_n, \quad \forall n \geq 0, \tag{iii'}$$

by the smoothing equality. For example, assuming (iii′)

$$E(X_{n+2}|\mathcal{B}_n) = E(E(X_{n+2}|\mathcal{B}_{n+1})|\mathcal{B}_n) = E(X_{n+1}|\mathcal{B}_n) = X_n.$$

(iv) If $\{X_n\}$ is a martingale, then $E(X_n)$ is constant. In the case of a submartingale, the mean increases and for a supermartingale, the mean decreases.

(v) If $\{(X_n, \mathcal{B}_n), n \geq 0\}$ is a (sub, super) martingale, then

$$\{X_n, \sigma(X_0, \ldots, X_n), n \geq 0\}$$

is also a (sub, super) martingale.

The reason for this is that since $X_n \in \mathcal{B}_n$,

$$\sigma(X_0, \ldots, X_n) \subset \mathcal{B}_n$$

and by smoothing

$$\begin{aligned} E(X_{n+1}|\sigma(X_0, \ldots, X_n)) &= E\left(E(X_{n+1}|\mathcal{B}_n)|X_0, \ldots, X_n\right) \\ &= E(X_n|X_0, \ldots, X_n) = X_n. \end{aligned}$$

Why do we need $\mathcal{B}_n$? Why not just condition on $\sigma(X_0, \ldots, X_n)$? Sometimes it is convenient to carry along auxiliary information.

(vi) *Martingale differences.* Call $\{(d_j, \mathcal{B}_j), j \geq 0\}$ a (sub, super) *martingale difference sequence* or a (sub, super) *fair sequence* if

(i) For $j \geq 0$, $\mathcal{B}_j \subset \mathcal{B}_{j+1}$.

(ii) For $j \geq 0$, $d_j \in L_1$, $d_j \in \mathcal{B}_j$.

(iii) For $j \geq 0$

$$\begin{aligned} E(d_{j+1}|\mathcal{B}_j) &= 0, && \text{(fair)} \\ &\geq 0, && \text{(subfair)} \\ &\leq 0, && \text{(superfair).} \end{aligned}$$

Here are the basic facts about martingale differences:

(a) If $\{(d_j, \mathcal{B}_j), j \geq 0\}$ is (sub, super) fair, then

$$\{(X_n := \sum_{j=0}^{n} d_j, \mathcal{B}_n), n \geq 0\}$$

is a (sub, super) martingale.

(b) Suppose $\{(X_n, \mathcal{B}_n), n \geq 0\}$ is a (sub, super) martingale. Define

$$d_0 = X_0 - E(X_0), \ d_j = X_j - X_{j-1}, \ j \geq 1.$$

Then $\{(d_j, \mathcal{B}_j), j \geq 0)$ is a (sub, super) fair sequence.

We now check facts (a) and (b). For (a), we have for instance in the case that $\{(d_j, \mathcal{B}_j), j \geq 0\}$ is assumed fair that

$$E\left(\sum_{j=0}^{n+1} d_j|\mathcal{B}_n\right) = E(d_{n+1}|\mathcal{B}_n) + E\left(\sum_{j=1}^{n} d_j|\mathcal{B}_n\right) = 0 + \sum_{j=1}^{n} d_j,$$

which verifies the martingale property.

For (b), observe that if $\{X_n\}$ is a martingale, then

$$E\big((X_j - X_{j-1})|\mathcal{B}_{j-1}\big) = E(X_j|\mathcal{B}_{j-1}) - X_{j-1} = X_{j-1} - X_{j-1} = 0.$$

(vii) *Orthogonality of martingale differences.* If $\{(X_n = \sum_{j=0}^{n} d_j, \mathcal{B}_n), n \geq 0\}$ is a martingale and $E(d_j^2) < \infty, j \geq 0$, then $\{d_j\}$ are orthogonal:

$$Ed_i d_j = 0, \quad i \neq j.$$

This is an easy verification: If $j > i$, then

$$\begin{aligned} E(d_i d_j) &= E\big(E(d_i d_j|\mathcal{B}_i)\big) \\ &= E\big(d_i E(d_j|\mathcal{B}_i)\big) = 0. \end{aligned}$$

A consequence is that

$$E(X_n^2) = E(\sum_{j=1}^{n} d_j^2) + 2 \sum_{0 \leq i < j \leq n} E(d_i d_j) = E(\sum_{j=1}^{n} d_j^2),$$

which is non-decreasing. From this, it seems likely (and turns out to be true) that $\{X_n^2\}$ is a sub-martingale.

Historical note: Volume 6 of the Oxford English Dictionary gives the following entries for the term *martingale* and comments it is a word of *obscure entymology*.

- A strap or arrangement of straps fastened at one end to the noseband, bit or reins and at the other to the girth to prevent a horse from rearing or throwing back his head.
- A rope for guying down the jib–boom to the dolphin–striker.
- A system in gambling which consists in doubling the stake when losing in the hope of eventually recouping oneself.

10.5 Examples of Martingales

Those most skilled in applying the economy and power of martingale theory in stochastic process modeling are those wizards able to find martingales in surprising circumstances. It is thus crucial for someone trying to master this subject to study as many examples as possible of where martingales arise. In this section, we list some of the common examples.

(1) *Martingales and smoothing.* Suppose $X \in L_1$ and $\{\mathcal{B}_n, n \geq 0\}$ is an increasing family of sub σ-fields of $\mathcal{B}$. Define for $n \geq 0$

$$X_n := E(X|\mathcal{B}_n).$$

Then

$$\{(X_n, \mathcal{B}_n), n \geq 0\}$$

is a martingale.

Verification is easy:

$$\begin{aligned} E(X_{n+1}|\mathcal{B}_n) &= E(E(X|\mathcal{B}_{n+1})|\mathcal{B}_n) \\ &= E(X|\mathcal{B}_n) \qquad \text{(smoothing)} \\ &= X_n. \end{aligned}$$

(2) *Martingales's and sums of independent random variables.* Suppose that $\{Z_n, n \geq 0\}$ is an independent sequence of integrable random variables satisfying for $n \geq 0$, $E(Z_n) = 0$. Set $X_0 = 0$, $X_n = \sum_{i=1}^n Z_i$, $n \geq 1$, and $\mathcal{B}_n := \sigma(Z_0, \ldots, Z_n)$. Then $\{(X_n, \mathcal{B}_n), n \geq 0\}$ is a martingale since $\{(Z_n, \mathcal{B}_n), n \geq 0\}$ is a fair sequence.

(3) *New martingale's from old, transforms, discrete stochastic integration.* Let $\{(d_j, \mathcal{B}_j), j \geq 0\}$ be martingale differences. Let $\{U_j\}$ be *predictable.* This means that U_j is measurable with respect to the prior σ-field; that is $U_0 \in \mathcal{B}_0$ and

$$U_j \in \mathcal{B}_{j-1}, \quad j \geq 1.$$

To avoid integrability problems, suppose $U_j \in L_\infty$ which means that U_j is bounded. Then $\{(U_j d_j, \mathcal{B}_j), n \geq 1\}$ is still a fair sequence since

$$\begin{aligned} E(U_j d_j|\mathcal{B}_{j-1}) &= U_j E(d_j|\mathcal{B}_{j-1}) \qquad \text{(since } U_j \in \mathcal{B}_{j-1}) \\ &= U_j \cdot 0 = 0. \end{aligned} \tag{10.22}$$

We conclude that $\{(\sum_{j=0}^n U_j d_j, \mathcal{B}_n), n \geq 0\}$ is a martingale.

In gambling models, d_j might be ± 1 and U_j is how much you gamble so that U_j is a strategy based on previous gambles. In investment models, d_j might be the change in price of a risky asset and U_j is the number of shares of the asset held by the investor. In stochastic integration, the $d_j' s$ are increments of Brownian motion.

The notion of the martingale transform is formalized in the following simple result.

Lemma 10.5.1 *Suppose* $\{(M_n, \mathcal{B}_n), n \in \mathbb{N}\}$ *is an adapted integrable sequence so that* $M_n \in \mathcal{B}_n$. *Define* $d_0 = M_0$, *and* $d_n = M_n - M_{n-1}$, $n \geq 1$. *Then* $\{(M_n, \mathcal{B}_n), n \in \mathbb{N}\}$ *is a martingale iff for every bounded predictable sequence* $\{U_n, n \in \mathbb{N}\}$ *we have*

$$E(\sum_{n=0}^{N} U_n d_n) = 0, \quad \forall N \geq 0. \tag{10.23}$$

Proof. If $\{(M_n, \mathcal{B}_n), n \in \mathbb{N}\}$ is a martingale, then (10.23) follows from (10.22). Conversely, suppose (10.23) holds. For $j \geq 0$, let $A \in \mathcal{B}_j$ and define $U_n = 0, n \neq j+1$, and $U_{j+1} = 1_{A_j}$. Then $\{U_n, n \in \mathbb{N}\}$ is bounded and predictable, and hence from (10.23) we get

$$0 = E(\sum_{n=1}^{N} U_n d_n) = E(U_{j+1} d_{j+1}) = E(1_{A_j} d_{j+1})$$

so that

$$0 = \int_{A_j} d_{j+1} dP = \int_{A_j} E(d_{j+1} | \mathcal{B}_j) dP.$$

Hence, from the Integral Comparison Lemma 10.1.1 we conclude that $E(d_{j+1}|\mathcal{B}_j) = 0$ almost surely. So $\{(d_n, \mathcal{B}_n), n \in \mathbb{N}\}$ is a martingale difference and the result follows. □

(4) *Generating functions, Laplace transforms, chf's etc.* Let $\{Z_n, n \geq 1\}$ be iid random variables. The construction we are about to describe works for a variety of transforms; for concreteness we suppose that Z_n has range $\{0, 1, 2, \dots\}$ and we use the generating function as our typical transform. Define

$$\mathcal{B}_0 = \{\emptyset, \Omega\}, \quad \mathcal{B}_n = \sigma(Z_1, \dots, Z_n), \quad n \geq 1$$

and let the generating function of the Z's be

$$\phi(s) = Es^{Z_1}, \quad 0 \leq s \leq 1.$$

Define $M_0 = 1$, fix $s \in (0, 1)$, set $S_0 = 0$, $S_n = \sum_{i=1}^{n} Z_i$, $n \geq 1$ and

$$M_n = \frac{s^{S_n}}{\phi^n(s)}, \quad n \geq 0.$$

Then $\{(M_n, \mathcal{B}_n), n \geq 0\}$ is a martingale. This is a straightforward verification:

$$\begin{aligned} E\left(s^{S_{n+1}}|\mathcal{B}_n\right) &= E\left(s^{Z_{n+1}} s^{S_n}|\mathcal{B}_n\right) \\ &= s^{S_n} E\left(s^{Z_{n+1}}|\mathcal{B}_n\right) \\ &= s^{S_n} E\left(s^{Z_{n+1}}\right) \qquad \text{(independence)} \\ &= s^{S_n} \phi(s). \end{aligned}$$

So therefore

$$E\left(\frac{s^{S_{n+1}}}{\phi^{n+1}(s)}|\mathcal{B}_n\right) = \frac{s^{S_n}}{\phi^n(s)}$$

which is equivalent to the assertion.

(5) *Method of centering by conditional means.* Let $\{\xi_n, n \geq 1\}$ be an arbitrary sequence of L_1 random variables. Define

$$\mathcal{B}_j = \sigma(\xi_1, \ldots, \xi_j),\ j \geq 1; \quad \mathcal{B}_0 = \{\emptyset, \Omega\}.$$

Then

$$\left\{\left((\xi_j - E(\xi_j|\mathcal{B}_{j-1})), \mathcal{B}_j\right) j \geq 1\right\}$$

is a fair sequence since

$$E\left((\xi_j - E(\xi_j|\mathcal{B}_{j-1}))|\mathcal{B}_{j-1}\right) = E(\xi_j|\mathcal{B}_{j-1}) - E(\xi_j|\mathcal{B}_{j-1}) = 0.$$

So

$$X_0 = 0, \quad X_n = \sum_{j=1}^{n} (\xi_j - E(\xi_j|\mathcal{B}_{j-1}))$$

is a martingale.

(6) *Connections with Markov chains.* Suppose $\{Y_n, n \geq 0\}$ is a Markov Chain whose state space is the integers with transition probability matrix $P = (p_{ij})$. Let f be an eigenvector corresponding to eigenvalue λ; that is, in matrix notation

$$Pf = \lambda f.$$

In component form, this is

$$\sum_j p_{ij} f(j) = \lambda f(i).$$

In terms of expectations, this is

$$E(f(Y_{n+1})|Y_n = i) = \lambda f(i)$$

or

$$E(f(Y_{n+1})|Y_n) = \lambda f(Y_n)$$

and by the Markov property this is

$$E(f(Y_{n+1})|Y_n) = E(f(Y_{n+1})|Y_0, \ldots, Y_n) = \lambda f(Y_n).$$

So we conclude that

$$\left\{\left(\frac{f(Y_n)}{\lambda^n}, \sigma(Y_0, \ldots, Y_n)\right), n \geq 0\right\}$$

is a martingale.

A special case is the *simple branching process*. Suppose $\{p_k, k \geq 0\}$ is the offspring distribution so that p_k represents the probability of k offspring per individual. Let $m = \sum_k kp_k$ be the mean number of offspring per individual. Let $\{Z^{(n)}(i), n \geq 0, i \geq 1\}$ be an iid sequence whose common mass function is the offspring distribution $\{p_k\}$ and define recursively $Z_0 = 1$ and

$$Z_{n+1} = \begin{cases} Z^{(n)}(1) + \cdots + Z^{(n)}(Z_n), & \text{if } Z_n > 0, \\ 0, & \text{if } Z_n = 0, \end{cases}$$

which represents the number in the $(n+1)$- generation. Then $\{Z_n\}$ is a Markov chain and

$$p_{ij} := P[Z_{n+1} = j | Z_n = i] = \begin{cases} \delta_{0j}, & \text{if } i = 0, \\ p_j^{*i}, & \text{if } i \geq 1, \end{cases}$$

where for $i \geq 1$, p_j^{*i} is the jth component of the i-fold convolution of the sequence $\{p_n\}$. Note for $i \geq 1$

$$\sum_{j=0}^{\infty} p_{ij} j = \sum_{j=1}^{\infty} p_j^{*i} j = im,$$

while for $i = 0$,

$$\sum_{j=0}^{\infty} p_{ij} j = p_{00} \cdot 0 + 0 = 0 = mi.$$

With $f(j) = j$ we have $Pf = mf$. This means that the process

$$\{(Z_n/m^n, \sigma(Z_0, \dots, Z_n)), n \geq 0\} \tag{10.24}$$

is a martingale.

(7) *Likelihood ratios.* Suppose $\{Y_n, n \geq 0\}$ are iid random variables and suppose the true density of Y_1 is f_0. (The word "density" can be understood with respect to some fixed reference measure μ.) Let f_1 be some other probability density. For simplicity suppose $f_0(y) > 0$, for all y. Then for $n \geq 0$

$$X_n = \frac{\prod_{i=0}^n f_1(Y_i)}{\prod_{i=0}^n f_0(Y_i)}$$

is a martingale since

$$\begin{aligned} E(X_{n+1}|Y_0, \dots, Y_n) &= E\left(\left(\frac{\prod_{i=0}^n f_1(Y_i)}{\prod_{i=0}^n f_0(Y_i)}\right) \frac{f_1(Y_{n+1})}{f_0(Y_{n+1})} |Y_0, \dots, Y_n\right) \\ &= X_n E\left(\frac{f_1(Y_{n+1})}{f_0(Y_{n+1})} |Y_0, \dots, Y_n\right). \end{aligned}$$

By independence this becomes

$$= X_n E\left(\frac{f_1(Y_{n+1})}{f_0(Y_{n+1})}\right) = X_n \int \frac{f_1(y)}{f_0(y)} f_0(y)\mu(dy)$$
$$= X_n \int f_1 d\mu = X_n \cdot 1 = X_n$$

since f_1 is a density.

10.6 Connections between Martingales and Submartingales

This section describes some connections between martingales and submartingales by means of what is called the *Doob decomposition* and also some simple results arising from Jensen's inequality.

10.6.1 Doob's Decomposition

The Doob decomposition expresses a submartingale as the sum of a martingale and an *increasing process*. This latter phrase has a precise meaning.

Definition. Given a process $\{U_n, n \geq 0\}$ and σ-fields $\{\mathcal{B}_n, n \geq 0\}$. We call $\{U_n, n \geq 0\}$ *predictable* if $U_0 \in \mathcal{B}_0$, and for $n \geq 0$, we have

$$U_{n+1} \in \mathcal{B}_n.$$

Call a process $\{A_n, n \geq 0\}$ an *increasing* process if $\{A_n\}$ is predictable and almost surely

$$0 = A_0 \leq A_1 \leq A_2 \leq \cdots.$$

Theorem 10.6.1 (Doob Decomposition) *Any submartingale*

$$\{(X_n, \mathcal{B}_n), n \geq 0\}$$

can be written in a unique way as the sum of a martingale

$$\{(M_n, \mathcal{B}_n), n \geq 0\}$$

and an increasing process $\{A_n, n \geq 0\}$*; that is*

$$X_n = M_n + A_n, \quad n \geq 0.$$

Proof. (a) Existence of such a decomposition: Define

$$d_0^{\#} = X_0, \quad d_j^{\#} = X_j - E(X_j|\mathcal{B}_{j-1}), \quad j \geq 1$$
$$M_n := \sum_{j=0}^{n} d_j^{\#}.$$

Then $\{M_n\}$ is a martingale since $\{d_j^{\#}\}$ is a fair sequence. Set $A_n = X_n - M_n$. Then $A_0 = X_0 - M_0 = X_0 - X_0 = 0$, and

$$\begin{aligned} A_{n+1} - A_n &= X_{n+1} - M_{n+1} - X_n + M_n \\ &= X_{n+1} - X_n - (M_{n+1} - M_n) \\ &= X_{n+1} - X_n - d_{n+1}^{\#} \\ &= X_{n+1} - X_n - X_{n+1} + E(X_{n+1}|\mathcal{B}_n) \\ &= E(X_{n+1}|\mathcal{B}_n) - X_n \geq 0 \end{aligned}$$

by the submartingale property. Since

$$A_{n+1} = \sum_{j=0}^{n} (A_{j+1} - A_j) = \sum_{j=0}^{n} (E(X_{j+1}|\mathcal{B}_j) - X_j) \in \mathcal{B}_n,$$

this shows $\{A_n\}$ is predictable and hence increasing.

(b) Uniqueness of the decomposition: Suppose

$$X_n = M_n + A_n,$$

and that there is also another decomposition

$$X_n = M_n' + A_n'$$

where $\{M_n'\}$ is a martingale and $\{A_n'\}$ is an increasing process. Then

$$A_n' = X_n - M_n', \quad A_n = X_n - M_n,$$

and

$$A_{n+1}' - A_n' = X_{n+1} - X_n - (M_{n+1}' - M_n').$$

Because $\{A_n'\}$ is predictable and $\{M_n'\}$ is a martingale,

$$A_{n+1}' - A_n' = E(A_{n+1}' - A_n'|\mathcal{B}_n) = E(X_{n+1}|\mathcal{B}_n) - X_n - 0$$

and

$$A_{n+1} - A_n = E(A_{n+1} - A_n|\mathcal{B}_n) = E(X_{n+1}|\mathcal{B}_n) - X_n.$$

Thus, remembering $A_0 = A_0' = 0$,

$$\begin{aligned} A_n &= A_0 + (A_1 - A_0) + \cdots + (A_n - A_{n-1}) \\ &= A_0' + (A_1' - A_0') + \cdots + (A_n' - A_{n-1}') = A_n', \end{aligned}$$

therefore also

$$M_n = X_n - A_n = X_n - A'_n = M'_n. \qquad \square$$

We now discuss some simple relations between martingales and submartingales which arise as applications of Jensen's inequality.

Proposition 10.6.2 (Relations from Jensen) *(a) Let*

$$\{(X_n, \mathcal{B}_n), n \geq 0\}$$

be a martingale and suppose ϕ is a convex function satisfying

$$E(|\phi(X_n)|) < \infty.$$

Then

$$\{(\phi(X_n), \mathcal{B}_n), n \geq 0\}$$

is a submartingale.

(b) Let $\{(X_n, \mathcal{B}_n), n \geq 0\}$ be a submartingale and suppose ϕ is convex and non-decreasing and $E(|\phi(X_n)|) < \infty$. Then

$$\{(\phi(X_n), \mathcal{B}_n), n \geq 0\}$$

is submartingale.

Proof. If $n < m$ and ϕ is non-decreasing and the process is a submartingale, then

$$\begin{aligned} \phi(X_n) \leq& \phi(E(X_m|\mathcal{B}_n)) && \text{(submartingale property)} \\ \leq& E(\phi(X_m)|\mathcal{B}_n) && \text{(Jensen).} \end{aligned}$$

The case where the process is a martingale is easier. $\square$

Example 10.6.1 Let $\{(X_n, \mathcal{B}_n), n \geq 0\}$ be martingale. Suppose ϕ is one of the following functions:

$$\phi(x) = |x|, x^2, x^+, x^-, x \vee a.$$

Then

$$\{|X_n|\}, \{X_n^2\}, \{X_n^+\}, \{X_n^-\}, \{X_n \vee a\}$$

are all submartingales, provided they satisfy $E(|\phi(X_n)|) < \infty$.

Example 10.6.2 (Doob decomposition of X_n^2) Let $\{(X_n, \mathcal{B}_n), n \geq 0\}$ be a martingale and suppose $E(X_n^2) < \infty$. Then $\{(X_n^2, \mathcal{B}_n), n \geq 0\}$ is a submartingale. What is its Doob decomposition?

Recall

$$d_0^\# = X_0^2, \quad d_j^\# = X_j^2 - E\left(X_j^2|\mathcal{B}_{j-1}\right), \quad M_n = \sum_{j=0}^{n} d_j^\#,$$

from the previous construction in Theorem 10.6.1 and

$$A_j = X_j^2 - M_j, \quad A_{n+1} - A_n = E(X_{n+1}^2 | \mathcal{B}_n) - X_n^2.$$

Write $X_n = \sum_{j=0}^n d_j$ (note the distinction between d_j and $d_j^\#$) and because $\{X_n\}$ is a martingale, $\{d_j\}$ is fair, so that

$$X_j^2 = (X_{j-1} + d_j)^2 = X_{j-1}^2 + 2X_{j-1}d_j + d_j^2$$

and

$$E(X_j^2 | \mathcal{B}_{j-1}) = X_{j-1}^2 + 2X_{j-1}E(d_j | \mathcal{B}_{j-1}) + E(d_j^2 | \mathcal{B}_{j-1}).$$

Remembering $E(d_j | \mathcal{B}_{j-1}) = 0$ yields

$$A_{n+1} - A_n = X_n^2 + E(d_{n+1}^2 | \mathcal{B}_n) - X_n^2 = E(d_{n+1}^2 | \mathcal{B}_n)$$

and

$$A_n = \sum_{j=1}^n E(d_j^2 | \mathcal{B}_{j-1}).$$

Therefore, the Doob Decomposition of the submartingale $\{X_n^2\}$ is

$$\begin{aligned} X_n^2 &= X_n^2 - \sum_{j=1}^n E(d_j^2 | \mathcal{B}_{j-1}) + \sum_{j=1}^n E(d_j^2 | \mathcal{B}_{j-1}) \\ &= M_n + A_n. \end{aligned}$$

Note $E(d_j^2 | \mathcal{B}_{j-1})$ is the conditional variance of the martingale increment. Also, if $\{d_j\}$ is an independent sequence, then

$$E(d_j^2 | \mathcal{B}_{j-1}) = E(d_j^2) = \mathrm{Var}(d_j^2),$$

and

$$\mathrm{Var}(X_n) = \mathrm{Var}(A_n).$$

10.7 Stopping Times

Let $\mathbb{N} = \{0, 1, 2, \dots\}$, $\overline{\mathbb{N}} = \{0, 1, 2, \dots, \infty\}$ and suppose $\mathcal{B}_n \subset \mathcal{B}_{n+1}, n \in \mathbb{N}$ is an increasing family of σ-fields.

Definition. A mapping $\nu : \Omega \mapsto \overline{\mathbb{N}}$ is a *stopping time* if

$$[\nu = n] \in \mathcal{B}_n, \quad \forall n \in \mathbb{N}.$$

To fix ideas, imagine a sequence of gambles. Then ν is the rule for when to stop and $\mathcal{B}_n$ is the information accumulated up to time n. You decide whether or not to stop after the nth gamble based on information available up to and including the nth gamble.

Note ν can be $+\infty$. If ν is a waiting time for an event and the event never happens, then it is natural to characterize the waiting time as infinite and hence $\nu = \infty$.

Define

$$\mathcal{B}_\infty = \bigvee_{n\in\mathbb{N}} \mathcal{B}_n = \sigma(\mathcal{B}_n, n \in \mathbb{N}),$$

so that $\mathcal{B}_\infty$ is the smallest σ-field containing all $\mathcal{B}_n$, $n \in \mathbb{N}$. Then

$$[\nu = \infty] = [\nu < \infty]^c = \left(\bigcup_{n\in\mathbb{N}} [\nu = n]\right)^c = \bigcap_{n\in\mathbb{N}} [\nu = n]^c \in \mathcal{B}_\infty.$$

Requiring

$$[\nu = n] \in \mathcal{B}_n, \quad n \in \mathbb{N}$$

implies

$$[\nu = n] \in \mathcal{B}_n, \quad n \in \overline{\mathbb{N}}.$$

Example: Hitting times. Let $\{(X_n, \mathcal{B}_n), n \in \mathbb{N}\}$ be any *adapted process*, meaning $\mathcal{B}_n \subset \mathcal{B}_{n+1}$ and $X_n \in \mathcal{B}_n$ for all $n \in \mathbb{N}$. For $A \in \mathcal{B}(\mathbb{R})$, define

$$\nu = \inf\{n \in \mathbb{N} : X_n \in A\},$$

with the convention that $\inf \emptyset = \infty$. Then ν is a stopping time since for $n \in \mathbb{N}$

$$[\nu = n] = [X_0 \notin A, \ldots, X_{n-1} \notin A, X_n \in A] \in \mathcal{B}_n.$$

If ν is a stopping time, define $\mathcal{B}_\nu$, the σ–field of information up to time ν as

$$\mathcal{B}_\nu = \{B \in \mathcal{B}_\infty : \forall n \in \mathbb{N}, \ [\nu = n] \cap B \in \mathcal{B}_n\}.$$

So $\mathcal{B}_\nu$ consists of all events that have the property that adding the information of when ν occurred, places the intersection in the appropriate σ–field. One can check that $\mathcal{B}_\nu$ is a σ-field. By definition $\mathcal{B}_\nu \subset \mathcal{B}_\infty$.

BASIC FACTS:

1. If $\nu \equiv k$, ν is a stopping time and $\mathcal{B}_\nu = \mathcal{B}_k$.
2. If ν is a stopping time and $B \in \mathcal{B}_\nu$, then $B \cap [\nu = \infty] \in \mathcal{B}_\infty$, and hence $B \cap [\nu = n] \in \mathcal{B}_n$ for $n \in \overline{\mathbb{N}}$. To see this, note

$$B \cap [\nu = \infty] = B \cap [\nu < \infty]^c = B \cap \left(\bigcup_{n \in \mathbb{N}} [\nu = n] \right)^c = \bigcap_{n \in \mathbb{N}} B[\nu \neq n].$$

Since $B \in \mathcal{B}_\nu \subset \mathcal{B}_\infty$ and $[\nu \neq n] = [\nu = n]^c \in \mathcal{B}_n \subset \mathcal{B}_\infty$, we have $B \cap [\nu = \infty] \in \mathcal{B}_\infty$.

3. We have $\nu \in \mathcal{B}_\infty$, and $\nu \in \mathcal{B}_\nu$.
4. ν is a stopping time iff $[\nu \leq n] \in \mathcal{B}_n$, $n \in \mathbb{N}$ iff $[\nu > n] \in \mathcal{B}_n$, $n \in \mathbb{N}$.

We verify this as follows: Observe that

$$[\nu \leq n] = \bigcup_{0 \leq j \leq n} [\nu = j],$$

so

$$[\nu = n] = [\nu \leq n] - [\nu \leq n - 1],$$

and

$$[\nu > n] = [\nu \leq n]^c.$$

5. If $B \in \mathcal{B}_\infty$, then $B \in \mathcal{B}_\nu$ iff

$$B \cap [\nu \leq n] \in \mathcal{B}_n, \quad \forall n \in \mathbb{N}.$$

(Warning: If ν is a stopping time, it is *false* in general that if $B \in \mathcal{B}_\nu$ then $B \cap [\nu > n] \in \mathcal{B}_n$.)

6. If $\{\nu_k\}$ are stopping times, then $\vee_k \nu_k$ and $\wedge_k \nu_k$ are stopping times.

This follows since

$$[\vee_k \nu_k \leq n] = \bigcap_k [\nu_k \leq n] \in \mathcal{B}_n, \quad \forall n \in \mathbb{N}$$

since $[\nu_k \leq n] \in \mathcal{B}_n$ for every k. Likewise

$$[\wedge_k \nu_k > n] = \bigcap_k [\nu_k > n] \in \mathcal{B}_n.$$

7. If $\{\nu_k\}$ is a monotone family of stopping times, $\lim_{k \to \infty} \nu_k$ is a stopping time, since the limit is $\vee_k \nu_k$ or $\wedge_k \nu_k$.
8. If $\nu_i, i = 1, 2$ are stopping times, so is $\nu_1 + \nu_2$.

Example. If ν is a stopping time $\nu_n = \nu \wedge n$ is a stopping time (which is bounded), since both ν and n are stopping times.

We now list some facts concerning the comparison of two stopping times ν and ν'.

1. Each of the events $[\nu < \nu']$, $[\nu = \nu']$, $[\nu \leq \nu']$ belong to $\mathcal{B}_\nu$ and $\mathcal{B}_{\nu'}$.

2. If $B \in \mathcal{B}_\nu$, then
$$B \cap [\nu \leq \nu'] \in \mathcal{B}_{\nu'}, \quad B \cap [\nu < \nu'] \in \mathcal{B}_{\nu'}.$$

3. If $\nu \leq \nu'$ on Ω, then $\mathcal{B}_\nu \subset \mathcal{B}_{\nu'}$.

To verify these facts, we first prove $[\nu < \nu'] \in \mathcal{B}_\nu$. We have that
$$[\nu < \nu'] \cap [\nu = n] = [n < \nu'] \cap [\nu = n] \in \mathcal{B}_n,$$
since $[n < \nu'] \in \mathcal{B}_n$ and $[\nu = n] \in \mathcal{B}_n$.

Next, we verify that $[\nu = \nu'] \in \mathcal{B}_\nu$. We have that
$$[\nu = \nu'] \cap [\nu = n] = [n = \nu'] \cap [\nu = n] \in \mathcal{B}_n$$
and therefore $[\nu \leq \nu'] = [\nu < \nu'] \cup [\nu = \nu'] \in \mathcal{B}_\nu$. The rest follows by symmetry or complementation.

Now we prove 2. For any n
$$B \cap [\nu \leq \nu'] \cap [\nu' = n] = (B \cap [\nu \leq n]) \cap [\nu' = n] \in \mathcal{B}_n,$$
since $B \cap [\nu \leq n] \in \mathcal{B}_n$ and $[\nu' = n] \in \mathcal{B}_n$.

Proof of 3. This follows from the second assertion since $[\nu \leq \nu'] = \Omega$. □

10.8 Positive Super Martingales

Suppose $\{(X_n, \mathcal{B}_n), n \geq 0\}$ is a positive supermartingale so that $X_n \geq 0$, $X_n \in \mathcal{B}_n$ and $E(X_{n+1}|\mathcal{B}_n) \leq X_n$. In this section we consider the following questions.

1. When does $\lim_{n\to\infty} X_n$ exist? In what sense does convergence take place if some form of convergence holds? Since supermartingales tend to decrease, at least on the average, one expects that under reasonable conditions, supermartingales bounded below by 0 should converge.

2. Is fairness preserved under random stopping? If $\{X_n\}$ is a martingale, we know that we have constant mean; that is $E(X_n) = E(X_0)$. Is $E(X_\nu) = E(X_0)$ for some reasonable class of stopping times ν?

When it holds, preservation of the mean under random stopping is quite useful. However, we can quickly see that preservation of the mean under random stopping does not always hold. Let $\{X_0 = 0, X_n = \sum_{i=1}^n Y_i, n \geq 1\}$. be the Bernoulli random walk so that $\{Y_i, i \geq 1\}$ are iid and

$$P[Y_i = \pm 1] = \frac{1}{2}, \quad i \geq 1.$$

Let

$$\nu = \inf\{n \geq 1 : X_n = 1\}$$

be the first time the random walks hits 1. Standard Markov chain analysis (for example, see Resnick, 1992, Chapter 1) asserts that $P[\nu < \infty] = 1$. But $X_\nu = 1$ so that $E(X_\nu) = 1 \neq E(X_0) = 0$ and therefore $E(X_\nu) \neq E(X_0)$. Thus, for random stopping to preserve the process mean, we need restrictions either on $\{X_n\}$ or on ν or both.

10.8.1 Operations on Supermartingales

We consider two transformations of supermartingales which yield supermartingales.

Proposition 10.8.1 (Pasting of supermartingales) *For $i = 1, 2$, let*

$$\{(X_n^{(i)}, \mathcal{B}_n), n \geq 0\}$$

be positive supermartingales. Let ν be a stopping time such that on $[\nu < \infty]$, we have $X_\nu^{(1)}(\omega) \geq X_\nu^{(2)}(\omega)$. Define

$$X_n(\omega) = \begin{cases} X_n^{(1)}(\omega), & \text{if } n < \nu(\omega) \\ X_n^{(2)}(\omega), & \text{if } n \geq \nu(\omega). \end{cases}$$

Then $\{(X_n, \mathcal{B}_n), n \geq 0\}$ is a new positive supermartingale, called the pasted supermartingale.

Remark. The idea is to construct something which tends to decrease. The segments before and after ν tend to decrease. Moving from the first process to the second at time ν causes a decrease as well.

Proof. Write

$$X_n = X_n^{(1)} 1_{[n<\nu]} + X_n^{(2)} 1_{[n \geq \nu]}.$$

From this we conclude $X_n \in \mathcal{B}_n$. Also, since each $\{(X_n^{(i)}, \mathcal{B}_n), n \geq 0\}$ is a supermartingale,

$$\begin{aligned} X_n &\geq E\left(X_{n+1}^{(1)} | \mathcal{B}_n\right) 1_{[n<\nu]} + E\left(X_{n+1}^{(2)} | \mathcal{B}_n\right) 1_{[n \geq \nu]} \\ &= E\left((X_{n+1}^{(1)} 1_{[n<\nu]} + X_{n+1}^{(2)} 1_{[n \geq \nu]}) | \mathcal{B}_n\right). \end{aligned} \tag{10.25}$$

However, $X_n^{(1)} \geq X_n^{(2)}$ on the set $[\nu = n]$ so

$$\begin{aligned} X_{n+1}^{(1)} 1_{[n<\nu]} + X_{n+1}^{(2)} 1_{[n\geq\nu]} &= X_{n+1}^{(1)} 1_{[\nu>n+1]} + X_{n+1}^{(1)} 1_{[\nu=n+1]} + X_{n+1}^{(2)} 1_{[n\geq\nu]} \\ &\geq X_{n+1}^{(1)} 1_{[\nu>n+1]} + X_{n+1}^{(2)} 1_{[\nu=n+1]} + X_{n+1}^{(2)} 1_{[\nu\leq n]} \\ &= X_{n+1}^{(1)} 1_{[\nu>n+1]} + X_{n+1}^{(2)} 1_{[\nu\leq n+1]} \\ &= X_{n+1}. \end{aligned}$$

From (10.25) $X_n \geq E(X_{n+1}|\mathcal{B}_n)$ which is the supermartingale property. □

Our second operation is to freeze the supermartingale after n steps. We show that if $\{X_n\}$ is a supermartingale (martingale), $\{X_{\nu\wedge n}\}$ is still a supermartingale (martingale). Note that

$$(X_{\nu\wedge n}, n \geq 0) = (X_0, X_1, \ldots, X_\nu, X_\nu, X_\nu, \ldots).$$

Proposition 10.8.2 *If $\{(X_n, \mathcal{B}_n), n \geq 0\}$ is a supermartingale (martingale), then $\{(X_{\nu\wedge n}, \mathcal{B}_n), n \geq 0\}$ is also a supermartingale (martingale).*

Proof. First of all, $X_{\nu\wedge n} \in \mathcal{B}_n$ since

$$\begin{aligned} X_{\nu\wedge n} &= X_\nu 1_{[n>\nu]} + X_n 1_{[\nu\geq n]} \\ &= \sum_{j=0}^{n-1} X_j 1_{[\nu=j]} + X_n 1_{[\nu\geq n]} \in \mathcal{B}_n, \end{aligned}$$

since $X_n \in \mathcal{B}_n$ and $1_{[\nu\geq n]} \in \mathcal{B}_{n-1}$. Also, if $\{(X_n, \mathcal{B}_n), n \in \mathbb{N}\}$ is a supermartingale,

$$\begin{aligned} E(X_{\nu\wedge n}|\mathcal{B}_{n-1}) &= \sum_{j=0}^{n-1} X_j 1_{[\nu=j]} + 1_{[\nu\geq n]} E(X_n|\mathcal{B}_{n-1}) \\ &\leq \sum_{j=0}^{n-1} X_j 1_{[\nu=j]} + 1_{[\nu\geq n]} X_{n-1} \\ &= X_\nu 1_{[\nu<n]} + X_{n-1} 1_{[\nu\geq n]} \\ &= X_{\nu\wedge(n-1)}. \end{aligned}$$

If $\{X_n\}$ is a martingale, equality prevails throughout, verifying the martingale property. □

10.8.2 Upcrossings

Let $\{x_n, n \geq 0\}$ be a sequence of numbers in $\bar{\mathbb{R}} = [-\infty, \infty]$. Let $-\infty < a < b < \infty$. Define the crossing times of $[a, b]$ by the sequence $\{x_n\}$ as

$$\begin{aligned}
\nu_1 &= \inf\{n \geq 0 : x_n \leq a\} \\
\nu_2 &= \inf\{n \geq \nu_1 : x_n \geq b\} \\
\nu_3 &= \inf\{n \geq \nu_2 : x_n \leq a\} \\
\nu_4 &= \inf\{n \geq \nu_3 : x_n \geq b\} \\
&\vdots \qquad\quad \vdots
\end{aligned}$$

and so on. It is useful and usual to adopt the convention that $\inf \emptyset = \infty$. Define

$$\beta_{a,b} = \max\{p : \nu_{2p} < \infty\}$$

(with the understanding that if $\nu_k < \infty$ for all k and we call $\beta_{a,b} = \infty$) the number of *upcrossings* of $[a, b]$ by $\{x_n\}$.

Lemma 10.8.1 (Upcrossings and Convergence) *The sequence $\{x_n\}$ is convergent in $\bar{\mathbb{R}}$ iff $\beta_{a,b} < \infty$ for all rational $a < b$ in $\mathbb{R}$.*

Proof. If $\liminf_{n\to\infty} x_n < \limsup_{n\to\infty} x_n$, then there exist rational numbers $a < b$ such that

$$\liminf_{n\to\infty} x_n < a < b < \limsup_{n\to\infty} x_n.$$

So $x_n < a$ for infinitely many n, and $x_n > b$ for infinitely many n, and therefore $\beta_{a,b} = \infty$.

Conversely, suppose for some rational $a < b$, we have $\beta_{a,b} = \infty$. Then the sequence $\{x_n\}$ is below a infinitely often and above b infinitely often so that

$$\liminf_{n\to\infty} x_n \leq a, \quad \limsup_{n\to\infty} x_n \geq b$$

and thus $\{x_n\}$ does not converge. □

10.8.3 Boundedness Properties

This section considers how to prove the following intuitive fact: A positive supermartingale tends to decrease but must stay non-negative, so the process should be bounded. This fact will lead in the next subsection to convergence.

Proposition 10.8.3 *Let $\{(X_n, \mathcal{B}_n), n \geq 0\}$ be a positive supermartingale. We have that*

$$\sup_{n\in\mathbb{N}} X_n < \infty \text{ a.s. on } [X_0 < \infty]. \tag{10.26}$$

$$P(\bigvee_{n\in\mathbb{N}} X_n \geq a | \mathcal{B}_0) \leq a^{-1} X_0 \wedge 1 \tag{10.27}$$

for all constants $a > 0$ or for all $\mathcal{B}_0$-measurable positive random variables a.

Proof. Consider two supermartingales $\{(X_n^{(i)}, \mathcal{B}_n), n \geq 0\}$, $i = 1, 2$, defined by $X_n^{(1)} = X_n$, and $X_n^{(2)} \equiv a$. Define a stopping time

$$\nu_a = \inf\{n : X_n \geq a\}.$$

Since

$$X_{\nu_a}^{(1)} \geq X_{\nu_a}^{(2)} \text{ on } [\nu_a < \infty],$$

we may paste the two supermartingales together to get via the Pastings Proposition 10.8.1 that

$$Y_n = \begin{cases} X_n, & \text{if } n < \nu_a, \\ a, & \text{if } n \geq \nu_a \end{cases}$$

is a positive supermartingale. Since $\{(Y_n, \mathcal{B}_n), n \geq 0\}$ is a supermartingale,

$$Y_0 \geq E(Y_n | \mathcal{B}_0), \quad n \geq 0. \tag{10.28}$$

But we also have

$$Y_n \geq a 1_{[\nu_a \leq n]} \tag{10.29}$$

and

$$\begin{aligned} Y_0 &= X_0 1_{[0 < \nu_a]} + a 1_{[0 = \nu_a]} \\ &= X_0 1_{[X_0 < a]} + a 1_{[X_0 \geq a]} = X_0 \wedge a. \end{aligned}$$

From (10.28)

$$\begin{aligned} X_0 \wedge a &\geq E(Y_n | \mathcal{B}_0) \\ &\geq E(a 1_{[\nu_a \leq n]} | \mathcal{B}_0) \qquad \text{(from (10.29))} \\ &= a P[\nu_a \leq n | \mathcal{B}_0]. \end{aligned}$$

Divide by a to get, as $n \to \infty$,

$$P[\nu_a \leq n | \mathcal{B}_0] \to P[\nu_a < \infty | \mathcal{B}_0] = P\Big[\bigvee_{n \in \mathbb{N}} X_n \geq a | \mathcal{B}_0\Big] \leq a^{-1} X_0 \wedge 1.$$

This is (10.27). To get (10.26), multiply (10.27) by $1_{[X_0 < \infty]}$ and integrate:

$$\begin{aligned} E 1_{[X_0 < \infty]} P(\bigvee_n X_n \geq a | \mathcal{B}_0) &= P[\bigvee_n X_n \geq a, X_0 < \infty] \\ &\leq E 1_{[X_0 < \infty]} (a^{-1} X_0 \wedge 1). \end{aligned}$$

Since

$$1_{[X_0 < \infty]} (a^{-1} X_0 \wedge 1) \leq 1$$

and

$$1_{[X_0 < \infty]} \left(a^{-1} X_0 \wedge 1\right) \to 0$$

as $a \to \infty$, we apply dominated convergence to get

$$P[\bigvee_n X_n = \infty, X_0 < \infty] = 0.$$

This is (10.26). □

10.8.4 Convergence of Positive Super Martingales

Positive supermartingales tend to decrease but are bounded below and hence can be expected to converge. This subsection makes this precise.

Given $\{X_n\}$ define the *upcrossing number* $\beta_{a,b}$ by

$$\beta_{a,b}(\omega) = \#\text{ upcrossings of } [a,b] \text{ by } \{X_n(\omega)\};$$

that is,

$$\begin{aligned}
\nu_1(\omega) &= \inf\{n \geq 0 : X_n(\omega) \leq a\} \\
\nu_2(\omega) &= \inf\{n \geq \nu_1(\omega) : X_n(\omega) \geq b\} \\
\nu_3(\omega) &= \inf\{n \geq \nu_2(\omega) : X_n(\omega) \leq a\} \\
&\vdots \qquad\qquad \vdots
\end{aligned}$$

and so on, and

$$\beta_{a,b}(\omega) = \sup\{p : \nu_{2p}(\omega) < \infty\}.$$

Then

$$\{\omega : \lim_{n\to\infty} X_n(\omega) \text{ exists }\} = \bigcap_{\substack{a<b \\ a,b \text{ rational}}} \{\omega : \beta_{a,b}(\omega) < \infty\}.$$

So $\lim_{n\to\infty} X_n$ exists a.s. iff $\beta_{a,b} < \infty$ a.s. for all rational $a < b$. To analyze when $\beta_{a,b} < \infty$, we need an inequality due to Dubins.

Proposition 10.8.4 (Dubins' inequality) *Let $\{(X_n, \mathcal{B}_n), n \geq 0\}$ be a positive supermartingale. Suppose $0 < a < b$. Then*

(1) $P(\beta_{a,b} \geq k | \mathcal{B}_0) \leq \left(\frac{a}{b}\right)^k (a^{-1}X_0 \wedge 1), \quad k \geq 1$

(2) $\beta_{a,b} < \infty$ *almost surely.*

Proof. We again apply the Pasting Proposition 10.8.1 using the ν_k's. Start by considering the supermartingales

$$X_n^{(1)} \equiv 1, \quad X_n^{(2)} = \frac{X_n}{a}$$

and paste at ν_1. Note on $[\nu_1 < \infty]$,

$$X_{\nu_1}^{(1)} \equiv 1 \geq X_{\nu_1}^{(2)}.$$

Thus

$$Y_n^{(1)} = \begin{cases} 1, & \text{if } n < \nu_1, \\ X_n/a, & \text{if } n \geq \nu_1 \end{cases}$$

is a supermartingale.

Now compare and paste $X_n^{(3)} = Y_n^{(1)}$ and $X_n^{(4)} = b/a$ at the stopping time ν_2. On $[\nu_2 < \infty]$

$$X_{\nu_2}^{(3)} = Y_{\nu_2}^{(1)} = \frac{X_{\nu_2}}{a} \geq \frac{b}{a} = X_{\nu_2}^{(4)}$$

so

$$Y_n^{(2)} = \begin{cases} Y_n^{(1)}, & \text{if } n < \nu_2 \\ b/a, & \text{if } n \geq \nu_2 \end{cases}$$

$$= \begin{cases} 1, & \text{if } n < \nu_1 \\ X_n/a, & \text{if } \nu_1 \leq n < \nu_2 \\ b/a, & \text{if } n \geq \nu_2 \end{cases}$$

is a supermartingale. Now compare $Y_n^{(2)}$ and $\frac{b}{a}\frac{X_n}{a}$. On $[\nu_3 < \infty]$,

$$Y_{\nu_3}^{(2)} = \frac{b}{a} \geq \frac{b}{a}\frac{X_{\nu_3}}{a}$$

and so

$$Y_n^{(3)} = \begin{cases} Y_n^{(2)}, & \text{if } n < \nu_3 \\ \left(\frac{b}{a}\right)\frac{X_n}{a}, & \text{if } n \geq \nu_3 \end{cases}$$

is a supermartingale. Continuing on in this manner we see that for any k, the following is a supermartingale:

$$\begin{aligned} Y_n \;=\; & 1, && n < \nu_1 \\ & X_n/a, && \nu_1 \leq n < \nu_2 \\ & b/a, && \nu_2 \leq n < \nu_3 \\ & \tfrac{b}{a}\cdot\tfrac{X_n}{a}, && \nu_3 \leq n < \nu_4 \\ & \vdots && \vdots \\ & \left(\tfrac{b}{a}\right)^{k-1}\tfrac{X_n}{a}, && \nu_{2k-1} \leq n < \nu_{2k} \\ & \left(\tfrac{b}{a}\right)^{k}, && \nu_{2k} \leq n. \end{aligned}$$

Note that

$$Y_0 = 1 1_{[0<\nu_1]} + \frac{X_0}{a} 1_{[\nu_1=0]} = 1 \wedge \frac{X_0}{a}. \tag{10.30}$$

Also

$$Y_n \geq \left(\frac{b}{a}\right)^k 1_{[n \geq \nu_{2k}]}. \tag{10.31}$$

From the definition of supermartingales

$$Y_0 \geq E(Y_n|\mathcal{B}_0); \tag{10.32}$$

that is, from (10.30), (10.31) and (10.32)

$$1 \wedge \frac{X_0}{a} \geq \left(\frac{b}{a}\right)^k P[\nu_{2k} \leq n | \mathcal{B}_0].$$

This translates to

$$P[\nu_{2k} \leq n | \mathcal{B}_0] \leq \left(\frac{a}{b}\right)^k \left(1 \wedge \frac{X_0}{a}\right).$$

Let $n \to \infty$ to get

$$P[\beta_{a,b} \geq k | \mathcal{B}_0] = P[\nu_{2k} < \infty | \mathcal{B}_0] \leq \left(\frac{a}{b}\right)^k \left(1 \wedge \frac{X_0}{a}\right).$$

Let $k \to \infty$ and we see

$$P[\beta_{a,b} = \infty | \mathcal{B}_0] = \lim_{k \to \infty} \left(\frac{a}{b}\right)^k \left(1 \wedge \frac{X_0}{a}\right) = 0.$$

We conclude $\beta_{a,b} < \infty$ almost surely and in fact $E(\beta_{a,b}) < \infty$ since

$$E(\beta_{a,b}) = \sum_{k=0}^{\infty} P[\beta_{a,b} \geq k] \leq \sum_k \left(\frac{a}{b}\right)^k < \infty.$$

□

Theorem 10.8.5 (Convergence Theorem) *If $\{(X_n, \mathcal{B}_n), n \in \mathbb{N}\}$ is a positive supermartingale, then*

$$\lim_{n \to \infty} X_n =: X_\infty \textit{ exists almost surely}$$

and

$$E(X_\infty | \mathcal{B}_n) \leq X_n, \quad n \in \mathbb{N}$$

so $\{(X_n, \mathcal{B}_n), n \in \overline{\mathbb{N}}\}$ is a positive supermartingale.

Remark. The last statement says we can add a last variable which preserves the supermartingale property. This is the *closure* property to be discussed in the next subsection and is an essential concept for the stopping theorems.

Proof. Since $\beta_{a,b} < \infty$ a.s. for all rational $a < b$, $\lim_{n \to \infty} X_n$ exists almost surely by Lemma 10.8.1. To prove X_∞ can be added to $\{X_n, n \in \mathbb{N}\}$ while preserving the supermartingale property, observe for $n \geq p$ that

$$E\left(\bigwedge_{m \geq n} X_m | \mathcal{B}_p\right) \leq E(X_n | \mathcal{B}_p) \qquad \text{(monotonicity)}$$
$$\leq X_p \qquad \text{(supermartingale property).}$$

As $n \to \infty$, $\wedge_{m \geq n} X_m \uparrow X_\infty$, so by monotone convergence for conditional expectations, letting $n \to \infty$, we get $E(X_\infty | \mathcal{B}_p) \leq X_p$. □

10.8.5 Closure

If $\{(X_n, \mathcal{B}_n), n \geq 0\}$ is positive martingale, then we know it is almost surely convergent. But when is it also the case that

(a) $X_n \stackrel{L_1}{\to} X_\infty$ and

(b) $E(X_\infty|\mathcal{B}_n) = X_n$ so that $\{(X_n, \mathcal{B}_n), n \in \overline{\mathbb{N}}\}$ is a positive martingale?

Even though it is true that $X_n \stackrel{a.s.}{\to} X_\infty$ and $E(X_m|\mathcal{B}_n) = X_n, \ \forall m > n$, it is not necessarily the case that $E(X_\infty|\mathcal{B}_n) = X_n$. Extra conditions are needed.

Consider, for instance, the example of the simple branching process in Section 10.5. (See also the fuller discussion in Subsection 10.9.2 on page 380 to come.) If $\{Z_n, n \geq 0\}$ is the process with $Z_0 = 1$ and Z_n representing the number of particles in the nth generation and $m = E(Z_1)$ is the mean offspring number per individual, then $\{Z_n/m^n\}$ is a non-negative martingale so the almost sure limit exists: $W_n := Z_n/m^n \stackrel{a.s.}{\to} W$. However, if $m \leq 1$, then extinction is sure so $W \equiv 0$ and we do not have $E(W|\mathcal{B}_n) = Z_n/m^n$.

This leads us to the topic of martingale closure.

Definition 10.8.1 (Closed Martingale) A martingale $\{(X_n, \mathcal{B}_n), n \in \mathbb{N}\}$ is *closed* (on the right) if there exists an integrable random variable $X_\infty \in \mathcal{B}_\infty$ such that for every $n \in \mathbb{N}$,

$$X_n = E(X_\infty|\mathcal{B}_n). \tag{10.33}$$

In this case $\{(X_n, \mathcal{B}_n), n \in \overline{\mathbb{N}}\}$ is a martingale.

In what follows, we write L_p^+ for the random variables $\xi \in L_p$ which are non-negative.

The next result gives a class of examples where closure can be assured.

Proposition 10.8.6 *Let $p \geq 1$, $X \in L_p^+$ and define*

$$X_n := E(X|\mathcal{B}_n), \quad n \in \mathbb{N} \tag{10.34}$$

and

$$X_\infty := E(X|\mathcal{B}_\infty). \tag{10.35}$$

Then $X_n \to X_\infty$ almost surely and in L_p and

$$\{(X_n, \mathcal{B}_n), n \in \mathbb{N}, (X_\infty, \mathcal{B}_\infty), (X, \mathcal{B})\} \tag{10.36}$$

is a closed martingale.

Remark 10.8.1 (i) For the martingale $\{(X_n, \mathcal{B}_n), n \in \overline{\mathbb{N}}\}$ given in (10.34) and (10.35), it is also the case that

$$X_n = E(X_\infty|\mathcal{B}_n),$$

since by smoothing and (10.35)

$$E(X_\infty|\mathcal{B}_n) = E\left(E(X|\mathcal{B}_\infty)|\mathcal{B}_n\right) = E(X|\mathcal{B}_n)$$

almost surely.

(ii) We can extend Proposition 10.8.6 to cases where the closing random variable is not necessarily non-negative by writing $X = X^+ - X^-$.

The proof of Proposition 10.8.6 is deferred until we state and discuss Corollary 10.8.1. The proof of Corollary 10.8.1 assumes the validity of Proposition 10.8.6.

Corollary 10.8.1 *For $p \geq 1$, the class of L_p convergent positive martingales is the class of the form*

$$\left\{\left(E(X|\mathcal{B}_n), \mathcal{B}_n\right), n \in \mathbb{N}\right\}$$

with $X \in L_p^+$.

Proof of Corollary 10.8.1 If $X \in L_p^+$, apply Proposition 10.8.6 to get that $\{E(X|\mathcal{B}_n)\}$ is L_p convergent. Conversely, suppose $\{X_n\}$ is a positive martingale and L_p convergent. For $n < r$, the martingale property asserts

$$E(X_r|\mathcal{B}_n) = X_n.$$

Now $X_r \stackrel{L_p}{\to} X_\infty$ as $r \to \infty$ and $E(\cdot|\mathcal{B}_n)$ is continuous in the L_p-metric (see (10.21)). Thus as $r \to \infty$

$$X_n = E(X_r|\mathcal{B}_n) \stackrel{L_p}{\to} E(X_\infty|\mathcal{B}_n)$$

by continuity. Therefore $X_n = E(X_\infty|\mathcal{B}_n)$ as asserted. □

Proof of Proposition 10.8.6. We know $\{(EX|\mathcal{B}_n), \mathcal{B}_n), n \in \mathbb{N}\}$ is a positive martingale and hence convergent by Theorem 10.8.5. Call the limit $X_\infty^\#$. Since $E(X|\mathcal{B}_n) \in \mathcal{B}_n \subset \mathcal{B}_\infty$ and $E(X|\mathcal{B}_n) \to X_\infty^\#$, we have $X_\infty^\# \in \mathcal{B}_\infty$. We consider two cases.

CASE 1: Suppose temporarily that $P[X \leq \lambda] = 1$ for some $\lambda < \infty$. We need to show that

$$X_\infty := E(X|\mathcal{B}_\infty) = X_\infty^\#.$$

Since $X \leq \lambda$, we have $E(X|\mathcal{B}_\infty) \leq \lambda$ and for all $A \in \mathcal{B}$, as $n \to \infty$

$$\int_A E(X|\mathcal{B}_n)dP \to \int_A X_\infty^\# dP,$$

by the dominated convergence theorem. Fix m, and let $A \in \mathcal{B}_m$. For $n > m$, we have $A \in \mathcal{B}_m \subset \mathcal{B}_n$ and

$$\int_A E(X|\mathcal{B}_n)dP = \int_A X dP$$

by the definition of conditional expectation. Since

$$E(X|\mathcal{B}_n) \to X_\infty^\#$$

almost surely, and in L_1 we get

$$\int_A E(X|\mathcal{B}_n)dP \to \int_A X_\infty^\# dP.$$

Thus

$$\int_A X_\infty^\# dP = \int_A XdP$$

for all $A \in \cup_m \mathcal{B}_m$.

Define

$$m_1(A) = \int_A X_\infty^\# dP, \quad m_2(A) = \int_A XdP.$$

Then we have two positive measures m_1 and m_2 satisfying

$$m_1(A) = m_2(A), \quad \forall A \in \bigcup_m \mathcal{B}_m.$$

But $\cup_m \mathcal{B}_m$ is a π-class, so Dynkin's theorem 2.2.2 implies that

$$m_1(A) = m_2(A) \quad \forall A \in \sigma\Big(\bigcup_m \mathcal{B}_m\Big) = \mathcal{B}_\infty.$$

We conclude

$$\int_A X_\infty^\# dP = \int_A XdP = \int_A E(X|\mathcal{B}_\infty)dP = \int_A X_\infty dP$$

and the Integral Comparison Lemma 10.1.1 implies $X_\infty^\# = E(X|\mathcal{B}_\infty)$.

L_p convergence is immediate since $E(X|\mathcal{B}_n) \le \lambda$, for all n, so that dominated convergence applies.

CASE 2: Now we remove the assumption that $X \le \lambda$. Only assume that $0 \le X \in L_p$, $p \ge 1$. Write

$$X = X \wedge \lambda + (X - \lambda)^+.$$

Since $E(\cdot|\mathcal{B}_n)$ is L_p-norm reducing (see (10.20)) we have

$$\begin{aligned} &\|E(X|\mathcal{B}_n) - E(X|\mathcal{B}_\infty)\|_p \\ &\qquad \le \|E\big((X \wedge \lambda)|\mathcal{B}_n\big) - E\big((X \wedge \lambda)|\mathcal{B}_\infty\big)\|_p + \|E\big((X - \lambda)^+|\mathcal{B}_n\big)\|_p \\ &\qquad\qquad + \|E(X - \lambda)^+|\mathcal{B}_\infty)\|_p \\ &\qquad \le \|E(X \wedge \lambda|\mathcal{B}_n) - E(X \wedge \lambda|\mathcal{B}_\infty)\|_p + 2\|(X - \lambda)^+\|_p \\ &\qquad = I + II. \end{aligned}$$

Since $0 \le X \wedge \lambda \le \lambda$, $I \to 0$ by Case 1. For II, note as $\lambda \to \infty$

$$(X - \lambda)^+ \to 0$$

and

$$(X - \lambda)^+ \le X \in L_p.$$

The dominated convergence theorem implies that $\|(X - \lambda)^+\|_p \to 0$ as $\lambda \to \infty$. We may conclude that

$$\limsup_{n\to\infty} \|E(X|\mathcal{B}_n) - E(X|\mathcal{B}_\infty)\|_p \le 2\|(X - \lambda)^+\|_p.$$

The left side is independent of λ, so let $\lambda \to \infty$ to get

$$\limsup_{n\to\infty} \|E(X|\mathcal{B}_n) - E(X|\mathcal{B}_\infty)\|_p = 0.$$

Thus

$$E(X|\mathcal{B}_n) \stackrel{L_p}{\to} E(X|\mathcal{B}_\infty)$$

and

$$E(X|\mathcal{B}_n) \stackrel{a.s.}{\to} X^\#_\infty$$

and therefore $X^\#_\infty = E(X|\mathcal{B}_\infty)$. □

10.8.6 Stopping Supermartingales

What happens to the supermartingale property if deterministic indices are replaced by stopping times?

Theorem 10.8.7 (Random Stopping) *Suppose $\{(X_n, \mathcal{B}_n), n \in \mathbb{N}\}$ is a positive supermartingale and also suppose $X_n \stackrel{a.s.}{\to} X_\infty$. Let ν_1, ν_2 be two stopping times. Then*

$$X_{\nu_1} \ge E(X_{\nu_2}|\mathcal{B}_{\nu_1}) \text{ a.s. on } [\nu_1 \le \nu_2]. \tag{10.37}$$

Some SPECIAL CASES:

(i) If $\nu_1 = 0$, then $\nu_2 \ge 0$ and

$$X_0 \ge E(X_{\nu_2}|\mathcal{B}_0)$$

and

$$E(X_0) \ge E(X_{\nu_2}).$$

(ii) If $\nu_1 \le \nu_2$ pointwise everywhere, then

$$X_{\nu_1} \ge E(X_{\nu_2}|\mathcal{B}_{\nu_1}), \quad E(X_{\nu_1}) \ge E(X_{\nu_2}).$$

The proof of Theorem 10.8.7 requires the following result.

Lemma 10.8.2 *If ν is a stopping time and $\xi \in L_1$, then*

$$E(\xi|\mathcal{B}_\nu) = \sum_{n\in\bar{\mathbb{N}}} E(\xi|\mathcal{B}_n)1_{[\nu=n]}. \tag{10.38}$$

Proof of Lemma 10.8.2: The right side of (10.38) is $\mathcal{B}_\nu$-measurable and for any $A \in \mathcal{B}_\nu$,

$$\begin{aligned}\int_A \sum_{n\in\bar{\mathbb{N}}} E(\xi|\mathcal{B}_n)1_{[\nu=n]}dP &= \sum_{n\in\bar{\mathbb{N}}} \int_{A\cap[\nu=n]} E(\xi|\mathcal{B}_n)dP \\ &= \sum_{n\in\bar{\mathbb{N}}} \int_{A\cap[\nu=n]} \xi dP\end{aligned}$$

(since $A \cap [\nu = n] \in \mathcal{B}_n$)

$$= \int_A \xi dP = \int_A E(\xi|\mathcal{B}_\nu)dP.$$

Finish with an application of Integral Comparison Lemma 10.1.1 or an appeal to the definition of conditional expectation. □

Proof of Theorem 10.8.7. Since Lemma 10.8.2 gives

$$E(X_{\nu_2}|\mathcal{B}_{\nu_1}) = \sum_{n\in\bar{\mathbb{N}}} E(X_{\nu_2}|\mathcal{B}_n)1_{[\nu_1=n]},$$

for (10.37) it suffices to prove for $n \in \overline{\mathbb{N}}$ that

$$X_n \ge E(X_{\nu_2}|\mathcal{B}_n) \text{ on } [n \le \nu_2]. \tag{10.39}$$

Set $Y_n = X_{\nu_2\wedge n}$. Then, first of all, $\{(Y_n, \mathcal{B}_n), n \ge 0\}$ is a positive supermartingale from Proposition 10.8.2 and secondly, from Theorem 10.8.5, it is almost surely convergent:

$$Y_n \stackrel{a.s.}{\to} Y_\infty = X_{\nu_2}.$$

To verify the form of the limit, note that if $\nu_2(\omega) < \infty$, then for n large, we have $n \wedge \nu_2(\omega) = \nu_2(\omega)$. On the other hand, if $\nu_2(\omega) = \infty$, then

$$Y_n(\omega) = X_n(\omega) \to X_\infty(\omega) = X_{\nu_2}(\omega).$$

Observe also that for $n \in \overline{\mathbb{N}}$, we get from Theorem 10.8.5

$$Y_n \ge E(Y_\infty|\mathcal{B}_n);$$

that is,

$$X_{\nu_2 \wedge n} \geq E(X_{\nu_2}|\mathcal{B}_n). \tag{10.40}$$

On $[\nu_2 \geq n]$ (10.40) says $X_n \geq E(X_{\nu_2}|\mathcal{B}_n)$ as required. □

For martingales, we will see that it is useful to know when equality holds in Theorem 10.8.7. Unfortunately, this does not always hold and conditions must be present to guarantee preservation of the martingale property under random stopping.

10.9 Examples

We collect some examples in this section.

10.9.1 Gambler's Ruin

Suppose $\{Z_n\}$ are iid Bernoulli random variables satisfying

$$P[Z_i = \pm 1] = \frac{1}{2}$$

and let

$$X_0 = j_0, \quad X_n = \sum_{i=1}^{n} Z_i + j_0, \quad n \geq 1$$

be the simple random walk starting from j_0. Assume $0 \leq j_0 \leq N$ and we ask: starting from j_0, will the random walk hit 0 or N first?

Define

$$\nu = \inf\{n : X_n = 0 \text{ or } N\},$$
$$[\text{ ruin }] = [X_\nu = 0],$$
$$p = P[X_\nu = 0] = P[\text{ ruin }].$$

If random stopping preserves the martingale property (to be verified later), then

$$j_0 = E(X_0) = E(X_\nu) = 0 \cdot P[X_\nu = 0] + NP[X_\nu = N]$$

and since

$$P[X_\nu = 0] = p, \quad P[X_\nu = N] = 1 - p,$$

we get

$$j_0 = N(1 - p)$$

and

$$p = 1 - \frac{j_0}{N}.$$

10.9.2 Branching Processes

Let $\{Z_n, n \geq 0\}$ be a simple branching process with *offspring distribution* $\{p_k, k \geq 0\}$ so that $\{Z_n\}$ is a Markov chain with transition probabilities

$$P[Z_{n+1} = j | Z_n = i] = \begin{cases} p_j^{*i}, & \text{if } i \geq 1, \\ \delta_{0j}, & \text{if } i = 0, \end{cases}$$

where $\{p_j^{*i}, j \geq 0\}$ is the i-fold convolution of the sequence $\{p_j, j \geq 0\}$. We can also represent $\{Z_n\}$ as

$$Z_{n+1} = Z^{(n)}(1) + \cdots + Z^{(n)}(Z_n), \tag{10.41}$$

where $\{Z^{(j)}(m), j \geq 0, m \geq 0\}$ are iid with distribution $\{p_k, k \geq 0\}$. Define the generating functions ($0 \leq s \leq 1$),

$$\begin{aligned} f(s) &= \sum_{k=0}^{\infty} p_k s^k = E(s^{Z_1}), \\ f_n(s) &= E(s^{Z_n}), \\ f_0(s) &= s, \quad f_1 = f \end{aligned}$$

so that from standard branching process theory

$$f_{n+1}(s) = f_n(f(s)) = f(f_n(s)).$$

Finally, set

$$m = E(Z_1) = f'(1).$$

We claim that the following elementary *facts* are true (cf. Resnick, 1992, Section 1.4):

(1) The extinction probability $q := P[Z_n \to 0] = P[\text{ extinction }] = P\left\{\bigcup_{n=1}^{\infty}[Z_n = 0]\right\}$ satisfies $f(s) = s$ and is the minimal solution in $[0, 1]$. If $m > 1$, then $q < 1$ while if $m \leq 1$, $q = 1$.

(2) Suppose $q < 1$. Then either $Z_n \to 0$ or $Z_n \to \infty$. We define the event

$$[\text{ explosion }] := [Z_n \to \infty]$$

and we have

$$1 = P[Z_n \to 0] + P[(Z_n \to \infty]$$

so that

$$q = P[\text{ extinction }], \quad 1 - q = P[\text{ explosion }].$$

We now verify fact (2) using martingale arguments. For $n \geq 0$, set $\mathcal{B}_n = \sigma(Z_0, \ldots, Z_n)$. We begin by observing that $\{(q^{Z_n}, \mathcal{B}_n), n \in \mathbb{N}\}$ is a positive martingale. We readily see this using (10.41) and (10.17):

$$\begin{aligned} E(s^{Z_{n+1}}|\mathcal{B}_n) &= E(s^{\sum_{i=1}^{Z_n} Z^n(i)}|\mathcal{B}_n) \\ &= \left(Es^{Z^{(n)}(1)}\right)^{Z_n} = f(s)^{Z_n}. \end{aligned}$$

Set $s = q$, and since $f(q) = q$, we get

$$E\left(q^{Z_{n+1}}|\mathcal{B}_n\right) = q^{Z_n}.$$

Since $\{(q^{Z_n}, \mathcal{B}_n), n \in \mathbb{N}\}$ is a positive martingale, it converges by Theorem 10.8.5. So $\lim_{n\to\infty} q^{Z_n}$ exists and therefore $\lim_{n\to\infty} Z_n =: Z_\infty$ also exists.

Let $\nu = \inf\{n : Z_n = 0\}$. Since $\lim_{n\to\infty} Z_n =: Z_\infty$, we also have $Z_{\nu\wedge n} \to Z_\nu$. From Proposition 10.8.2 $\{(q^{Z_{\nu\wedge n}}, \mathcal{B}_n), n \in \mathbb{N}\}$ is a positive martingale, which satisfies

$$1 \geq q^{Z_{\nu\wedge n}} \to q^{Z_\nu};$$

and because a martingale has a constant mean, $E(q^{Z_{\nu\wedge n}}) = E(q^{Z_{\nu\wedge 0}}) = E(q^{Z_0}) = q$. Applying dominated convergence

$$q = E(q^{Z_{\nu\wedge n}}) \to E(q^{Z_\nu}),$$

that is,

$$q = E(q^{Z_\nu}) = E(q^{Z_\infty} 1_{[\nu=\infty]}) + E(q^{Z_\nu} 1_{[\nu<\infty]}).$$

On $[\nu < \infty]$, $Z_\nu = 0$ and recall $q = P[\nu < \infty] = P[\text{extinction}]$ so

$$q = E(q^{Z_\infty} 1_{\nu=\infty}) + q,$$

and therefore

$$E(q^{Z_\infty} 1_{[\nu=\infty]}) = 0.$$

This implies that on $[\nu = \infty]$, $q^{Z_\infty} = 0$, and thus $Z_\infty = \infty$. So on $[\nu = \infty] =$ [non-extinction], $Z_\infty = \infty$ as claimed in fact (2).

We next recall that $\{(W_n := \frac{Z_n}{m^n}, \mathcal{B}_n), n \in \mathbb{N}\}$ is a non-negative martingale. An almost sure limit exists, namely,

$$W_n = \frac{Z_n}{m^n} \stackrel{a.s.}{\to} W.$$

On the event [extinction], $Z_n \to 0$, so $W(\omega) = 0$ for $\omega \in$ [extinction]. Also $E(W) \leq 1$, since by Fatou's lemma

$$E(W) = E(\liminf_{n\to\infty} \frac{Z_n}{m^n}) \leq \liminf_{n\to\infty} \frac{E(Z_n)}{m^n} = 1.$$

Consider the special case that $q = 1$. Then $Z_n \to 0$ almost surely and $P[W = 0] = 1$. So $\{W_n := \frac{Z_n}{m^n}\}$ is a positive martingale such that $W_n \to 0 = W$. We have $E(W_n) = 1$, but $E(W) = 0$. So this martingale is **NOT** closable. There is no hope that

$$W_n = E(W|\mathcal{B}_n)$$

since $W = 0$. For later reference, note that in this case $\{Z_n/m^n, n \geq 0\}$ is **NOT** uniformly integrable since if it were, $E(Z_n/m^n) = 1$ would imply $E(W) = 1$, which is false.

10.9.3 Some Differentiation Theory

Recall the Lebesgue decomposition of two measures and the Radon–Nikodym theorem of Section 10.1. We are going to consider these results when the σ-fields are allowed to vary.

Suppose Q is a finite measure on $\mathcal{B}$. Let the restriction of Q to a sub σ-field $\mathcal{G}$ be denoted $Q|_{\mathcal{G}}$. Suppose we are given a family of σ-fields $\mathcal{B}_n, n \in \mathbb{N}$, $\mathcal{B}_\infty = \vee_n \mathcal{B}_n$ and $\mathcal{B}_n \subset \mathcal{B}_{n+1}$. Write the Lebesgue decomposition of $Q|_{\mathcal{B}_n}$ with respect to $P|_{\mathcal{B}_n}$ as

$$Q|_{\mathcal{B}_n} = f_n dP|_{\mathcal{B}_n} + Q|_{\mathcal{B}_n}(\cdot \cap N_n), \quad n \in \overline{\mathbb{N}} \tag{10.42}$$

where $P(N_n) = 0$ for $n \in \overline{\mathbb{N}}$.

Proposition 10.9.1 *The family $\{(f_n, \mathcal{B}_n), n \geq 0\}$ is a positive supermartingale and $f_n \stackrel{a.s.}{\to} f_\infty$ where f_∞ is given by (10.42) with $n = \infty$.*

The proof requires the following characterization of the density appearing in the Lebesgue decomposition.

Lemma 10.9.1 *Suppose Q is a finite measure on $(\Omega, \mathcal{G})$ whose Lebesgue decomposition with respect to the probability measure P is*

$$Q(A) = \int_A X dP + Q(A \cap N), \quad A \in \mathcal{G},$$

where $P(N) = 0$. Then X is determined up to P-sets of measure 0 as the largest $\mathcal{G}$-measurable function such that $XdP \leq Q$ on $\mathcal{G}$.

Proof of Lemma 10.9.1. We assume the Lebesgue decomposition is known. If Y is a non-negative $\mathcal{G}$-measurable and integrable function such that $YdP \leq Q$ on $\mathcal{G}$, then for any $A \in \mathcal{G}$,

$$\begin{aligned}\int_A YdP &= \int_{AN^c} YdP \leq Q(AN^c)\\ &= \int_{AN^c} XdP + Q(AN^cN)\\ &= \int_{AN^c} XdP = \int_A XdP.\end{aligned}$$

Hence by the Integral Comparison Lemma 10.1.1, we have $X \geq Y$ almost surely. □

Proof of Proposition 10.9.1. We have from (10.42)

$$f_{n+1} dP|_{\mathcal{B}_{n+1}} + Q|_{\mathcal{B}_{n+1}}(\cdot \cap N_{n+1}) = Q|_{\mathcal{B}_{n+1}},$$

so that

$$f_{n+1} dP|_{\mathcal{B}_{n+1}} \leq Q|_{\mathcal{B}_{n+1}}.$$

Hence for all $A \in \mathcal{B}_n$, we get, by the definition of conditional expectation, that

$$\int_A E(f_{n+1}|\mathcal{B}_n) dP = \int_A f_{n+1} dP \leq Q(A).$$

So $E(f_{n+1}|\mathcal{B}_n)$ is a function in $L_1(\mathcal{B}_n)$ such that for all $A \in \mathcal{B}_n$

$$\int_A E(f_{n+1}|\mathcal{B}_n) dP \leq Q(A);$$

that is,

$$E(f_{n+1}|\mathcal{B}_n) dP|_{\mathcal{B}_n} \leq Q|_{\mathcal{B}_n}.$$

Since f_n is the maximal $\mathcal{B}_n$-measurable function with this property, we have

$$E(f_{n+1}|\mathcal{B}_n) \leq f_n,$$

which is the supermartingale property. It therefore follows that $\lim_{n\to\infty} f_n$ exists almost surely. Call the limit f and we show $f = f_\infty$. Since

$$Q|_{\mathcal{B}_\infty} \geq f_\infty dP|_{\mathcal{B}_\infty},$$

we have for all $A \in \mathcal{B}_n$,

$$\int_A E(f_\infty|\mathcal{B}_n) dP = \int_A f_\infty dP \leq Q(A).$$

Thus, since f_n is the maximal $\mathcal{B}_n$-measurable function satisfying

$$f_n dP|_{\mathcal{B}_n} \leq Q|_{\mathcal{B}_n},$$

we have

$$E(f_\infty|\mathcal{B}_n) \leq f_n.$$

Let $n \to \infty$ and use Proposition 10.8.6 to get

$$f_\infty = E(f_\infty|\mathcal{B}_\infty) = \lim_{n\to\infty} E(f_\infty|\mathcal{B}_n) \leq \lim_{n\to\infty} f_n = f.$$

We conclude $f_\infty \leq f$. Also, by Fatou's lemma, for all $A \in \cup_n \mathcal{B}_n$,

$$\int_A f dP = \int_A \liminf_{n\to\infty} f_n dP \leq \liminf_{n\to\infty} \int_A f_n dP \leq Q(A). \tag{10.43}$$

This statement may be extended to $A \in \mathcal{B}_\infty$ by Dynkin's theorem (see, for example Corollary 2.2.1 on page 38). Since $f \in \mathcal{B}_\infty$ and f_∞ is the maximal $\mathcal{B}_\infty$-measurable function satisfying (10.43), we get $f \leq f_\infty$. Therefore $f = f_\infty$. □

Proposition 10.9.2 *Suppose $Q|_{\mathcal{B}_n} << P|_{\mathcal{B}_n}$ for all $n \in \mathbb{N}$; that is, there exists $f_n \in L_1(\Omega, \mathcal{B}_n, P)$ such that*

$$Q|_{\mathcal{B}_n} = f_n dP|_{\mathcal{B}_n}.$$

Then (a) the family $\{(f_n, \mathcal{B}_n), n \in \mathbb{N}\}$ is positive martingale and (b) we have $f_n \to f_\infty$ almost surely and in L_1 iff $Q|_{\mathcal{B}_\infty} << P|_{\mathcal{B}_\infty}$.

Proof. (a) Since $Q|_{\mathcal{B}_n} << P|_{\mathcal{B}_n}$ for all $n \in \mathbb{N}$, for any $A \in \mathcal{B}_n$

$$\begin{aligned} Q|_{\mathcal{B}_n}(A) &= \int_A f_n dP = Q|_{\mathcal{B}_{n+1}}(A) \\ &= \int_A f_{n+1} dP = \int_A E(f_{n+1}|\mathcal{B}_n) dP, \end{aligned}$$

and therefore we have $f_n = E(f_{n+1}|\mathcal{B}_n)$ by Lemma 10.1.1.

(b) Given $f_n \to f_\infty$ almost surely and in L_1, we have by Corollary 10.8.1 that

$$f_n = E(f_\infty|\mathcal{B}_n).$$

For all $A \in \mathcal{B}_n$ and using the definition of conditional expectation and the martingale property, we get

$$\begin{aligned} \int_A f_\infty dP &= \int_A E(f_\infty|\mathcal{B}_n) dP = \int_A f_n dP \\ &= Q|_{\mathcal{B}_n}(A) = Q(A). \end{aligned}$$

So for $A \in \cup_n \mathcal{B}_n$

$$\int_A f_\infty dP = Q(A).$$

Extend this by Dynkin's theorem to $A \in \mathcal{B}_\infty$ to get

$$Q(A) = \int_A f_\infty dP,$$

that is,

$$Q|_{\mathcal{B}_\infty} << f_\infty dP|_{\mathcal{B}_\infty}.$$

Conversely, suppose $Q|_{\mathcal{B}_\infty} << P|_{\mathcal{B}_\infty}$. Then from the previous proposition

$$f_n \overset{a.s.}{\to} f_\infty.$$

Densities of $Q|_{\mathcal{B}_n}$ converge and Scheffé's lemma 8.2.1 on page 253 implies L_1 convergence:

$$\int |f_n - f| dP \to 0.$$

Note that Scheffé's lemma applies since each f_n is a density with the same total mass

$$\int_\Omega f_n dP = Q(\omega). \qquad \square$$

Example 10.9.1 We give a special case of the previous Proposition 10.9.2. Suppose $\Omega = [0, 1)$, and P is Lebesgue measure. Let

$$\mathcal{B}_n = \sigma\left\{[\frac{k}{2^n}, \frac{k+1}{2^n}), k = 0, 1, \ldots, 2^n - 1\right\},$$

so that $\mathcal{B}_n \uparrow \mathcal{B}_\infty = \mathcal{B}([0, 1))$.

Let Q be a finite, positive measure on $\mathcal{B}([0, 1))$. Then trivially

$$Q|_{\mathcal{B}_n} << P|_{\mathcal{B}_n},$$

since if $A \in \mathcal{B}_n$ and $P(A) = 0$, then $A = \emptyset$ and $Q(A) = 0$. What is

$$f_n = \frac{dQ|_{\mathcal{B}_n}}{dP|_{\mathcal{B}_n}}?$$

We claim

$$f_n(x) = \sum_{i=0}^{2^n-1} \frac{Q([\frac{i}{2^n}, \frac{i+1}{2^n}))}{P([\frac{i}{2^n}, \frac{i+1}{2^n}))} 1_{[\frac{i}{2^n}, \frac{i+1}{2^n})}(x).$$

The reason is that for all j

$$Q([\frac{j}{2^n}, \frac{j+1}{2^n})) = \int_{[\frac{j}{2^n}, \frac{j+1}{2^n})} f_n dP.$$

We also know

$$f_n \stackrel{a.s.}{\to} f_\infty;$$

that is,

$$f_n(x) = \frac{Q(I_n(x))}{P(I_n(x))} \to f_\infty(\omega),$$

where $I_n(x)$ is the interval containing x and f_∞ satisfies

$$Q = f_\infty dP + Q(\cdot \cap N_\infty).$$

Since $\mathcal{B}_\infty = \mathcal{B}([0, 1))$, we conclude that $Q << P$ iff $f_n \to f$ almost surely and in L_1 where f satisfies $dP/dQ = f$. $\square$

10.10 Martingale and Submartingale Convergence

We have already seen some relations between martingales and submartingales, for instance Doob's decomposition. This section begins by discussing another relation between martingales and submartingales called the *Krickeberg decomposition*. This decomposition is used to extend convergence properties of positive supermartingales to more general martingale structures.

10.10.1 Krickeberg Decomposition

Krickeberg's decomposition takes a submartingale and expresses it as the difference between a positive martingale and a positive supermartingale.

Theorem 10.10.1 (Krickeberg Decomposition) *If $\{(X_n, \mathcal{B}_n), n \geq 0\}$ is a submartingale such that*

$$\sup_n E(X_n^+) < \infty,$$

then there exists a positive martingale $\{(M_n, \mathcal{B}_n), n \geq 0\}$ and a positive supermartingale $\{(Y_n, \mathcal{B}_n), n \geq 0\}$ and

$$X_n = M_n - Y_n.$$

Proof. If $\{X_n\}$ is a submartingale, then also $\{X_n^+\}$ is a submartingale. (See Example 10.6.1.) Additionally, $\{E(X_p^+|\mathcal{B}_n), p \geq n\}$ is monotone non-decreasing in p. To check this, note that by smoothing,

$$E(X_{p+1}^+|\mathcal{B}_n) = E(E(X_{p+1}^+|\mathcal{B}_p)|\mathcal{B}_n) \geq E(X_p^+|\mathcal{B}_n)$$

where the last inequality follows from the submartingale property. Monotonicity in p implies

$$\lim_{p\to\infty} \uparrow E(X_p^+|\mathcal{B}_n) =: M_n$$

exists.

We claim that $\{(M_n, \mathcal{B}_n), n \geq 0\}$ is a positive martingale. To see this, observe that

(a) $M_n \in \mathcal{B}_n$, and $M_n \geq 0$.

(b) The expectation of M_n is finite and constant in n since

$$\begin{aligned} E(M_n) &= E\big(\lim_{p\to\infty} \uparrow E(X_p^+|\mathcal{B}_n)\big) \\ &= \lim_{p\to\infty} \uparrow E(E(X_p^+|\mathcal{B}_n)) \qquad \text{(monotone convergence)} \\ &= \lim_{p\to\infty} \uparrow EX_p^+ \\ &= \sup_{p\geq 0} EX_p^+ < \infty, \end{aligned}$$

since expectations of submartingales increase. Thus $E(M_n) < \infty$.

(c) The martingale property holds since

$$\begin{aligned} E(M_{n+1}|\mathcal{B}_n) &= E\big(\lim_{p\to\infty} \uparrow E(X_p^+|\mathcal{B}_{n+1})|\mathcal{B}_n\big) \\ &= \lim_{p\to\infty} \uparrow E\big(E(X_p^+|\mathcal{B}_{n+1})|\mathcal{B}_n\big) \qquad \text{(monotone convergence)} \\ &= \lim_{p\to\infty} \uparrow E(X_p^+|\mathcal{B}_n) = M_n. \qquad \text{(smoothing)} \end{aligned}$$

We now show that

$$\{(Y_n = M_n - X_n, \mathcal{B}_n), n \ge 0\}$$

is a positive supermartingale. Obviously, $Y_n \in \mathcal{B}_n$. Why is $Y_n \ge 0$? Since $M_n = \lim_{p\to\infty} \uparrow E(X_p^+|\mathcal{B}_n)$, if we take $p = n$, we get

$$M_n \ge E(X_n^+|\mathcal{B}_n) = X_n^+ \ge X_n^+ - X_n^- = X_n.$$

To verify the supermartingale property note that

$$\begin{aligned} E(Y_{n+1}|\mathcal{B}_n) &= E(M_{n+1}|\mathcal{B}_n) - E(X_{n+1}|\mathcal{B}_n) \\ &\le M_n - X_n = Y_n \end{aligned}$$

since $E(M_{n+1}|\mathcal{B}_n) = M_n$ and $E(X_{n+1}|\mathcal{B}_n) \ge X_n$. □

10.10.2 Doob's (Sub)martingale Convergence Theorem

Krickeberg's decomposition leads to the Doob submartingale convergence theorem.

Theorem 10.10.2 (Submartingale Convergence) *If $\{(X_n, \mathcal{B}_n), n \ge 0\}$ is a (sub)-martingale satisfying*

$$\sup_{n\in\mathbb{N}} E(X_n^+) < \infty,$$

then there exists $X_\infty \in L_1$ such that

$$X_n \stackrel{a.s.}{\to} X_\infty.$$

Remark. If $\{X_n\}$ is a martingale

$$\sup_{n\in\mathbb{N}} E(X_n^+) < \infty \text{ iff } \sup_{n\in\mathbb{N}} E(|X_n|) < \infty$$

in which case the martingale is called L_1-bounded. To see this equivalence, observe that if $\{(X_n, \mathcal{B}_n), n \in \mathbb{N}\}$ is a martingale then

$$\begin{aligned} E(|X_n|) &= E(X_n^+) + E(X_n^-) = 2E(X_n^+) - E(X_n) \\ &= 2EX_n^+ - \text{const.} \end{aligned}$$

Proof. From the Krickberg decomposition, there exist a positive martingale $\{M_n\}$ and a positive supermartingale $\{Y_n\}$ such that

$$X_n = M_n - Y_n.$$

From Theorem 10.8.5, the following are true:

$$M_n \stackrel{a.s.}{\to} M_\infty, \quad Y_n \stackrel{a.s.}{\to} Y_\infty$$
$$E(M_\infty|\mathcal{B}_n) \le M_n, \quad E(Y_\infty|\mathcal{B}_n) \le Y_n,$$

so

$$E(M_\infty) \le E(M_n), \quad E(Y_\infty) \le E(Y_n)$$

and M_∞ and Y_∞ are integrable. Hence M_∞ and Y_∞ are finite almost surely, $X_\infty = M_\infty - Y_\infty$ exists, and $X_n \to X_\infty$. □

10.11 Regularity and Closure

We begin this section with two reminders and a recalled fact.

Reminder 1. (See Subsection 10.9.2.) Let $\{Z_n\}$ be a simple branching process with $P(\text{ extinction }) = 1 =: q$. Then with $Z_0 = 1$, $E(Z_1) = m$

$$W_n := Z_n/m^n \to 0 \text{ a.s.}$$

So the martingale $\{W_n\}$ satisfies

$$E(W_n) = 1 \not\to E(0) = 0$$

so $\{W_n\}$ does NOT converge in L_1. Also, there does NOT exist a random variable W_∞ such that $W_n = E(W_\infty|\mathcal{B}_n)$ and $\{W_n\}$ is NOT uniformly integrable (ui).

Reminder 2. Recall the definition of uniform integrability and its characterizations from Subsection 6.5.1 of Chapter 6. A family of random variables $\{X_t, t \in I\}$ is ui if $X_t \in L_1$ for all $t \in I$ and

$$\lim_{b\to\infty} \sup_{t\in I} \int_{|X_t|>b} |X_t| dP = 0.$$

Review Subsection 6.5.1 for full discussion and characterizations and also review Theorem 6.6.1 on page 191 for the following FACT: If $\{X_n\}$ converges a.s. and $\{X_n\}$ is ui, then $\{X_n\}$ converges in L_1.

Here is an example relevant to our development.

Proposition 10.11.1 *Let $X \in L_1$. Let $\mathcal{G}$ vary over all sub σ-fields of $\mathcal{B}$. The family $\{E(X|\mathcal{G}) : \mathcal{G} \subset \mathcal{B}\}$ is a ui family of random variables.*

Proof. For any $\mathcal{G} \subset \mathcal{B}$

$$\begin{aligned}\int_{[|E(X|\mathcal{G})|>b]} |E(X|\mathcal{G})| dP &\le \int_{[E(|X|\,|\mathcal{G})>b]} E(|X|\,|\mathcal{G}) dP \\ &= \int_{[E(|X|\,|\mathcal{G})>b]} |X| dP \qquad \text{(definition)} \\ &= \int_{[E(|X|\,|\mathcal{G})>b]\cap[|X|\le K]} |X| dP \\ &\quad + \int_{[E(|X|\,|\mathcal{G})>b]\cap[|X|>K]} |X| dP \\ &\le KP[E(|X|\,|\mathcal{G}) > b] + \int_{[|X|>K]} |X| dP,\end{aligned}$$

and applying Markov's inequality yields a bound

$$\begin{aligned}&\le \frac{K}{b} E(E(|X|\,|\mathcal{G})) + \int_{[|X|>K]} |X| dP \\ &= \frac{K}{b} E(|X|) + \int_{[|X|>K]} |X| dP;\end{aligned}$$

that is,

$$\begin{aligned}\limsup_{b\to\infty} \sup_{\mathcal{G}} \int_{[|E(X|\mathcal{G})|>b]} |E(X|\mathcal{G})| dP &\le \limsup_{b\to\infty} \left(\frac{K}{b} E(|X|) + \int_{|X|>K} |X| dP \right) \\ &= \int_{|X|>K} |X| dP \to 0\end{aligned}$$

as $K \to \infty$ since $X \in L_1$. □

We now characterize ui martingales. Compare this result to Proposition 10.8.6 on page 374.

Proposition 10.11.2 (Uniformly Integrable Martingales) *Suppose that* $\{(X_n, \mathcal{B}_n), n \ge 0\}$ *is a martingale. The following are equivalent:*

(a) $\{X_n\}$ *is* L_1*-convergent.*

(b) $\{X_n\}$ *is* L_1*-bounded and the almost sure limit is a closing random variable; that is,*

$$\sup_n E(|X_n|) < \infty.$$

There exists a random variable X_∞ *such that* $X_n \stackrel{a.s.}{\to} X_\infty$ *(guaranteed by the Martingale Convergence Theorem 10.10.2) which satisfies*

$$X_n = E(X_\infty|\mathcal{B}_n), \quad \forall n \in \mathbb{N}.$$

(c) *The martingale is closed on the right; that is, there exists $X \in L_1$ such that*

$$X_n = E(X|\mathcal{B}_n), \quad \forall n \in \mathbb{N}.$$

(d) *The sequence $\{X_n\}$ is ui.*

If any one of (a)-(d) is satisfied, the martingale is called regular or closable.

Proof. (a)→(b). If $\{X_n\}$ is L_1-convergent, $\lim_{n\to\infty} E(|X_n|)$ exists, so $\{E(|X_n|)\}$ is bounded and thus $\sup_n E(|X_n|) < \infty$. Hence the martingale is L_1-bounded and by the martingale convergence theorem 10.10.2, $X_n \stackrel{a.s.}{\to} X_\infty$. Since conditional expectations preserve L_1 convergence (cf (10.21)) we have as a consequence of $X_n \stackrel{L_1}{\to} X_\infty$ that as $j \to \infty$

$$X_n = E(X_j|\mathcal{B}_n) \stackrel{L_1}{\to} E(X_\infty|\mathcal{B}_n).$$

Thus, X_∞ is a closing random variable.

(b)→(c). We must find a closing random variable satisfying (c). The random variable $X = X_\infty$ serves the purpose and $X_\infty \in L_1$ since from (b)

$$E(|X_\infty|) = E(\liminf_{n\to\infty} |X_n|) \leq \liminf_{n\to\infty} E(|X_n|) \leq \sup_{n\in\bar{\mathbb{N}}} E(|X_n|) < \infty.$$

(c)→(d). The family $\{E(X|\mathcal{B}_n), n \in \mathbb{N}\}$ is ui by Proposition 10.11.1.

(d)→(a). If $\{X_n\}$ is ui, $\sup_n E(|X_n|) < \infty$ by the characterization of uniform integrability , so $\{X_n\}$ is L_1-bounded and therefore $X_n \to X_\infty$ a.s. by the martingale convergence theorem 10.10.2). But uniform integrability and almost sure convergence imply L_1 convergence. □

10.12 Regularity and Stopping

We now discuss when a stopped martingale retains the martingale characteristics. We begin with a simple but important case.

Theorem 10.12.1 *Let $\{(X_n, \mathcal{B}_n), n \geq 0\}$ be a regular martingale.*

(a) *If ν is a stopping time, then $X_\nu \in L_1$.*

(b) *If ν_1 and ν_2 are stopping times and $\nu_1 \leq \nu_2$, then*

$$\{(X_{\nu_1}, \mathcal{B}_{\nu_1}), (X_{\nu_2}, \mathcal{B}_{\nu_2})\}$$

is a two term martingale and

$$X_{\nu_1} = E(X_{\nu_2}|\mathcal{B}_{\nu_1});$$

therefore

$$E(X_{\nu_1}) = E(X_{\nu_2}) = E(X_0).$$

For regular martingales, random stopping preserves fairness and for a stopping time ν, we have $E(X_\nu) = E(X_0)$ since we may take $\nu = \nu_2$ and $\nu_1 \equiv 0$.

Proof. The martingale is assumed regular so that we can suppose

$$X_n = E(X_\infty | \mathcal{B}_n),$$

where $X_n \to X_\infty$ a.s. and in L_1. Hence when $\nu = \infty$, we may interpret $X_\nu = X_\infty$.

For any stopping time ν

$$E(X_\infty | \mathcal{B}_\nu) = X_\nu \tag{10.44}$$

since by Lemma 10.8.2

$$\begin{aligned} E(X_\infty | \mathcal{B}_\nu) &= \sum_{n \in \bar{\mathbb{N}}} E(X_\infty | \mathcal{B}_n) 1_{[\nu = n]} \\ &= \sum_{n \in \bar{\mathbb{N}}} X_n 1_{[\nu = n]} = X_\nu. \end{aligned}$$

Since $X_\infty \in L_1$,

$$\begin{aligned} E(|X_\nu|) \le& E\big(|E(X_\infty | \mathcal{B}_\nu)|\big) \le E\big(E(|X_\infty|)|\mathcal{B}_\nu\big) \\ =& E(|X_\infty|) < \infty. \end{aligned}$$

Thus $X_\nu \in L_1$.

If $\nu_1 \le \nu_2$, then $\mathcal{B}_{\nu_1} \subset \mathcal{B}_{\nu_2}$ and

$$\begin{aligned} E(X_{\nu_2} | \mathcal{B}_{\nu_1}) &= E(E(X_\infty | \mathcal{B}_{\nu_2}) | \mathcal{B}_{\nu_1}) \quad \text{(by (10.44))} \\ &= E(X_\infty | \mathcal{B}_{\nu_1}) \quad \text{(smoothing)} \\ &= X_{\nu_1}. \quad \text{(by (10.44)).} \end{aligned}$$

□

Remark. A criterion for regularity is L_p-boundedness: If $\{(X_n, \mathcal{B}_n), n \ge 0\}$ is a martingale and

$$\sup_n E(|X_n|^p) < \infty, \quad p > 1,$$

then $\{X_n\}$ is ui and hence regular. See (6.13) of Chapter 6 on page 184. The result is false for $p = 1$. Take the branching process $\{(W_n = Z_n/m^n, \mathcal{B}_n), n \in \mathbb{N}\}$. Then $\sup_n E(|W_n|) = 1$, but as noted in *Reminder 1* of Section 10.11, $\{W_n\}$ is NOT ui.

Example 10.12.1 (An L_2-bounded martingale) An example of an L_2-bounded martingale can easily be constructed from the simple branching process martingale $W_n = Z_n/m^n$ with $m > 1$. Let

$$\sigma^2 = \text{Var}(Z_1) = \sum_{k=0}^{\infty} k^2 p_k - (\sum_{k=0}^{\infty} k p_k)^2 < \infty.$$

The martingale $\{W_n\}$ is L_2 bounded and

$$W_n \to W, \quad \text{almost surely and in } L_1,$$

and $E(W) = 1$, $\mathrm{Var}(W) = \frac{\sigma^2}{m^2-m}$.

Proof. Standard facts arising from solving difference equations (cf. Resnick (1994)) yield

$$\mathrm{Var}(Z_n) = \sigma^2 \frac{m^n(m^n-1)}{m^2-m}$$

so

$$\mathrm{Var}(W_n) = \frac{1}{m^{2n}} \left\{ \frac{\sigma^2 m^n (m^n - 1)}{m^2 - m} \right\} = \frac{\sigma^2 (1 - \frac{1}{m^n})}{m^2 - m}$$

and

$$EW_n^2 = \mathrm{Var}(W_n) + \underbrace{(EW_n)^2}_{1} = 1 + \frac{\sigma^2 (1 - \frac{1}{m^n})}{m^2 - m}.$$

For $m > 1$

$$EW_n^2 \to 1 + \frac{\sigma^2}{m^2 - m}.$$

Thus, $\sup_n E(W_n^2) < \infty$, so that $\{W_n\}$ is L_2 bounded and

$$1 = E(W_n) \to E(W),$$

$$E(W_n^2) \to E(W^2) = 1 + \frac{\sigma^2}{m^2 - m},$$

and

$$\mathrm{Var}(W) = \frac{\sigma^2}{m^2 - m}.$$

□

10.13 Stopping Theorems

We now examine more flexible conditions for a stopped martingale to retain martingale characteristics. In order for this to be the case, either one must impose conditions on the sequence (such as the ui condition discussed in the last section) or on the stopping time or both. We begin with a reminder of Proposition 10.8.2 on page 368 which says that if $\{(X_n, \mathcal{B}_n), n \in \mathbb{N}\}$ is a martingale and ν is a stopping time, then $\{(X_{\nu \wedge n}, \mathcal{B}_n), n \in \mathbb{N}\}$ is still a martingale.

With this reminder in mind, we call a stopping time ν *regular* for the martingale $\{(X_n, \mathcal{B}_n), n \in \mathbb{N}\}$ if $\{(X_{\nu \wedge n}, \mathcal{B}_n), n \geq 0\}$ is a regular martingale. The next result presents necessary and sufficient conditions for a stopping time to be regular.

Proposition 10.13.1 (Regularity) *Let $\{(X_n, \mathcal{B}_n), n \in \mathbb{N}\}$ be a martingale and suppose ν is a stopping time. Then ν is regular for $\{X_n\}$ iff the following three conditions hold.*

(i) $X_\infty := \lim_{n\to\infty} X_n$ exists a.s. on $[\nu = \infty]$ which means $\lim_{n\to\infty} X_{\nu\wedge n}$ exists a.s. on Ω.

(ii) $X_\nu \in L_1$. Note from (i) we know X_ν is defined a.s. on Ω.

(iii) $X_{\nu\wedge n} = E(X_\nu|\mathcal{B}_n), \quad n \in \mathbb{N}$.

Proof. Suppose ν is regular. Then $\{(Y_n = X_{\nu\wedge n}, \mathcal{B}_n), n \geq 0\}$ is a regular martingale. Thus from Proposition 10.11.2 page 389

(i) $Y_n \to Y_\infty$ a.s. and in L_1 and on the set $[\nu = \infty]$, $Y_n = X_{\nu\wedge n} = X_n$, and so $\lim_{n\to\infty} X_n$ exists a.s. on $[\nu = \infty]$.

(ii) $Y_\infty \in L_1$. But $Y_\infty = X_\nu$.

(iii) We have $E(Y_\infty|\mathcal{B}_n) = Y_n$; that is, $E(X_\nu|\mathcal{B}_n) = X_{\nu\wedge n}$.

Conversely, suppose (i), (ii) and (iii) from the statement of the proposition hold. From (i), we get X_ν is defined a.s. on Ω. From (ii), we learn $X_\nu \in L_1$ and from (iii), we get that X_ν is a closing random variable for the martingale $\{X_{\nu\wedge n}\}$. So $\{X_{\nu\wedge n}\}$ is regular from Proposition 10.11.2. □

Here are two circumstances which guarantee that ν is regular.

(i) If $\nu \leq M$ a.s., then ν is regular since

$$\{X_{\nu\wedge n}, n \in \mathbb{N}\} = \{X_0, X_1, \dots, X_\nu, X_\nu, \dots\}$$

is ui. To check uniform integrability, note that

$$|X_{\nu\wedge n}| \leq \sup_m |X_{\nu\wedge m}| = \sup_{m\leq M} |X_m| \in L_1.$$

Recall from Subsection 6.5.1 that domination by an integrable random variable is sufficient for uniform integrability.

(ii) If $\{X_n\}$ is regular, then any stopping time ν is regular. (See Corollary 10.13.1 below.)

The relevance of Proposition 10.13.1 is shown in the next result which yields the same information as Theorem 10.12.1 but under somewhat weaker, more flexible conditions.

Theorem 10.13.2 *If ν is regular and $\nu_1 \leq \nu_2 \leq \nu$ for stopping times ν_1 and ν_2, then for $i = 1, 2$, X_{ν_i} exists, $X_{\nu_i} \in L_1$ and*

$$\{(X_{\nu_1}, \mathcal{B}_{\nu_1}), (X_{\nu_2}, \mathcal{B}_{\nu_2})\}$$

is a two term martingale; that is,

$$E(X_{\nu_2}|\mathcal{B}_{\nu_1}) = X_{\nu_1}.$$

Note the following conclusion from Theorem 10.13.2. Suppose ν is regular, $\nu_1 = 0$ and $\nu_2 = \nu$. Then

$$E(X_\nu|\mathcal{B}_0) = X_0 \text{ and } E(X_\nu) = E(X_0).$$

Proof of Theorem 10.13.2. Let $Y_n = X_{\nu\wedge n}$. So $\{(Y_n, \mathcal{B}_n), n \in \mathbb{N}\}$ is a regular martingale. For regular martingales, Theorem 10.12.1 implies whenever $\nu_1 \leq \nu_2$ that

$$Y_{\nu_1} = E(Y_{\nu_2}|\mathcal{B}_{\nu_1}).$$

But if $\nu_1 \leq \nu_2 \leq \nu$, then $Y_{\nu_1} = X_{\nu_1\wedge\nu} = X_{\nu_1}$ and $Y_{\nu_2} = X_{\nu\wedge\nu_2} = X_{\nu_2}$. □

Corollary 10.13.1 *(a) Suppose ν_1 and ν_2 are stopping times and $\nu_1 \leq \nu_2$. If ν_2 is regular for the martingale $\{(X_n, \mathcal{B}_n), n \geq 0\}$, so is ν_1.*

(b) If $\{(X_n, \mathcal{B}_n), n \geq 0\}$ is a regular martingale, every stopping time ν is regular.

Proof. (b) Set $\nu_2 \equiv \infty$. Then

$$\{X_{\nu_2\wedge n}\} = \{X_{\infty\wedge n}\} = \{X_n\}$$

is regular so ν_2 is regular for $\{X_n\}$. If we assume (a) is true, we conclude ν is also regular.

(a) In the Theorem 10.13.2, put $\nu_2 = \nu$ to get $X_{\nu_1} \in L_1$. It suffices to show $\{X_{\nu_1\wedge n}\}$ is ui. We have

$$\begin{aligned}\int_{[|X_{\nu_1\wedge n}|>b]} |X_{\nu_1\wedge n}|dP &= \int_{[|X_{\nu_1\wedge n}|>b,\nu_1\leq n]} |X_{\nu_1\wedge n}|dP \\ &\quad + \int_{[|X_{\nu_1\wedge n}|>b,\nu_1>n]} |X_{\nu_1\wedge n}|dP \\ &= A + B.\end{aligned}$$

Now for B we have

$$\begin{aligned}B &\leq \int_{[|X_n|>b,\nu_1>n]} |X_n|dP \leq \int_{[|X_n|>b,\nu_2>n]} |X_n|dP \\ &\leq \int_{[|X_{\nu_2\wedge n}|>b]} |X_{\nu_2\wedge n}|dP \overset{b\to\infty}{\to} 0,\end{aligned}$$

since ν_2 regular implies $\{X_{\nu_2\wedge n}\}$ is ui.

For the term A we have

$$A = \int_{[|X_{\nu_1}|>b,\nu_1\leq n]} |X_{\nu_1}|dP \leq \int_{[|X_{\nu_1}|>b]} |X_{\nu_1}|dP \overset{b\to\infty}{\to} 0,$$

since $X_{\nu_1} \in L_1$. □

Here is another characterization of a regular stopping time ν.

Theorem 10.13.3 *In order for the stopping time ν to be regular for the martingale $\{(X_n, \mathcal{B}_n), n \geq 0\}$, it is necessary and sufficient that*

$$(a) \quad \int_{[\nu<\infty]} |X_\nu| dP < \infty, \tag{10.45}$$

and

$$(b) \quad \{X_n 1_{[\nu>n]}, n \in \mathbb{N}\} \text{ is ui.} \tag{10.46}$$

Proof. Sufficiency: We show that (a) and (b) imply that $\{X_{\nu\wedge n}\}$ is ui and therefore that ν is regular. To prove uniform integrability, note that

$$\begin{aligned}
\int_{[|X_{\nu\wedge n}|>b]} |X_{\nu\wedge n}| dP &= \int_{[\nu\leq n, |X_\nu|>b]} |X_\nu| dP + \int_{[\nu>n, |X_n|>b]} |X_n| dP \\
&\leq \int_{[\nu<\infty]\cap[|X_\nu|>b]} |X_\nu| dP \\
&\quad + \int_{[|X_n|1_{[\nu>n]}>b]} |X_n| 1_{[\nu>n]} dP \\
&= A + B.
\end{aligned}$$

For A we have that

$$A = \int_{[|X_\nu|1_{[\nu<\infty]}>b]} |X_\nu| 1_{[\nu<\infty]} dP \stackrel{b\to\infty}{\to} 0,$$

since $X_\nu 1_{[\nu<\infty]} \in L_1$ by (a). For B we have $B \to 0$ as $b \to \infty$ since $\{X_n 1_{[\nu>n]}\}$ is assumed ui by assumption (b).

Converse: Suppose ν is regular. We show that (a) and (b) are necessary.

Necessity of (a):

$$\begin{aligned}
\int_{[\nu<\infty]} |X_\nu| dP &= \lim_{n\to\infty} \uparrow \int_{[\nu\leq n]} |X_\nu| dP \\
&= \lim_{n\to\infty} \uparrow \int_{[\nu\leq n]} |X_{\nu\wedge n}| dP \\
&\leq \sup_n E(|X_{\nu\wedge n}|) < \infty
\end{aligned}$$

since $\{X_{\nu\wedge n}\}$ is ui.

Necessity of (b): If ν is regular, then

$$|X_n 1_{[\nu>n]}| \leq |X_{\nu\wedge n}|,$$

and $\{X_{\nu\wedge n}\}$ is ui implies that $\{X_n 1_{[\nu>n]}\}$ is ui since a smaller sequence is ui if a dominating one is ui. (See Subsection 6.5.1.) □

Remark 10.13.1 A sufficient condition for (10.45) is that $\{X_n\}$ be an L_1-bounded martingale. To see this, recall that L_1-bounded martingales converge almost surely so that

$$X_n \stackrel{a.s.}{\to} X_\infty$$

and X_ν is thus defined almost everywhere. We claim $X_\nu \in L_1$, and therefore $X_\nu 1_{[\nu<\infty]} \in L_1$. To verify the claim, observe that $X_{\nu\wedge n} \stackrel{a.s.}{\to} X_\nu$, and so by Fatou's lemma

$$E(|X_\nu|) = E\big(\lim_{n\to\infty} |X_{\nu\wedge n}|\big) \le \liminf_{n\to\infty} E(|X_{\nu\wedge n}|). \tag{10.47}$$

Also,

$$E(X_n|\mathcal{B}_{\nu\wedge n}) = X_{\nu\wedge n}, \tag{10.48}$$

since by Lemma 10.8.2

$$\begin{aligned} E(X_n|\mathcal{B}_{\nu\wedge n}) &= \sum_{j\in\mathbb{N}} E(X_n|\mathcal{B}_j)1_{[\nu\wedge n=j]} \\ &= \sum_{j\le n} E(X_n|\mathcal{B}_j)1_{[\nu\wedge n=j]} + \sum_{j>n} E(X_n|\mathcal{B}_j)1_{[\nu\wedge n=j]} \end{aligned}$$

and since $[\nu\wedge n = j] = \emptyset$ when $j > n$, on applying the martingale property to the first sum, we get

$$= \sum_{j\le n} X_j 1_{[\nu\wedge n=j]} = X_{\nu\wedge n},$$

as claimed in (10.48). Thus

$$\begin{aligned} E(|X_{\nu\wedge n}|) &\le E|E(X_n|\mathcal{B}_{\nu\wedge n})| \\ &\le E\big(E(|X_n|\,|\mathcal{B}_{\nu\wedge n})\big) = E(|X_n|). \end{aligned} \tag{10.49}$$

From (10.47) and (10.49)

$$\begin{aligned} E(|X_\nu|) &\le \liminf_{n\to\infty} E(|X_{\nu\wedge n}|) \le \lim_{n\to\infty} E(|X_n|) \\ &= \sup_n E(|X_n|) < \infty. \end{aligned}$$

□

Additional remark. If the martingale $\{X_n\}$ is non-negative, then it is automatically L_1-bounded since $\sup_n E(|X_n|) = \sup_n E(X_n) = E(X_0)$. Thus (10.45) holds.

We now apply Theorem 10.13.3 to study the first escape times from a strip.

Corollary 10.13.2 *Let $\{(X_n, \mathcal{B}_n), n \ge 0\}$ be an L_1-bounded martingale.*

(a) *For any level $a > 0$, the escape time $\nu_a := \inf\{n : |X_n| > a\}$ is regular. In particular, this holds if $\{X_n\}$ is a positive martingale.*

(b) *For any $b < 0 < a$, the time $\nu_{a,b} = \inf\{n : X_n > a \text{ or } X_n < b\}$ is regular.*

Proof. (a) We apply Theorem 10.13.3. Since $\{X_n\}$ is L_1-bounded (10.45) is immediate from the previous remark. For (10.46) we need that $\{X_n 1_{\nu>n}\}$ is ui but since $|X_n 1_{\nu>n}| \leq a$, uniform integrability is automatic.

(b) We have

$$\nu_{a,b} \leq \nu_{|a|\vee b} = \inf\{n : |X_n| > |a| \vee |b|\}$$

and $\nu_{|a|\vee|b|}$ is regular from part (a) so $\nu_{a,b}$, being dominated by a regular stopping time, is regular from Corollary 10.13.1. □

We end this section with an additional regularity criterion.

Proposition 10.13.4 *Suppose $\{(X_n, \mathcal{B}_n), n \geq 0\}$ is a martingale. Then*

$$(i) \quad \nu \text{ is regular for } \{X_n\} \text{ and}$$
$$(ii) \quad X_n \stackrel{a.s.}{\to} 0 \text{ on } [\nu = \infty]$$

is equivalent to

$$(iii) \quad \int_{[\nu<\infty]} |X_\nu| dP < \infty \text{ and}$$
$$(iv) \quad \int_{[\nu>n]} |X_n| dP \to 0.$$

Proof. Assume (i) and (ii). Then $\{(X_{\nu\wedge n}, \mathcal{B}_n), n \in \mathbb{N}\}$ is a regular martingale, and hence $X_{\nu\wedge n} \to X_\nu \in L_1$ almost surely and in L_1 and from (ii), $X_\nu = 0$ on $[\nu = \infty]$. Then

$$\infty > \int_\Omega |X_\nu| dP = \int_{[\nu<\infty]} |X_\nu| dP$$

since $X_\nu = 0$ on $[\nu = \infty]$. This is (iii). Also

$$\int_{[\nu>n]} |X_n| dP = \int_{[\nu>n]} |X_{\nu\wedge n}| dP \to \int_{[\nu=\infty]} |X_\nu| dP = 0$$

since $X_{\nu\wedge n} \stackrel{L_1}{\to} X_\nu$ entails

$$\sup_A \left| \int_A (X_{\nu\wedge n}| - \int_A |X_\nu| \right| \to 0.$$

Thus (i)+(ii) ⇒ (iii)+(iv).

Given (iii) and (iv): Recall from Theorem 10.13.3 that ν is regular for $\{X_n\}$ iff

(a) $\int_{[\nu<\infty]} |X_\nu| dP < \infty$

(b) $\{X_n 1_{\nu>n}\}$ ui.

If we show (iv) implies (b), then we get ν is regular. Note (iv) implies

$$\xi_n := |X_n| 1_{[\nu>n]} \xrightarrow{L_1} 0.$$

We show this implies $\{\xi_n\}$ is ui. Observe

$$\sup_n \int_{[\xi_n>b]} \xi_n dP \leq \bigvee_{n\leq n_0} \int_{[\xi_n>b]} \xi_n dP \bigvee \bigvee_{n>n_0} E(\xi_n).$$

Choose n_0 so large that $\vee_{n>n_0} E(\xi_n) < \epsilon$ (since $E(\xi_n) \to 0$, we can do this) and choose b so large that

$$\bigvee_{n\leq n_0} \int_{[\xi_n>b]} \xi_n dP < \epsilon,$$

for a total bound of 2ϵ.

So we get (i). Since ν is regular, X_ν is defined on Ω and $X_\nu \in L_1$ (recall Theorem 10.13.2) and $X_{\nu\wedge n} \to X_\nu$ a.s. and in L_1. But, as before

$$\begin{aligned} 0 = \lim_{n\to\infty} E(\xi_n) &= \lim_{n\to\infty} \int_{[\nu>n]} |X_n| dP \\ &= \lim_{n\to\infty} \int_{[\nu>n]} |X_{\nu\wedge n}| dP = \int_{[\nu=\infty]} |X_\nu| dP. \end{aligned}$$

So $X_\nu 1_{\nu=\infty} = 0$ almost surely; that is, $X_n \to 0$ on $[\nu = \infty]$. Hence (ii) holds. □

10.14 Wald's Identity and Random Walks

This section discusses a martingale approach to some facts about the random walk. Consider a sequence of iid random variables $\{Y_n, n \geq 1\}$ which are not almost surely constant and define the *random walk* $\{X_n, n \geq 0\}$ by

$$X_0 = 0, \quad X_n = \sum_1^n Y_i, n \geq 1,$$

with associated σ-fields

$$\mathcal{B}_0 = \{\emptyset, \Omega\}, \quad \mathcal{B}_n = \sigma(Y_1, \ldots, Y_n) = \sigma(X_0, \ldots, X_n).$$

Define the cumulant generating function by

$$\phi(u) = \log E(\exp\{uY_1\}), \quad u \in \mathbb{R}.$$

We recall the following facts about cumulant generating functions.

1. ϕ is convex.

   Let $\alpha \in [0,1]$. Recall Hölder's inequality from Subsection 6.5.2: If $p > 0$, $q > 0$, $p^{-1} + q^{-1} = 1$, then

$$E(\xi\eta) \leq (E|\xi|^p)^{1/p}(E|\eta|^q)^{1/q}.$$

   Set $p = 1/\alpha$, and $q = 1/(1-\alpha)$ and we have

$$\begin{aligned}\phi(\alpha u_1 + (1-\alpha)u_2) &= \log E\left(e^{\alpha u_1 Y_1} e^{(1-\alpha)u_2 Y_1}\right)\\ &\leq \log\left(E(e^{u_1 Y_1})\right)^{\alpha}\left(E(e^{u_2 Y_1})\right)^{1-\alpha}\\ &= \alpha\phi(u_1) + (1-\alpha)\phi(u_2).\end{aligned}$$

2. The set $\{u : \phi(u) < \infty\} =: [\phi < \infty]$ is an interval containing 0. (This interval might be $[0,0] = \{0\}$, as would be the case if Y_1 were Cauchy distributed).

   If $u_1 < u_2$ and $\phi(u_i) < \infty$, $i = 1, 2$, then for $0 \leq \alpha \leq 1$,

$$\phi(\alpha u_1 + (1-\alpha)u_2) \leq \alpha\phi(u_1) + (1-\alpha)\phi(u_2) < \infty.$$

   So if $u_i \in [\phi < \infty]$, $i = 1, 2$, then

$$[u_1, u_2] \subset [\phi < \infty].$$

   Note $\phi(0) = \log E(e^{0 Y_1}) = \log 1 = 0$.

3. If the interior of $[\phi < \infty]$ is non-empty, ϕ is analytic there, hence infinitely differentiable there, and

$$\phi'(u) = E\left(Y_1 \exp\{uY_1 - \phi(u)\}\right),$$

   so

$$\phi'(0) = E(Y_1).$$

   One may also check that

$$\phi''(0) = \mathrm{Var}(Y_1).$$

4. On $[\phi < \infty]$, ϕ is strictly convex and ϕ' is strictly increasing.

10.14.1 The Basic Martingales

Here is a basic connection between martingales and the random walk.

Proposition 10.14.1 *For any $u \in [\phi < \infty]$, define*

$$M_n(u) = \exp\{uX_n - n\phi(u)\} = \frac{e^{uX_n}}{\left(E(e^{uY_1})\right)^n}.$$

Then $\{(M_n(u), \mathcal{B}_n), n \in \mathbb{N}\}$ is a positive martingale with $E(M_n(u)) = 1$. Also, $M_n(u) \to 0$ a.s. as $n \to \infty$ and hence $\{M_n(u)\}$ is a non-regular martingale.

Proof. The fact that $\{M_n(u)\}$ is a martingale was discussed earlier. See item 4 of Section 10.5.

Now we check that $M_n(u) \to 0$ almost surely as $n \to \infty$. We have that $u \in [\phi < \infty]$ and $0 \in [\phi < \infty]$ implies $\frac{u}{2} \in [\phi < \infty]$, and by strict convexity

$$\phi(u/2) = \phi(\frac{1}{2}0 + \frac{1}{2}u) < \frac{1}{2}\phi(0) + \frac{1}{2}\phi(u) = \frac{1}{2}\phi(u). \tag{10.50}$$

Also $\{M_n(\frac{u}{2})\}$ is a positive martingale and L_1-bounded, so there exists a random variable Z such that

$$M_n(\frac{u}{2}) = \exp\{\frac{u}{2}X_n - n\phi(\frac{u}{2})\} \to Z < \infty$$

and

$$M_n^2(\frac{u}{2}) = \exp\{uX_n - 2n\phi(\frac{u}{2})\} \to Z^2 < \infty.$$

Therefore

$$\begin{aligned} M_n(u) &= \exp\{uX_n - n\phi(u)\} \\ &= \exp\{uX_n - 2n\phi(\frac{u}{2}) + n[2\phi(\frac{u}{2}) - \phi(u)]\} \\ &= (X^2 + o(1))\exp\{n[2(\phi(\frac{u}{2}) - \frac{1}{2}\phi(u))]\} \\ &\to 0 \end{aligned}$$

since $\phi(\frac{u}{2}) - \frac{1}{2}\phi(u) < 0$ from (10.50). □

MORE MARTINGALES. From Proposition 10.14.1 we can get many other martingales. Since ϕ is analytic on $[\phi < \infty]$, it is also true that

$$u \mapsto \exp\{ux - n\phi(u)\}$$

is analytic. We may expand in a power series to get

$$\exp\{ux - n\phi(u)\} = \sum_{k=0}^{\infty} \frac{u^k}{k!} f_k(n, x). \tag{10.51}$$

If we expand the left side of (10.51) as a power series in u, the coefficient of u^k is $f_k(n, x)/k!$. As an example of this procedure, we note the first few terms:

$$\begin{aligned} f_0(n, x) &= \exp\{ux - n\phi(u)\}|_{u=0} = 1 \\ f_1(n, x) &= \frac{\partial}{\partial u} \exp\{ux - n\phi(u)\}|_{u=0} \\ &= \exp\{ux - n\phi(u)\}(x - n\phi'(u))|_{u=0} \\ &= 1 \cdot (x - nEY_1) \end{aligned}$$

and

$$\begin{aligned} f_2(n, x) &= \frac{\partial^2}{\partial u^2} \exp\{ux - n\phi(u)\}|_{u=0} \\ &= \frac{\partial}{\partial u}\{e^{ux - n\phi(u)}(x - n\phi'(u))\}|_{u=0} \\ &= e^{ux-n\phi(u)}(-n\phi''(u)) + \\ &\qquad e^{ux-n\phi(u)}(x - n\phi'(u))^2|_{u=0} \\ &= \left(x - nE(Y_1)\right)^2 - n\mathrm{Var}(Y_1). \end{aligned}$$

Each of these coefficients can be used to generate a martingale.

Proposition 10.14.2 *For each $k \geq 1, \{(f_k(n, X_n), \mathcal{B}_n), n \geq 0\}$ is a martingale. In particular*

$$\begin{aligned} k = 1, &\quad \{(f_1(n, X_n) = X_n - nE(Y_1) = X_n - E(X_n), \mathcal{B}_n), n \in \mathbb{N}\} \\ k = 2, &\quad \{((X_n - E(X_n))^2 - Var(X_n), \mathcal{B}_n), n \in \mathbb{N}\} \end{aligned}$$

are martingales.
(If $Var(Y_1) = \sigma^2$ and $E(Y_1) = 0$, then $\{X_n^2 - n\sigma^2\}$ is a martingale.)

For the most important cases where $k = 1$ or $k = 2$, one can verify directly the martingale property (see Section 10.5). Thus we do not give a formal proof of the general result but only give the following heuristic derivation. From the martingale property of $\{M_n(u)\}$ we have for $m < n$

$$E(M_n(u)|\mathcal{B}_m) = M_m(u);$$

that is,

$$E\left(\exp\{uX_n - n\phi(u)\}|\mathcal{B}_m\right) = e^{uX_m - m\phi(u)}$$

so that

$$E\left(\sum_{k=0}^{\infty} \frac{u^k}{k!} f_k(n, X_n)|\mathcal{B}_m\right) = \sum_{k=0}^{\infty} \frac{u^k}{k!} f_k(m, X_m).$$

Now differentiate inside $E(\ |\mathcal{B}_m)$ k times and set $u = 0$. This needs justification which can be obtained from the Dominated Convergence Theorem. The differentiation yields

$$E(f_k(n, X_n)|\mathcal{B}_m) = f_k(m, X_m). \qquad \square$$

10.14.2 Regular Stopping Times

We will call the martingale $\{(M_n(u), \mathcal{B}_n), n \in \mathbb{N}\}$, where

$$M_n(u) = \exp\{uX_n - u\phi(u)\} = e^{uX_n}/(Ee^{uY_1})^n$$

the *exponential* martingale. Recall from Proposition 10.14.1 that if $u \neq 0$, and $u \in [\phi < \infty]$, then $M_n(u) \to 0$, almost surely. Here is Wald's Identity for the exponential martingale.

Proposition 10.14.3 (Wald Identity) *Let $u \in [\phi < \infty]$ and suppose $\phi'(u) \geq 0$. Then for $a > 0$.*

$$\nu_a^+ = \inf\{n : X_n \geq a\}$$

is regular for the martingale $\{(M_n(u), \mathcal{B}_n), n \in \mathbb{N}\}$. Consequently, by Corollary 10.13.1, any stopping time $\nu \leq \nu_a^+$ is regular and hence Wald's identity holds

$$\begin{aligned}1 = E(M_0(u)) &= E(M_\nu) \\ &= \int \exp\{uX_\nu - \nu\phi(u)\}dP \\ &= \int_{[\nu<\infty]} exp\{uX_\nu - \nu\phi(u)\}dP.\end{aligned}$$

Proof. Recall from Proposition 10.13.4, that for a stopping time ν and a martingale $\{\xi_n\}$

$$\begin{cases} (i) & \nu \text{ is regular for } \{\xi_n\} \\ (ii) & \xi_n \to 0 \text{ on } [\nu = \infty] \end{cases} \Leftrightarrow \begin{cases} (iii) & \int_{[\nu<\infty]} |\xi_\nu| dP < \infty \\ (iv) & \int_{[\nu>n]} |\xi_n| dP \to 0. \end{cases}$$

Recall from Remark 10.13.1 that (iii) automatically holds when the martingale is L_1-bounded which is implied by the martingale being positive.

So we need check that

$$\int_{[\nu_a^+>n]} M_n(u)dP = \int_{[\nu_u^+>n]} e^{uX_n - n\phi(u)}dP \to 0. \tag{10.52}$$

For the proof of (10.52) we need the following random walk *fact*. Let $\{\xi_i, i \geq 1\}$ be iid, $E(\xi_i) \geq 0$. Then

$$\limsup_{n\to\infty} \sum_{i=1}^n \xi_i = +\infty, \tag{10.53}$$

and so if

$$\nu_a^\xi := \inf\{n : \sum_{i=1}^{n} \xi_i \geq a\},$$

we have $\nu_a^\xi < \infty$ a.s. and $P[\nu_a^\xi > n] \to 0$. For the proof of (10.53), note that if $E(\xi_1) > 0$, then almost surely, by the strong law of large numbers $\sum_{i=1}^{n} \xi_i \sim nE(\xi_i) \to \infty$. If $E(\xi_i) = 0$, the result is still true but one must use standard random walk theory as discussed in, for example, Chung (1974), Feller (1971), Resnick (1992).

We now verify (10.52). We use a technique called exponential tilting. Suppose the step random variables Y_i have distribution F. On a space $(\Omega^\#, \mathcal{B}^\#, P^\#)$, define $\{Y_i^\#, i \geq 1\}$ to be iid with distribution $F^\#$ defined by

$$F^\#(dy) = e^{uy-\phi(u)}F(dy).$$

Note $F^\#$ is a probability distribution since

$$\begin{aligned} F^\#(R) &= \int_R e^{uy-\phi(u)}F(dy) = \int_\Omega e^{uY_1-\phi(u)}dP \\ &= Ee^{uY_1}/e^{\phi(u)} = 1. \end{aligned}$$

$F^\#$ is sometimes called the *Esscher transform* of F. Also

$$E^\#(Y_1^\#) = \int_R yF^\#(dy) = \int ye^{uy-\phi(u)}F(dy) = \frac{m'(u)}{m(u)} = \phi'(u)$$

where $m(u) = E(e^{uY_1})$, and by assumption

$$E^\#(Y_1^\#) = \phi'(u) \geq 0.$$

Note the joint distribution of $Y_1^\#, \dots, Y_n^\#$ is

$$\begin{aligned} P^\#[Y_1^\# \in dy_1, \dots, Y_n^\# \in dy_n] &= \prod_{i=1}^{n} e^{uy_i-\phi(u)}F(dy_i) \\ &= \exp\{u\sum_{i=1}^{n} y_i - n\phi(n)\}\prod_{i=1}^{n} F(dy_i). \end{aligned} \tag{10.54}$$

Now in order to verify (10.52), observe

$$\begin{aligned} &\int_{[\nu_a^+ > n]} e^{uX_n - n\phi(u)}dP \\ &= \int_{\{(y_1,\dots,y_n):\sum_{i=1}^{j} y_i < a, j=1,\dots,n\}} \exp\{u\sum_{i=1}^{n} y_i - n\phi(u)\}\prod_{i=1}^{n} F(dy_i) \\ &= P^\#[\sum_{i=1}^{j} Y_i^\# < a, j = 1, \dots, n] \qquad \text{(from (10.54))} \\ &= P^\#[\nu_a^{Y_1^\#} > n] \to 0. \end{aligned}$$

□

Corollary 10.14.1 *Let $-b < 0 < a$, $u \in [\phi < \infty]$ and*

$$\nu_{a,b} = \inf\{n : X_n \geq a \text{ or } X_n \leq -b\}.$$

Then $\nu_{a,b}$ is regular for $\{M_n(u)\}$ and thus satisfies Wald identity.

Proof. Note $\nu_{a,b}$ is not defined directly in terms of $\{M_n(u)\}$ and therefore Corollary 10.13.2 is not directly applicable. If $\phi'(u) \geq 0$, Proposition 10.14.3 applies, then ν_a^+ is regular for $\{M_n(u)\}$, and hence $\nu_{a,b} \leq \nu_a^+$ is regular by Corollary 10.13.1. If $\phi'(u) \leq 0$, check the previous Proposition 10.14.3 to convince yourself that

$$\nu_b^- := \inf\{n : X_n \leq b\}$$

is regular and hence $\nu_{a,b} \leq \nu_b^-$ is also regular . □

Example 10.14.1 (Skip free random walks) Suppose the step random variable Y_1 has range $\{1, 0, -1, -2, \dots\}$ and that $P[Y_1 = 1] > 0$. Then the random walk $\{X_n\}$ with steps $\{Y_j\}$ is *skip free positive* since it cannot jump over states in the upward direction.

Let $a > 0$ be an integer. Because $\{X_n\}$ is skip free positive,

$$X_{\nu_a^+} = a \text{ on } [\nu_a^+ < \infty].$$

Note

$$\phi(u) = \log\left(e^u P[Y_1 = 1] + \sum_{j=0}^{\infty} e^{-uj} P[Y_1 = -j]\right)$$

so $[0, \infty) \subset [\phi < \infty]$ and $\phi(\infty) = \infty$. By convexity, there exists $u^* \in [0, \infty)$ such that

$$\inf_{u \in [0,\infty)} \phi(u) = \phi(u^*).$$

On the interval $[u^*, \infty)$, ϕ increases continuously from the minimum $\phi(u^*)$ to ∞. Thus for $u \geq u^*$, we have $\phi'(u) \geq 0$.

For $u \geq u^*$, Wald's identity is

$$\begin{aligned}1 &= \int_{[\nu_a^+ < \infty]} \exp\{uX_{\nu_a^+} - \nu_a^+ \phi(u)\} dP \\ &= \int_{[\nu_a^+ < \infty]} \exp\{ua - \nu_a^+ \phi(u)\} dP,\end{aligned}$$

so that

$$\int_{[\nu_a^+ < \infty]} e^{-\phi(u)\nu_a^+} dP = e^{-ua}. \tag{10.55}$$

This holds for $u \in [\phi < \infty]$ and in particular it holds for $u \in [u^*, \infty)$.

Consider the following cases.

CASE (I) Suppose $\phi'(0) = E(Y_1) \geq 0$. Then $u^* = 0$ and since $E(Y_1) \geq 0$ implies $\nu_a^+ < \infty$ almost surely, we have

$$\int e^{-\phi(u)\nu_a^+} dP = e^{-ua}, \quad u \geq 0.$$

Setting $\lambda = \phi(u)$ gives

$$E\big(e^{-\lambda \nu_a^+}\big) = \int e^{-\lambda \nu_a^+} dP = e^{-\phi^{\leftarrow}(\lambda)a}.$$

In this case, Wald's identity gives a formula for the Laplace transform of ν_a^+.

CASE (II) Suppose $E(Y_1) = \phi'(0) < 0$. Since $\phi(0) = 0$, convexity requires $\phi(u^*) < 0$ and there exists a unique $u_0 > u^* > 0$ such that $\phi(u_0) = 0$. Thus if we substitute u_0 in (10.55) we get

$$e^{-u_0 a} = \int_{[\nu_a^+ < \infty]} e^{-\phi(u_0)\nu_a^+} dP = \int_{[\nu_a^+ < \infty]} e^0 dP = P[\nu_a^+ < \infty] < 1.$$

In this case, Wald's identity gives a formula for $P[\nu_a^+ < \infty]$. □

We now examine the following martingales:

$$\{X_n - nE(Y_1), n \geq 0\}, \quad \{(X_n - nE(Y_1))^2 - n\mathrm{Var}(Y_1), n \geq 0\}.$$

Neither is regular but we can find regular stopping times.

Proposition 10.14.4 *Let ν be a stopping time which satisifes $E(\nu) < \infty$. Then*

(a) ν is regular for $\{X_n - nE(Y_1)\}$ assuming $E(|Y_1|) < \infty$.

(b) ν is regular for $\{(X_n - nE(Y_1))^2 - n\mathit{Var}(Y_1)\}$ assuming $E(Y_1^2) < \infty$.

From (a) we get

$$E(X_\nu) = E(\nu)E(Y_1).$$

From (b) we get

$$E\big(X_\nu - \nu E(Y_1)\big)^2 = E(\nu)\mathit{Var}(Y_1).$$

Proof. (a) Since $\{X_n - nEY_1\}$ has mean 0, without loss of generality we can suppose $E(Y_1) = 0$. If $E(\nu) < \infty$, then $P[\nu < \infty] = 1$ and so $X_{\nu\wedge n} \stackrel{as}{\to} X_\nu$. We show, in fact, this convergence is also L_1 and thus $\{X_{\nu\wedge n}\}$ is a regular martingale and ν is a regular stopping time.

Note that

$$|X_{\nu\wedge n} - X_\nu| = \begin{cases} 0, & \text{if } \nu < n \\ |\sum_{i=1}^n Y_i - \sum_{i=1}^\nu Y_i| = |\sum_{j=n+1}^\nu Y_i|, & \text{if } \nu \geq n, \end{cases}$$

so that

$$|X_{\nu\wedge n} - X_\nu| = |\sum_{j=n+1}^{\infty} Y_j 1_{[\nu\geq j]}| \leq \sum_{j=n+1}^{\infty} |Y_j| 1_{[\nu\geq j]} =: \xi_{n+1}.$$

Note $\xi_{n+1} \leq \xi_1$ and

$$E(\xi_1) = E(\sum_{j=1}^{\infty} |Y_j| 1_{[\nu\geq j]}) = \sum_{j=1}^{\infty} E|Y_j| 1_{[\nu\geq j]}$$

and because $[\nu \geq j] \in \mathcal{B}_{j-1}$ we get by independence that this equals

$$= \sum_{j=1}^{\infty} E(|Y_1|)P[\nu \geq j] = E(|Y_1|)E(\nu) < \infty.$$

Now

$$\xi_{n+1} = \sum_{j=n+1}^{\infty} |Y_j| 1_{[\nu\geq j]} \to 0$$

as $n \to \infty$, since the series is zero when $n + 1 > \nu$. Furthermore

$$\xi_{n+1} \leq \xi_1 \in L_1,$$

and so by the dominated convergence theorem

$$0 \leftarrow E(\xi_{n+1}) \geq E\big(|X_{\nu\wedge n} - X_\nu|\big)$$

which means

$$X_{\nu\wedge n} \xrightarrow{L_1} X_\nu.$$

(b) Now suppose $E(Y_1) = 0$, $E(Y_1^2) < \infty$. We first check $X_{\nu\wedge n} \xrightarrow{L_2} X_\nu$. Note that $1_{[\nu\geq m]} \in \mathcal{B}_{m-1}$ is predictable so that $\{Y_m 1_{\nu\geq m}\}$ is a fair (martingale difference) sequence and hence orthogonal. Also,

$$\begin{aligned}\sum_{m=1}^{\infty} E(Y_m 1_{[\nu\geq m]})^2 &= \sum_{m=1}^{\infty} E(Y_m^2)P[\nu \geq m] \\ &= E(Y_1^2) \sum_{m=1}^{\infty} P[\nu \geq m] = E(Y_1^2)E(\nu) < \infty.\end{aligned}$$

As in (a), we get using orthogonality, that as $n \to \infty$

$$E(X_{\nu\wedge n} - X_\nu)^2 = E(\sum_{j=n+1}^{\infty} Y_j 1_{\nu\geq j})^2 = \sum_{j=n+1}^{\infty} E(Y_j 1_{\nu\geq j})^2 \to 0$$

since we already checked that $\sum_{m=1}^{\infty} E(Y_m 1_{\nu\geq m})^2 < \infty$. So $X_{\nu\wedge n} \xrightarrow{L_2} X_\nu$.

It follows that $X^2_{\nu\wedge n} \xrightarrow{L_1} X^2_\nu$. Furthermore

$$X^2_{\nu\wedge n} - (\nu\wedge n)\mathrm{Var}(Y_1) \xrightarrow{L_1} X^2_\nu - \nu\mathrm{Var}(Y_1),$$

since

$$\begin{aligned}
&E|X^2_{\nu\wedge n} - (\nu\wedge n)\mathrm{Var}(Y_1) - (X^2_\nu - \nu\mathrm{Var}(Y_1))| \\
&\quad \leq E\big(|X^2_{\nu\wedge n} - X^2_\nu|\big) + E\big(|\nu\wedge n - \nu|\big)\mathrm{Var}(Y_1) \\
&\quad = o(1) + \mathrm{Var}(Y_1)E\big(|\nu - n|1_{[\nu>n]}\big) \\
&\quad \leq o(1) + \mathrm{Var}(Y_1)E(\nu 1_{[\nu>n]}) \to 0.
\end{aligned}$$

Thus $\{X^2_{\nu\wedge n} - (\nu\wedge n)\mathrm{Var}Y_1\}$ is regular by Proposition 10.11.2. □

10.14.3 Examples of Integrable Stopping Times

Proposition 10.14.4 has a hypothesis that the stopping time be integrable. In this subsection, we give sufficient conditions for first passage times and first escape times from strips to be integrable.

Proposition 10.14.5 *Consider the random walk with steps $\{Y_j\}$.*

(i) If $E(Y_1) > 0$, then for $a > 0$

$$\nu_a^+ = \inf\{n : X_n \geq a\} \in L_1.$$

(ii) If $E(Y_1) < 0$, then for $b > 0$

$$\nu_b^- = \inf\{n : X_n \leq -b\} \in L_1.$$

(iii) If $E(Y_1) \neq 0$, and $Y_1 \in L_1$, then

$$\nu_{a,b} = \inf\{n : X_n \geq a \text{ or } X_n \leq -b\} \in L_1.$$

Proof. Observe that (i) implies (ii) since given (i), we can replace Y_i by $-Y_i$ to get (ii). Also (i) and (ii) imply (iii) since

$$\nu_{a,b} \leq \nu_a^+ \wedge \nu_b^- \leq \nu_a^+.$$

It suffices to show (i) and we now suppose $E(Y_1) > 0$. Then

$$\{X_{\nu_a^+\wedge n} - (\nu_a^+ \wedge n)E(Y_1), n \geq 0\}$$

is a zero mean martingale so

$$0 = E(X_0) - OE(Y_1) = E(X_{\nu_a^+\wedge n} - (\nu_a^+ \wedge n)E(Y_1)),$$

which translates to

$$E(X_{\nu_a^+\wedge n}) = E(Y_1)E(\nu_a^+ \wedge n). \tag{10.56}$$

Since

$$\nu_a^+ \wedge n \nearrow \nu_a^+$$

we get by the monotone convergence theorem

$$E(\nu_a^+ \wedge n) \nearrow E(\nu_a^+).$$

From (10.56), we need a bound on $EX_{\nu_a^+\wedge n}$.

We consider two cases:

CASE 1. Suppose that Y_1 is bounded above; that is, suppose there exists c and $Y_1 \le c$ with probability 1. On $[\nu_a^+ < \infty]$ we have $X_{\nu_a^+-1} \le a$ and $Y_{\nu_a^+} \le c$ so that

$$X_{\nu_a^+} = X_{\nu_a^+-1} + Y_{\nu_a^+} \le a + c$$

and

$$X_{\nu_a^+\wedge n} \le a \text{ if } n < \nu_a^+.$$

In any case

$$X_{\nu_a^+\wedge n} \le a + c.$$

Thus (10.56) and $E(Y_1) > 0$ imply

$$\frac{a+c}{E(Y_1)} \ge E(\nu_a^+ \wedge n) \nearrow E(\nu_a^+),$$

so $\nu_a^+ \in L_1$.

CASE 2. If Y_1 is not bounded above by c, we proceed as follows. Note as $c \uparrow \infty$, $Y_1 \wedge c \uparrow Y_1$ and $|Y_1 \wedge c| \le |Y_1| \in L_1$. By Dominated Convergence $E(Y_1 \wedge c) \to E(Y_1) > 0$. Thus, there exists $c > 0$ such that $E(Y_1 \wedge c) > 0$. Then for $n \ge 0$

$$X_n^{(c)} := \sum_{i=1}^{n}(Y_i \wedge c) \le \sum_{i=1}^{n} Y_i =: X_n$$

and

$$\nu_a^{+(c)} = \inf\{n : X_n^{(c)} \ge a\} \ge \nu_a^+ = \inf\{n : X_n \ge a\}.$$

From Case 1, $\nu_a^{+(c)} \in L_1$, so $\nu_a^+ \in L_1$. □

10.14.4 The Simple Random Walk

Suppose $\{Y_n, n \geq 1\}$ are iid random variables with range $\{\pm 1\}$ and

$$P[Y_1 = \pm 1] = \frac{1}{2}.$$

Then $E(Y_1) = 0$. As usual, define

$$X_0 = 0, \quad X_n = \sum_{i=1}^{n} Y_i, \quad n \geq 1,$$

and think of X_n as your fortune after the nth gamble. For a positive integer $a > 0$, define

$$\nu_a^+ = \inf\{n : X_n = a\}.$$

Then $P[\nu_a^+ < \infty] = 1$. This follows either from the standard random walk result (Resnick, 1994)

$$\limsup_{n\to\infty} X_n = +\infty,$$

or from the following argument. We have $\nu_a^+ < \infty$ a.s. iff $\nu_1^+ < \infty$ a.s. since if the random walk can reach state 1 in finite time, then it can start afresh and advance to state 2 with the same probability that governed its transition from 0 to 1. Suppose

$$p := P[\nu_a^+ = \infty].$$

Then

$$\begin{aligned} 1 - p &= P[\nu_1^+ < \infty] \\ &= P[\nu_1^+ < \infty, X_1 = -1] + P[\nu_1^+ < \infty, X_1 = 1] \\ &= \frac{1}{2}(1-p)(1-p) + \frac{1}{2} \end{aligned}$$

since $(1-p)(1-p)$ is the probability the random walk starts from -1, ultimately hits 0, and then starting from 0 ultimately hits 1. Therefore

$$1 - p = \frac{1}{2}(1-p)^2 + \frac{1}{2}$$

so

$$2 - 2p = 2 - 2p + p^2,$$

and $p^2 = 0$ which implies $p = 0$. Notice that even though $P[\nu_u^+ < \infty] = 1$, $E(\nu_a^+) = \infty$ since otherwise, by Wald's equation

$$a = E(X_{\nu_a^+}) = E(Y_1)E(\nu_a^+) = 0,$$

a contradiction. □

GAMBLER'S RUIN. Starting from 0, the gambling game ends when the random walk hits either a or $-b$. From Theorem 10.13.3, $\nu_{a,b}$ is regular since

$$\int_{[\nu_{a,b}<\infty]} |X_{\nu_{a,b}}|dP \leq (|a| \vee |b|)P[\nu_{a,b} < \infty] < \infty,$$

and

$$|X_n 1_{[\nu_{a,b}>n]}| \leq |a| \bigvee |b|$$

so that $\{X_n 1_{[\nu_{a,b}>n]}\}$ is ui.

Now regularity of the stopping time allows optimal stopping

$$\begin{aligned}0 = E(X_0) = E(X_{\nu_{a,b}}) &= -bP[\nu_b^- < \nu_a^+] + aP[\nu_a^+ < \nu_b^-]\\ &= -bP[\nu_b^- < \nu_a^+] + a(1 - P[\nu_b^- < \nu_a^+]).\end{aligned}$$

We solve for the probability to get

$$P[\nu_b^- < \nu_a^+] = P[\text{ hit } -b \text{ before hit } a] = \frac{a}{a+b}. \tag{10.57}$$

We now compute the expected duration of the game $E(\nu_{a,b})$. Continue to assume $P[Y_1 = \pm 1] = \frac{1}{2}$. Recall $\{X_n^2 - n, n \geq 0\}$ is a martingale and $E(X_n^2 - n) = 0$. Also $\{(X^2_{\nu_{a,b}\wedge n} - (\nu_{a,b} \wedge n), \mathcal{B}_n), n \in \mathbb{N}\}$ is a zero mean martingale so that

$$0 = E(X^2_{\nu_{a,b}\wedge n} - (\nu_{a,b} \wedge n));$$

that is,

$$E(X^2_{\nu_{a,b}\wedge n}) = E(\nu_{a,b} \wedge n). \tag{10.58}$$

As $n \to \infty$,

$$\nu_{a,b} \wedge n \uparrow \nu_{a,b}$$

so the monotone convergence theorem implies that

$$E(\nu_{a,b} \wedge n) \uparrow E(\nu_{a,b}).$$

Also

$$|X^2_{\nu_{a,b}\wedge n}| \leq |a|^2 \bigvee |b|^2$$

and

$$X_{\nu_{a,b}\wedge n} \to X_{\nu_{a,b}}$$

implies by the dominated convergence theorem that

$$E(X^2_{\nu_{a,b}\wedge n}) \to E(X^2_{\nu_{a,b}}).$$

From (10.58) and (10.57)

$$E(\nu_{a,b}) = E(X^2_{\nu_{a,b}})$$

(so $\nu_{a,b} \in L_1$ and is therefore regular by Proposition 10.14.4)

$$= a^2 P[X^2_{\nu_{a,b}} = a^2] + b^2 P[X^2_{\nu_{a,b}} = b^2]$$
$$= a^2(1 - \frac{a}{a+b}) + b^2(\frac{a}{a+b})$$
$$= a^2 \frac{b}{a+b} + \frac{b^2 a}{a+b}$$
$$= \frac{ab(a+b)}{a+b} = ab.$$

□

GAMBLER'S RUIN IN THE ASYMMETRIC CASE. Suppose now that

$$P[Y_1 = 1] = p, \quad P[Y_1 = -1] = 1 - p =: q$$

for $p \neq \frac{1}{2}$ and $0 < p < 1$. Then for $u \in \mathbb{R}$,

$$Ee^{uY_1} = e^u p + e^{-u} q$$

and from Corollary 10.14.1, $\nu_{a,b}$ is regular for the martingale

$$\{M_n(u) = \frac{e^{uX_n}}{(e^u p + e^{-u} q)^n}, n \geq 0\}$$

and Wald's identity becomes

$$1 = \int_{[\nu_{a,b} < \infty]} M_{\nu_{a,b}}(u) dP = \int \frac{e^{uX_{\nu_{a,b}}}}{(e^u p + e^{-u} q)^{\nu_{a,b}}} dP.$$

To get rid of the denominator, substitute $u = \log q/p$ so that

$$e^u p + e^{-u} q = \frac{q}{p} \cdot p + \frac{p}{q} \cdot q = q + p = 1.$$

Then with $e^u = q/p$ we have

$$1 = E\left(\exp\{uX_{\nu_{a,b}}\}\right) = e^{ua} P[\nu_a^+ < \nu_b^-] + e^{-ub} P[\nu_b^- < \nu_a^+].$$

Solving we get

$$P[\nu_b^- < \nu_a^+] = P[\text{ exit the strip at } -b]$$
$$= \frac{1 - (\frac{q}{p})^a}{(\frac{p}{q})^b - (\frac{q}{p})^a}.$$

□

10.15 Reversed Martingales

Suppose that $\{\mathcal{B}_n, n \geq 0\}$ is a *decreasing* family of σ-fields; that is, $\mathcal{B}_n \supset \mathcal{B}_{n+1}$. Call $\{(X_n, \mathcal{B}_n), n \geq 0\}$ a *reversed martingale* if $X_n \in \mathcal{B}_n$ and

$$E(X_n|\mathcal{B}_{n+1}) = X_{n+1}, \quad n \geq 0.$$

This says the index set has been reversed. For $n \leq 0$, set

$$\mathcal{B}'_n = \mathcal{B}_{-n}, \quad X'_n = X_{-n}.$$

Then $\mathcal{B}'_n \subset \mathcal{B}'_m$ if $n < m < 0$ and $\{(X'_n, \mathcal{B}'_n), n \leq 0\}$ is a martingale with index set $\{\dots, -2, -1, 0\}$ with time flowing as usual from left to right. Note this martingale is closed on the right by X'_0 and for $n < 0$

$$E(X'_0|\mathcal{B}'_n) = X'_n.$$

So the martingale $\{(X'_n, \mathcal{B}'_n), n \leq 0\}$ is ui and as we will see, this implies the original sequence is convergent a.s. and in L_1.

Example. Let $\{\xi_k, k \geq 1\}$ be iid, L_1 random variables. For $n \geq 1$ define $S_n = \sum_{i=1}^{n} \xi_i$ and $\mathcal{B}_n = \sigma(S_n, S_{n+1}, \dots)$. Hence $\mathcal{B}_n$ is a decreasing family. For $1 \leq k \leq n$, $\mathcal{B}_n = \sigma(S_n, \xi_{n+1}, \xi_{n+2}, \dots)$. Furthermore, by symmetry

$$E(\xi_k|\mathcal{B}_n) = E(\xi_1|\mathcal{B}_n), \quad 1 \leq k \leq n.$$

Adding over $k = 1, 2, \dots, n$, we get

$$S_n = E(S_n|\mathcal{B}_n) = \sum_{k=1}^{n} E(\xi_k|\mathcal{B}_n) = nE(\xi_1|\mathcal{B}_n),$$

and thus

$$\frac{S_n}{n} = E(S_1|\mathcal{B}_n)$$

which is a reversed martingale sequence and thus uniformly integrable. From Theorem 10.15.1 below, this sequence is almost surely convergent. The Kolmogorov 0-1 law gives $\lim_{n\to\infty} \frac{S_n}{n}$ is a constant, say c. But this means

$$c = \frac{1}{n}E(S_n) = E(\xi_1).$$

Thus, the Reversed Martingale Convergence Theorem 10.15.1 provides a very short proof of the strong law of large numbers.

Here are the basic convergence properties of reversed martingales.

Theorem 10.15.1 (Reversed Martingale Convergence Theorem) *Suppose that $\{\mathcal{B}_n, n \geq 0\}$ is a decreasing family of σ-fields and suppose*

$$\{(X_n, \mathcal{B}_n), n \geq 0\}$$

is a positive reversed martingale. Set

$$\mathcal{B}_\infty = \bigcap_{n\geq 0} \mathcal{B}_n.$$

(i) There exists $X_\infty \in \mathcal{B}_\infty$ and $X_n \stackrel{as}{\to} X_\infty$.

(ii) $E(X_n|\mathcal{B}_\infty) = X_\infty$ almost surely.

(iii) $\{X_n\}$ is ui and $X_n \stackrel{L_1}{\to} X_\infty$.

Proof. Recall $X'_n = X_{-n}$, $\mathcal{B}'_n = \mathcal{B}_{-n}$, $n \leq 0$ defines a martingale on the index set $\{\dots, -2, -1, 0\}$. Define

$$\begin{aligned}\delta_{a,b}^{(n)} &= \#\text{ downcrossings of } [a,b] \text{ by } X_0, \dots, X_n \\ &= \#\text{ upcrossings of } [a,b] \text{ by } X_n, X_{n-1}, \dots, X_0 \\ &= \#\text{ upcrossings of } [a,b] \text{ by } X'_{-n}, X'_{-n+1}, \dots, X'_0 \\ &= \gamma_{ab}^{(n)}.\end{aligned}$$

Now apply Dubins' inequality 10.8.4 to the positive martingale $X'_{-n}, \dots, X'_0$ to get for $n \geq 1$,

$$\begin{aligned}P[\gamma_{a,b}^{(n)} \geq k|\mathcal{B}'_{-n}] &= P[\delta_{a,b}^{(n)} \geq k|\mathcal{B}_n] \\ &\leq (\frac{a}{b})^k(\frac{X'_{-n}}{a} \wedge 1) \\ &= (\frac{a}{b})^k(\frac{X_n}{a} \wedge 1).\end{aligned}$$

Taking $E(\cdot|\mathcal{B}_\infty)$ on both sides yields

$$P[\delta_{a,b}^n \geq k|\mathcal{B}_\infty) \leq (\frac{a}{b})^k E\big((\frac{X_n}{a} \wedge 1)|\mathcal{B}_\infty\big).$$

As $n \uparrow \infty$,

$$\delta_{ab}^{(n)} \uparrow \delta_{a,b} = \#\text{ downcrossings of } [a,b] \text{ by } \{X_0, X_1, \dots\},$$

and

$$P(\delta_{ab} \geq k|\mathcal{B}_\infty) \leq (\frac{a}{b})^k \sup_n E\big((\frac{X_n}{a} \wedge 1)|\mathcal{B}_\infty\big) \leq (\frac{a}{b})^k.$$

Thus $\delta_{a,b} < \infty$ almost surely for all $a < b$ and therefore $\lim_{n\to\infty} X_n$ exists almost surely. Set $X_\infty = \limsup_{n\to\infty} X_n$ so that X_∞ exists everywhere, and $X_n \to X_\infty$ a.s. Since $X_n \in \mathcal{B}_n$, and $\{\mathcal{B}_n\}$ is decreasing, we have for $n \geq p$ that $X_n \in \mathcal{B}_p$ so $X_\infty \in \mathcal{B}_p$ for all p. Thus

$$X_\infty \in \bigcap_p \mathcal{B}_p = \mathcal{B}_\infty.$$

Now for all $n \geq 0$, $X_n = E(X_0|\mathcal{B}_n)$ so $\{X_n\}$ is ui by Proposition 10.11.1. Uniform integrability and almost sure convergence imply L_1 convergence. (See Theorem 6.6.1 on page 191. This gives (iii).

Also we have

$$E(X_n|\mathcal{B}_\infty) = E(E(X_n|\mathcal{B}_{n+1})|\mathcal{B}_\infty) = E(X_{n+1}|\mathcal{B}_\infty). \tag{10.59}$$

Now let $n \to \infty$ and use the fact that $X_n \stackrel{L_1}{\to} X_\infty$ implies that the conditional expectations are L_1-convergent. We get from (10.59)

$$X_\infty = E(X_\infty|\mathcal{B}_\infty) = \lim_{n\to\infty} E(X_n|\mathcal{B}_\infty) = E(X_n|\mathcal{B}_\infty)$$

for any $n \geq 0$. This concludes the proof. □

These results are easily extended when we drop the assumption of positivity which was only assumed in order to be able to apply Dubins' inequality 10.8.4.

Corollary 10.15.1 *Suppose $\{\mathcal{B}_n\}$ is a decreasing family and $X \in L_1$. Then*

$$E(X|\mathcal{B}_n) \to E(X|\mathcal{B}_\infty)$$

almost surely and in L_1. (The result also holds if $\{\mathcal{B}_n\}$ is an increasing family. See Proposition 10.11.2.)

Proof. Observe that if we define $\{X_n\}$ by $X_n := E(X|\mathcal{B}_n)$, then this sequence is a reversed martingale from smoothing. From the previous theorem, we know

$$X_n \to X_\infty \in \mathcal{B}_\infty$$

a.s. and in L_1. We must identify X_∞. From L_1-convergence we have that for all $A \in \mathcal{B}$,

$$\int_A E(X|\mathcal{B}_n)dP \to \int_A X_\infty dP. \tag{10.60}$$

Thus for all $A \in \mathcal{B}_\infty$

$$\begin{aligned}\int_A E(X|\mathcal{B}_n)dP &= \int_A XdP \qquad \text{(definition)}\\ &= \int_A E(X|\mathcal{B}_\infty)dP \qquad \text{(definition)}\\ &\to \int_A X_\infty dP \qquad \text{(from (10.60))}.\end{aligned}$$

So by the Integral Comparison Lemma 10.1.1

$$X_\infty = E(X|\mathcal{B}_\infty).$$

□

Example 10.15.1 (Dubins and Freedman) Let $\{X_n\}$ be some sequence of random elements of a metric space $(\mathbb{S}, \mathcal{S})$ defined on the probability space $(\Omega, \mathcal{B}, P)$ and define

$$\mathcal{B}_n = \sigma(X_n, X_{n+1}, \dots).$$

Define the *tail* σ-field

$$\mathcal{T} = \bigcap_n \mathcal{B}_n.$$

Proposition 10.15.2 *$\mathcal{T}$ is a.s. trivial (that is, $\Lambda \in \mathcal{T}$ implies $P(\Lambda) = 0$ or 1) iff*

$$\forall A \in \mathcal{B}: \sup_{B \in \mathcal{B}_n} |P(AB) - P(A)P(B)| \to 0.$$

Proof. $\to$. If $\mathcal{T}$ is a.s. trivial, then

$$P(A|\mathcal{B}_n) \to P(A|\mathcal{B}_\infty) = P(A|\mathcal{T}) = P(A|\{\emptyset, \Omega\}) = P(A) \tag{10.61}$$

a.s. and in L_1. Therefore,

$$\begin{aligned}\sup_{B \in \mathcal{B}_n} |P(AB) - P(A)P(B)| &= \sup_{B \in \mathcal{B}_n} |E\big(P(AB|\mathcal{B}_n)\big) - P(A)E(1_B)| \\ &= \sup_{B \in \mathcal{B}_n} |E\,(1_B\{P(A|\mathcal{B}_n) - P(A)\})| \\ &\le \sup_{B \in \mathcal{B}_n} E\,|P(A|\mathcal{B}_n) - P(A)| \to 0\end{aligned}$$

from (10.61).

$\leftarrow$. If $\Lambda \in \mathcal{T}$, then $\Lambda \in \mathcal{B}_n$ and therefore

$$P(\Lambda \cap \Lambda) = P(\Lambda)P(\Lambda)$$

which yields $P(\Lambda) = (P(\Lambda))^2$. □

Call a sequence $\{X_n\}$ of random elements of $(\mathbb{S}, \mathcal{S})$ *mixing* if there exists a probability measure F on $\mathcal{S}$ such that for all $A \in \mathcal{S}$

$$P[X_n \in A] \to F(A)$$

and

$$P([X_n \in \cdot] \cap A) \to F(\cdot)P(A).$$

So $\{X_n\}$ possesses a form of asymptotic independence.

Corollary 10.15.2 *If the tail σ-field $\mathcal{T}$ of $\{X_n\}$ is a.s. trivial, and*

$$P[X_n \in \cdot] \to F(\cdot),$$

then $\{X_n\}$ is mixing.

10.16 Fundamental Theorems of Mathematical Finance

This section briefly shows the influence and prominence of martingale theory in mathematical finance. It is based on the seminal papers by Harrison and Pliska (1981), Harrison and Krebs (1979) and an account in the book by Lamberton and Lapeyre (1996).

10.16.1 A Simple Market Model

The probability setup is the following. We have a probability space $(\Omega, \mathcal{B}, P)$ where Ω is finite and $\mathcal{B}$ is the set of all subsets. We assume

$$P(\{\omega\}) > 0, \quad \forall \omega \in \Omega. \tag{10.62}$$

We think of ω as a state of nature and (10.62) corresponds to the idea that all investors agree on the possible states of nature but may not agree on probability forecasts.

There is a finite time horizon $0, 1, \dots, N$ and N is the terminal date for economic activity under consideration. There is a family of σ-fields $\mathcal{B}_0 \subset \mathcal{B}_1 \subset \cdots \subset \mathcal{B}_N = \mathcal{B}$. Securities are traded at times $0, 1, \dots, N$ and we think of $\mathcal{B}_n$ as the information available to the investor at time n. We assume $\mathcal{B}_0 = \{\Omega, \emptyset\}$.

Investors trade $d + 1$ assets ($d \geq 1$) and the price of the ith asset at time n is $S_n^{(i)}$ for $i = 0, 1, \dots, d$. Assets labelled $1, \dots, d$ are risky and their prices change randomly. The asset labelled 0 is a *riskless* asset with price at time n given by $S_n^{(0)}$, and we assume as a normalization $S_0^{(0)} = 1$. The riskless asset may be thought of as a money market or savings account or as a bond growing deterministically. For instance, one model for $\{S_n^{(0)}, 0 \leq n \leq N\}$ if there is a constant interest rate r, is $S_n^{(0)} = (1+r)^n$. We assume each stochastic process $\{S_n^{(i)}, 0 \leq n \leq N\}$ is non-negative and adapted so that $0 \leq S_n^{(i)} \in \mathcal{B}_n$ for $i = 0, \dots, d$. Assume $S_n^{(0)} > 0$, $n = 0, \dots, N$. We write

$$\{\mathbf{S}_n = (S_n^{(0)}, S_n^{(1)}, \dots, S_n^{(d)}),\ 0 \leq n \leq N\}$$

for the $\mathbb{R}^{d+1}$-valued price process.

Since the money market account is risk free, we often wish to judge the quality of our investments in terms of how they compare to the riskfree asset. We can apply the discount factor $\beta_n = 1/S_n^{(0)}$ to our price process and get the discounted price process

$$\{\overline{\mathbf{S}}_n = \mathbf{S}_n / S_n^{(0)},\ 0 \leq n \leq N\}$$

which may be thought of as the original price process denominated in units of the current price of the riskless asset. Note that $\overline{S}_n^{(0)} \equiv 1$.

The change in the prices from period to period is given by the $\mathbb{R}^{d+1}$-valued process

$$\mathbf{d}_0 = \mathbf{S}_0, \quad \mathbf{d}_n = \mathbf{S}_n - \mathbf{S}_{n-1},\ n = 1, \dots, N$$

and the change in the discounted prices is

$$\overline{\mathbf{d}}_0 = \overline{\mathbf{S}}_0, \quad \overline{\mathbf{d}}_n = \overline{\mathbf{S}}_n - \overline{\mathbf{S}}_{n-1}, \ n = 1, \dots, N.$$

Note that $\overline{d}_n^{(0)} \equiv 0$ for $n = 1, \dots, N$.

A TRADING STRATEGY is an $\mathbb{R}^{d+1}$-valued stochastic process

$$\{\phi_n = (\phi_n^{(0)}, \phi_n^{(1)}, \dots, \phi_n^{(d)}), \ 0 \le n \le N\}$$

which is predictable, so that for each $i = 0, \dots, d$, we have $\phi_n^{(i)} \in \mathcal{B}_{n-1}$ for $n \ge 1$.

Note that since Ω is finite, each random variable $|\phi_n^{(i)}|, 0 \le n \le N, \ 0 \le i \le d$ is bounded. Think of the vector ϕ_n as the number of shares of each asset held in the investors portfolio between times $n-1$ and n based on information that was available up to time $n-1$. At time n, when new prices $\mathbf{S}_n$ are announced, a repositioning of the investor's portfolio is enacted leading to a position where ϕ_{n+1} shares of each asset are held. When the prices $\mathbf{S}_n$ are announced and just prior to the rebalancing of the portfolio, the *value* of the portfolio is

$$V_n(\phi) = (\phi_n, \mathbf{S}_n) = \phi_n' \mathbf{S}_n = \sum_{i=0}^{d} \phi_n^{(i)} S_n^{(i)}.$$

The discounted value process is

$$\overline{V}_n(\phi) = \beta_n V_n(\phi) = (\phi_n, \overline{\mathbf{S}}_n).$$

To summarize: we start with value $V_0(\phi) = (\phi_0, \mathbf{S}_0)$. Now $\mathbf{S}_0$ is known so we can rebalance the portfolio with ϕ_1. The current value $(\phi_1, \mathbf{S}_0)$ persists until prices $\mathbf{S}_1$ are announced. When this announcement is made, the value of the portfolio is $V_1(\phi) = (\phi_1, \mathbf{S}_1)$. Then since $\mathbf{S}_1$ is known, we rebalance the portfolio using ϕ_2 and the value is $(\phi_2, \mathbf{S}_1)$ until $\mathbf{S}_2$ is announced when the value is $(\phi_2, \mathbf{S}_2)$ and so on.

A trading strategy ϕ is called *self-financing* if we have

$$(\phi_n, \mathbf{S}_n) = (\phi_{n+1}, \mathbf{S}_n), \quad 0 \le n \le N-1. \tag{10.63}$$

This means that at time n, just after prices $\mathbf{S}_n$ are announced, the value of the portfolio is $(\phi_n, \mathbf{S}_n)$. Then using the new information of the current prices, the portfolio is rebalanced using ϕ_{n+1} yielding new value $(\phi_{n+1}, \mathbf{S}_n)$. The equality in (10.63) means that the portfolio adjustment is done without infusing new capital into the portfolio and without consuming wealth from the portfolio.

Here are some simple characterizations of when a strategy is self-financing.

Lemma 10.16.1 *If ϕ is a trading strategy, then the following are equivalent:*

(i) ϕ is self-financing.

(ii) For $1 \le n \le N$

$$V_n(\phi) = V_0(\phi) + \sum_{j=1}^{n} (\phi_j, \mathbf{d}_j). \tag{10.64}$$

(iii) For $1 \le n \le N$

$$\overline{V}_n(\phi) = V_0(\phi) + \sum_{j=1}^{n} (\phi_j, \overline{\mathbf{d}}_j). \tag{10.65}$$

Proof. Observe that the self-financing condition (10.63) is equivalent to

$$(\phi_{j+1}, \mathbf{d}_{j+1}) = (\phi_{j+1}, \mathbf{S}_{j+1}) - (\phi_j, \mathbf{S}_j) = V_{j+1}(\phi) - V_j(\phi), \tag{10.66}$$

which says that for a self-financing strategy, changes in the value function are due to price moves. Summing (10.66) over $j = 0, \dots, n$ gives (10.64). Conversely, differencing (10.64) yields (10.66) and hence (10.63). Next, in (10.63) multiply through by the discount factor β_n to get $(\phi_n, \overline{\mathbf{S}}_n) = (\phi_{n+1}, \overline{\mathbf{S}}_n)$ or

$$(\phi_{n+1}, \overline{\mathbf{d}}_{n+1}) = (\phi_{n+1}, \overline{\mathbf{S}}_{n+1}) - (\phi_n, \overline{\mathbf{S}}_n) = \overline{V}_{n+1}(\phi) - \overline{V}_n(\phi). \tag{10.67}$$

Proceed as in (ii). □

Note that since $\overline{d}_j^{(0)} = 0$, for $j = 1, \dots, N$ we can rewrite (10.65) as

$$\overline{V}_n(\phi) = V_0(\phi) + \sum_{j=1}^{n} \sum_{i=1}^{d} \phi_j^{(i)} \overline{d}_j^{(i)} \tag{10.68}$$

showing that the discounted wealth at n from a self-financing strategy is only dependent on $V_0(\phi)$ and $\{\phi_j^{(i)}, i = 1, \dots, d;\ \ j = 1, \dots, n\}$. The next result shows that if a predictable process $\{(\phi_j^{(1)}, \phi_j^{(2)}, \dots, \phi_j^{(d)}), 1 \le j \le N\}$ and an initial value $V_0 \in \mathcal{B}_0$ is given, one may always find a unique predictable process $\{\phi_j^{(0)}, 0 \le j \le N\}$ such that $\phi = \{(\phi_j^{(0)}, \phi_j^{(1)}, \dots, \phi_j^{(d)}), 0 \le j \le N\}$ is self-financing.

Lemma 10.16.2 *Given $\{(\phi_j^{(1)}, \phi_j^{(2)}, \dots, \phi_j^{(d)}), 1 \le j \le N\}$, a predictable process, and a non-negative random variable $V_0 \in \mathcal{B}_0$, there exists a unique predictable process $\{\phi_j^{(0)}, 0 \le j \le N\}$ such that*

$$\phi = \{(\phi_j^{(0)}, \phi_j^{(1)}, \dots, \phi_j^{(d)}), 0 \le j \le N\}$$

is self-financing with initial value V_0.

Proof. Suppose $(\phi_0, \ldots, \phi_N)$ is a self-financing strategy and that V_0 is the initial wealth. On the one hand, we have (10.68) with V_0 replacing $V_0(\phi)$ and on the other hand, we have from the definition that

$$\begin{aligned}\overline{V}_n(\phi) =&(\phi_n, \overline{\mathbf{S}}_n)\\ =&\phi_n^{(0)}1 + \sum_{i=1}^{d} \phi_n^{(i)}\overline{S}_n^{(i)}.\end{aligned} \tag{10.69}$$

Now equate (10.69) and (10.68) and solving for $\phi_n^{(0)}$, we get

$$\begin{aligned}\phi_n^{(0)} =&V_0 + \sum_{j=1}^{n}\sum_{i=1}^{d} \phi_j^{(i)}\overline{d}_j^{(i)} - \sum_{i=1}^{d} \phi_n^{(i)}\overline{S}_n^{(i)}\\ =&V_0 + \sum_{j=1}^{n-1}\sum_{i=1}^{d} \phi_j^{(i)}\overline{d}_j^{(i)} + \sum_{i=1}^{d} \phi_n^{(i)}\left((\overline{S}_n^{(i)} - \overline{S}_{n-1}^{(i)}) - \overline{S}_n^{(i)}\right)\\ =&V_0 + \sum_{j=1}^{n-1}\sum_{i=1}^{d} \phi_j^{(i)}\overline{d}_j^{(i)} + \sum_{i=1}^{d} \phi_n^{(i)}(-\overline{S}_{n-1}^{(i)}) \in \mathcal{B}_{n-1},\end{aligned}$$

since $\phi_{n-1}^{(i)} \in \mathcal{B}_{n-1}$ for $i = 1, \ldots, d$.

This shows how $\phi_n^{(0)}$ is determined if the strategy is self-financing, but it also shows how to pick $\phi_n^{(0)}$, given V_0 and $\{(\phi_n^{(i)}, 1 \le i \le d),\ n = 1, \ldots, N\}$ to make the strategy self-financing. □

10.16.2 Admissible Strategies and Arbitrage

There is nothing in our definitions that requires $\phi \ge \mathbf{0}$. If $\phi_n^{(i)} < 0$ for some $i = 0, 1, \ldots, d$, then we are *short* $|\phi_n^{(i)}|$ shares of the asset. We imagine borrowing $|\phi_n^{(i)}|$ shares to produce capital to invest in the other hot assets. We want to restrict the risk inherent in short sales.

We call ϕ an *admissible strategy* if ϕ is self-financing and

$$V_n(\phi) \ge 0, \quad n = 0, \ldots, N.$$

This is a kind of margin requirement which increases the likelihood of being able to pay back what was borrowed should it be necessary to do so at any time. Not only do we require that the initial value $V_0(\phi)$ be non-negative, but we require that the investor never be in a position of debt.

In fair markets in economic equilibrium, we usually assume there are no *arbitrage opportunities*. An arbitrage strategy ϕ is an admissible strategy satisfying

$$V_0(\phi) = 0 \text{ and } V_N(\phi)(\omega_0) > 0,$$

for some $\omega_0 \in \Omega$. Equivalently, we require

$$V_0(\phi) = 0 \text{ and } E(V_N(\phi)) > 0. \tag{10.70}$$

Such arbitrage strategies, if they exist, represent riskless stategies which produce positive expected profit. No initial funds are required, the investor can never come out behind at time N, and yet the investor can under certain circumstances make a positive profit. Markets that contain arbitrage opportunities are not consistent with economic equilibrium. Markets without arbitrage opportunities are called *viable*.

The next subsection characterizes markets without arbitrage opportunities.

10.16.3 Arbitrage and Martingales

There is a fundamental connection between absence of arbitrage opportunities and martingales.

Recall the given probability measure is P. For another probability measure P^*, we write $P^* \equiv P$ if $P \ll P^*$ and $P^* \ll P$ so that P and P^* have the same null sets.

Theorem 10.16.1 *The market is viable iff there exists a probability measure $P^* \equiv P$ such that with respect to P^*, $\{(\overline{\mathbf{S}}_n, \mathcal{B}_n), 0 \le n \le N\}$ is a P^*-martingale.*

Remark. Since $\overline{\mathbf{S}}_n$ is $\mathbb{R}^{d+1}$-valued, what does it mean for this quantity to be a martingale? One simple explanation is to interpret the statement component-wise so that the theorem statement asserts that for each $i = 0, \ldots, N$, $\{(\overline{S}_n^{(i)}, \mathcal{B}_n), 0 \le n \le N\}$ is a P^*-martingale.

A measure P^* which makes $\{(\overline{\mathbf{S}}_n, \mathcal{B}_n), 0 \le n \le N\}$ a P^*-martingale is called an *equivalent martingale measure* or a *risk neutral measure*.

Proof. Suppose first that $P^* \equiv P$ and $\{(\overline{\mathbf{S}}_n, \mathcal{B}_n), 0 \le n \le N\}$ is a P^*-martingale. Then $\{(\overline{\mathbf{d}}_n, \mathcal{B}_n), 0 \le n \le N\}$ is a P^*-martingale difference and thus

$$E^*(\overline{\mathbf{d}}_{j+1} | \mathcal{B}_j) = \mathbf{0}, \quad j = 0, \ldots, N-1.$$

From (10.65) we see that $\{\overline{V}_n(\phi), 0 \le n \le N\}$ is a P^*-martingale transform, and hence a martingale. Thus

$$E^*\big(\overline{V}_n(\phi)\big) = E^*\big(\overline{V}_0(\phi)\big), \quad 0 \le n \le N. \tag{10.71}$$

Suppose now that an arbitrage strategy ϕ exists. Then by definition $\overline{V}_0(\phi) = 0$ and from (10.71), we get $E^*(\overline{V}_N(\phi)) = 0$. Since ϕ is admissible, $V_N(\phi) \ge 0$ and this coupled with a zero E^*-expectation means $V_N(\phi) = 0$ P^*-almost surely, and thus (since $P^* \equiv P$) also P-almost surely. Since $P(\{\omega\}) > 0$ for all $\omega \in \Omega$, we have no exception sets and hence $V_N(\phi) \equiv 0$. This contradicts the definition of an arbitrage strategy, and we thus conclude that they do not exist.

We now consider the converse and for this we need the following lemma.

Lemma 10.16.3 *Suppose there exists a self-financing strategy ϕ which, while not necessarily admissible, nonetheless satisfies $V_0(\phi) = 0$, $V_N(\phi) \geq 0$ and $E(V_N(\phi)) > 0$. Then there exists an arbitrage strategy and the market is not viable.*

Proof of Lemma 10.16.3. If $V_n(\phi) \geq 0$, $n = 0, \ldots, N$, then ϕ is admissible and hence an arbitrage strategy.

Otherwise, there exists

$$n_0 = \sup\{k : P[V_k(\phi) < 0] > 0\},$$

and since $V_N(\phi) \geq 0$ we have $1 \leq n_0 \leq N - 1$ and thus

$$\text{(a) } P[V_{n_0}(\phi) < 0] > 0, \tag{10.72}$$

$$\text{(b) } V_n(\phi) \geq 0, \quad n_0 < n \leq N. \tag{10.73}$$

We now construct a new strategy $\psi = (\psi_0, \ldots, \psi_N)$. To do this define

$$\mathbf{e}_0 = (1, 0, \ldots, 0) \in \mathbb{R}^{d+1}.$$

The definition of ψ is

$$\psi_k = \begin{cases} \mathbf{0}, & \text{if } k \leq n_0, \\ 1_{[V_{n_0}(\phi)<0]} \left(\phi_k - \frac{V_{n_0}(\phi)}{S_{n_0}^{(0)}} \mathbf{e}_0 \right), & \text{if } k > n_0. \end{cases} \tag{10.74}$$

We observe from its definition that ψ is predictable, and in a series of steps we will show

(i) ψ is self-financing,

(ii) ψ is admissible,

(iii) $E(V_N(\psi)) > 0$,

and hence ψ is an arbitrage strategy and thus the market is not viable.

We now take up (i), (ii) and (iii) in turn.

STEP (I). To show ψ is self-financing, we need to show

$$(\psi_k, \mathbf{S}_k) = (\psi_{k+1}, \mathbf{S}_k) \tag{10.75}$$

for $k = 0, \ldots, N-1$. First of all, for $k + 1 \leq n_0$ both sides of (10.75) are 0. Now consider the case $k > n_0$. We have

$$\begin{aligned} V_k(\psi) = (\psi_k, \mathbf{S}_k) &= \left(1_{[V_{n_0}(\phi)<0]} \left(\phi_k - \frac{V_{n_0}(\phi)}{S_{n_0}^{(0)}} \mathbf{e}_0\right), \mathbf{S}_k \right) \\ &= 1_{[V_{n_0}(\phi)<0]} \left((\phi_k, \mathbf{S}_k) - \frac{V_{n_0}(\phi)}{S_{n_0}^{(0)}} S_k^{(0)} \right) \end{aligned} \tag{10.76}$$

and because ϕ is self-financing we get

$$
\begin{aligned}
&= 1_{[V_{n_0}(\phi)<0]}\left((\phi_{k+1}, \mathbf{S}_k) - \frac{V_{n_0}(\phi)}{S_{n_0}^{(0)}}(\mathbf{e}_0, \mathbf{S}_k)\right) \\
&= \left(1_{[V_{n_0}(\phi)<0]}\left(\phi_{k+1} - \frac{V_{n_0}(\phi)}{S_{n_0}^{(0)}}\mathbf{e}_0\right), \mathbf{S}_k\right) \\
&= (\psi_{k+1}, \mathbf{S}_k),
\end{aligned}
$$

as required.

The last case to consider is $k = n_0$. In this case $(\psi_k, \mathbf{S}_k) = 0$ and we need to check that $(\psi_{k+1}, \mathbf{S}_k) = 0$. This follows:

$$
\begin{aligned}
(\psi_{k+1}, \mathbf{S}_k) =&(\psi_{n_0+1}, \mathbf{S}_{n_0}) \\
=&\left(1_{[V_{n_0}(\phi)<0]}\left(\phi_{n_0+1} - \frac{V_{n_0}(\phi)}{S_{n_0}^{(0)}}\mathbf{e}_0\right), \mathbf{S}_{n_0}\right) \\
=&1_{[V_{n_0}(\phi)<0]}\left((\phi_{n_0+1}, \mathbf{S}_{n_0}) - \frac{V_{n_0}(\phi)}{S_{n_0}^{(0)}}S_{n_0}^{(0)}\right)
\end{aligned}
$$

and again using the fact that ϕ is self-financing this is

$$
=1_{[V_{n_0}(\phi)<0]}\left((\phi_{n_0}, \mathbf{S}_{n_0}) - V_{n_0}(\phi)\right) = 0,
$$

since $(\phi_{n_0}, \mathbf{S}_{n_0}) = V_{n_0}(\phi)$. Thus ψ is self-financing.

STEP (II). Next we show ψ is admissible. First of all, for $k \leq n_0$ we have $V_k(\psi) = 0$ so we focus on $k > n_0$. For such k we have from (10.76) that $V_k(\psi) \geq 0$ since on $[V_{n_0}(\phi) < 0]$ the term $V_{n_0}(\phi)S_k^{(0)}/S_{n_0}^{(0)} < 0$. This verifies the admissibility of ψ.

STEP (III). Now we finish the verification that ψ is an arbitrage strategy. This follows directly from (10.76) with N substituted for k since $V_N(\psi) > 0$ on $[V_{n_0}(\phi) < 0]$. So an arbitrage strategy exists and the market is not viable. □

We now return to the proof of the converse of Theorem 10.16.1. Suppose we have a viable market so no arbitrage strategies exist. We need to find an equivalent martingale measure. Begin by defining two sets of random variables

$$
\begin{aligned}
\Gamma :=&\{X : \Omega \mapsto \mathbb{R} : X \geq 0,\ E(X) > 0\} \\
\mathcal{V} :=&\{V_N(\phi) : V_0(\phi) = 0,\ \phi \text{ is self-financing and predictable}\}.
\end{aligned}
$$

Lemma 10.16.3 implies that $\Gamma \cap \mathcal{V} = \emptyset$. (Otherwise, there would be a strategy ϕ such that $V_N(\phi) \geq 0$, $E(V_N(\phi)) > 0$ and $V_0(\phi = 0$ and Lemma 10.16.3 would imply the existence of an arbitrage strategy in violation of our assumption that the market is viable.)

We now think of Γ and $\mathcal{V}$ as subsets of the Euclidean space $\mathbb{R}^\Omega$, the set of all functions with domain Ω and range $\mathbb{R}$. (For example, if Ω has m elements $\omega_1, \dots, \omega_m$, we can identify $\mathbb{R}^\Omega$ with $\mathbb{R}^m$.) The set $\mathcal{V}$ is a vector space. To see this, suppose $\phi(1)$ and $\phi(2)$ are two self-financing predictable strategies such that $V_N(\phi(i)) \in \mathcal{V}$ for $i = 1, 2$. For real numbers a, b we have

$$aV_N(\phi(1)) + bV_N(\phi(2)) = V_N(a\phi(1) + b\phi(2))$$

and

$$0 = aV_0(\phi(1)) + bV_0(\phi(2)) = V_0(a\phi(1) + b\phi(2))$$

and $aV_N(\phi(1)) + bV_N(\phi(2))$ is the value function corresponding to the self-financing predictable strategy $a\phi(1) + b\phi(2)$; this value function is 0 at time 0. Thus $aV_N(\phi(1)) + bV_N(\phi(2)) \in \mathcal{V}$.

Next define

$$\mathcal{K} := \{X \in \Gamma : \sum_{\omega \in \Omega} X(\omega) = 1\}$$

so that $\mathcal{K} \subset \Gamma$. Observe that $\mathcal{K}$ is closed in the Euclidean space $\mathbb{R}^\Omega$ and is compact and convex. (If $X, Y \in \mathcal{K}$, then we have $\sum_\omega \alpha X(\omega) + (1-\alpha)Y(\omega) = \alpha + 1 - \alpha = 1$ for $0 \le \alpha \le 1$.) Furthermore, since $\mathcal{V} \cap \Gamma = \emptyset$, we have $\mathcal{V} \cap \mathcal{K} = \emptyset$.

Now we apply the separating hyperplane theorem. There exists a linear function $\lambda : \mathbb{R}^\Omega \mapsto \mathbb{R}$ such that

(i) $\lambda(X) > 0$, for $X \in \mathcal{K}$,

(ii) $\lambda(X) = 0$, for $X \in \mathcal{V}$.

We represent the linear functional λ as the vector

$$\lambda = (\lambda(\omega), \omega \in \Omega)$$

and rewrite (i) and (ii) as

(i′) $\sum_{\omega \in \Omega} \lambda(\omega)X(\omega) > 0$, for $X \in \mathcal{K}$,

(ii′) $\sum_{\omega \in \Omega} V_N(\phi)(\omega) = 0$, for $V_N(\phi) \in \mathcal{V}$, so that ϕ is self-financing and predictable.

From (i′) we claim $\lambda(\omega) > 0$ for all $\omega \in \Omega$. The reason for the claim, is that if $\lambda(\omega_0) = 0$ for some ω_0, then $X = 1_{\{\omega_0\}}$ satisfies $\sum_{\omega \in \Omega} X(\omega) = 1$ so $X \in \mathcal{K}$ but

$$\sum_{\omega \in \Omega} \lambda(\omega)X(\omega) = \lambda(\omega_0) = 0$$

violates (i′).

Define P^* by

$$P^*(\omega) = \frac{\lambda(\omega)}{\sum_{\omega' \in \Omega} \lambda(\omega')}.$$

Then $P^*(\omega) > 0$ for all ω so that $P^* \equiv P$. It remains to check that P^* is that which we desire: an equivalent martingale measure.

For any $V_N(\phi) \in \mathcal{V}$, (ii′) gives

$$\sum_{\omega\in\Omega} \lambda(\omega) V_N(\phi)(\omega) = 0,$$

so that, since $V_0(\phi) = 0$, we get from (10.65)

$$E^*\big(\sum_{j=1}^{N} (\phi_j, \overline{\mathbf{d}}_j)\big) = 0, \tag{10.77}$$

for any ϕ which is predictable and self-financing.

Now pick $1 \le i \le d$ and suppose $(\phi_n^{(i)}, 0 \le n \le N)$ is any predictable process. Using Lemma 10.16.2 with $V_0 = 0$ we know there exists a predictable process $(\phi_n^{(0)}, 0 \le n \le N)$ such that

$$\phi^\# := \{(\phi_n^{(0)}, 0, \dots, 0, \phi_n^{(i)}, 0, \dots, 0), 0 \le n \le N\}$$

is predictable and self-financing. So applying (10.77), we have

$$\begin{aligned} 0 = E^*\big(\sum_{j=1}^{N} (\phi_j^\#, \overline{\mathbf{d}}_j)\big) &= E^*\big(\sum_{j=1}^{N}\sum_{l=1}^{d} \phi_j^{\#(l)} \overline{d}_j^{(l)}\big) \\ &= E^*\big(\sum_{j=1}^{N} \phi_j^{(i)} \overline{d}_j^{(i)}\big). \end{aligned} \tag{10.78}$$

Since (10.78) holds for an arbitrary predictable process $\{\phi_j^{(i)}, 0 \le j \le N\}$, we conclude from Lemma 10.5.1 that $\{(\overline{S}_n^{(i)}, \mathcal{B}_n), 0 \le n \le N\}$ is a P^*-martingale. □

Corollary 10.16.1 *Suppose the market is viable and P^* is an equivalent martingale measure making $\{(\overline{\mathbf{S}}_n, \mathcal{B}_n), 0 \le n \le N\}$ a P^*-martingale. Then*

$$\{(\overline{V}_n(\phi), \mathcal{B}_n), 0 \le n \le N\}$$

is a P^*-martingale for any self-financing strategy ϕ.

Proof. If ϕ is self-financing, then by (10.65)

$$\overline{V}_n(\phi) = V_0(\phi) + \sum_{j=1}^{n} (\phi_j, \overline{\mathbf{d}}_j)$$

so that $\{(\overline{V}_n(\phi), \mathcal{B}_n), 0 \le n \le N\}$ is a P^*-martingale transform and hence a martingale. □

10.16.4 Complete Markets

A *contingent claim* or *European option of maturity N* is a non-negative $\mathcal{B}_N = \mathcal{B}$-measurable random variable X. We think of the contingent claim X paying $X(\omega) \geq 0$ at time N if the state of nature is ω. An investor may choose to buy or sell contingent claims. The seller has to pay the buyer of the option $X(\omega)$ dollars at time N. In this section and the next we will see how an investor who sells a contingent claim can perfectly protect himself or *hedge* his move by selling the option at the correct price.

Some examples of contingent claims include the following.

- *European call with strike price K*. A call on (for example) the first asset with strike price K is a contingent claim or random variable of the form $X = (S_N^{(1)} - K)^+$. If the market moves so that $S_N^{(1)} > K$, then the holder of the claim receives the difference. In reality, the call gives the holder the right (but not the obligation) to buy the asset at price K which can then be sold for profit $(S_N^{(1)} - K)^+$.

- *European put with strike price K*. A put on (for example) the first asset with strike price K is a contingent claim or random variable of the form $X = (K - S_N^{(1)})^+$. The buyer of the option makes a profit if $S_N^{(1)} < K$. The holder of the option has the right to sell the asset at price K even if the market price is lower.

There are many other types of options, some with exotic names: Asian options, Russian options, American options and passport options are some examples.

A contingent claim X is *attainable* if there exists an admissible strategy ϕ such that

$$X = V_N(\phi).$$

The market is *complete* if every contingent claim is attainable. Completeness will be shown to yield a simple theory of contingent claim pricing and hedging.

Remark 10.16.1 Suppose the market is viable. Then if X is a contingent claim such that for a self-financing strategy ϕ we have $X = V_N(\phi)$, then ϕ is admissible and hence X is attainable.

In a viable market, if a contingent claim is attainable with a self-financing strategy, this strategy is automatically admissible. To see this, suppose P^* is an equivalent martingale measure making $\{(\overline{\mathbf{S}}_n, \mathcal{B}_n), 0 \leq n \leq N\}$ a vector martingale. Then if ϕ is self-financing, Lemma 10.16.1 implies that $\{(\overline{V}_n(\phi), \mathcal{B}_n), 0 \leq n \leq N\}$ is a P^*-martingale. By the martingale property

$$\overline{V}_n(\phi) = E^*(\overline{V}_N(\phi)|\mathcal{B}_n), \quad 0 \leq n \leq N.$$

However, $0 \leq X = V_N(\phi)$ and hence $\overline{V}_N(\phi) \geq 0$ so that $\overline{V}_n(\phi) \geq 0$ for $0 \leq n \leq N$, and hence ϕ is admissible. □

Theorem 10.16.2 *Suppose the market is viable so an equivalent martingale measure P^* exists. The market is also complete iff there is a unique equivalent martingale measure.*

Proof. Suppose the market is viable and complete. Given a contingent claim X, there is a strategy ϕ which realizes X in the sense that $X = V_N(\phi)$. Thus we get by applying Lemma 10.16.1 that

$$\frac{X}{S_N^{(0)}} = \frac{V_N(\phi)}{S_N^{(0)}} = \overline{V}_N(\phi) = V_0(\phi) + \sum_{j=1}^{n}(\phi_j, \overline{\mathbf{d}}_j).$$

Suppose P_1^*, P_2^* are two equivalent martingale measures. Then by Corollary 10.16.1 $\{(\overline{V}_n(\phi), \mathcal{B}_n), 0 \le n \le N\}$ is a P_i^*-martingale ($i = 1, 2$). So

$$E_i^*(\overline{V}_N(\phi)) = E_i^*(\overline{V}_0(\phi)) = V_0(\phi),$$

since $\mathcal{B}_0 = \{\emptyset, \Omega\}$. We conclude that

$$E_1^*(X/S_N^{(0)}) = E_2^*(X/S_N^{(0)})$$

for any non-negative $X \in \mathcal{B}_N$. Let $X = 1_A S_N^{(0)}$ for arbitrary $A \in \mathcal{B}_N = \mathcal{B}$ and we get

$$E_1^*(1_A) = P_1^*(A) = P_2^*(A) = E_2^*(1_A),$$

and thus $P_1^* = P_2^*$ and the equivalent martingale measures are equal.

Conversely, suppose the market is viable but not complete. Then there exists a contingent claim $0 \le X$ such that, for no admissible strategy ϕ, is it true that $X = V_N(\phi)$.

Now we employ some elementary L_2-theory. Define

$$\mathcal{H} := \{U_0 + \sum_{n=1}^{N}\sum_{i=1}^{d}\phi_n^{(i)} d_n^{(i)} : U_0 \in \mathcal{B}_0,\ \phi_n^{(1)}, \dots, \phi_n^{(d)} \text{ are predictable.}\}$$

By Lemma 10.16.2 there exists $\phi_n^{(0)}, 1 \le n \le N\}$ such that

$$\phi := \{(\phi_n^{(0)}, \dots, \phi_n^{(d)}, 0 \le n \le N\}$$

is self-financing. Then with such $\{\phi_n^{(0)}\}$, since $\overline{d}_n^{(0)} = 0,\ 1 \le n \le N$, we have

$$U_0 + \sum_{n=1}^{N}\sum_{i=1}^{d}\phi_n^{(i)}\overline{d}_n^{(i)} = U_0 + \sum_{n=1}^{N}(\phi_n, \overline{\mathbf{d}}_n).$$

Thus, $X/S_N^{(0)} \notin \mathcal{H}$ since if it were, there would be a self-financing (and hence by Remark 10.16.1 also admissible) strategy such that

$$\frac{X}{S_N^{(0)}} = U_0 + \sum_{n=1}^{N}(\phi_n, \overline{\mathbf{d}}_n) = \overline{V}_N(\phi) = \frac{V_N(\phi)}{S_N^{(0)}}.$$

Since the market is viable, there exists an equivalent martingale measure P^*. All random variables are bounded since Ω is finite. Now consider $L_2(P^*)$ with inner product $\langle X, Y \rangle = E^*(XY)$. Then $\mathcal{H}$ is a subspace of $L_2(P^*)$ since it is clearly a vector space and it is closed in $L_2(P^*)$. So $\mathcal{H} \neq L_2(P^*)$ since $X/S_N^{(0)} \notin \mathcal{H}$ means that there exists $\xi \neq 0$ with $\xi \in \mathcal{H}^\perp$, the orthogonal complement of $\mathcal{H}$. Define

$$\|\xi\|^\vee = \sup_{\omega\in\Omega} |\xi(\omega)|$$

and

$$P^{**}(\{\omega\}) = \left(1 + \frac{\xi(\omega)}{2\|\xi\|^\vee}\right) P^*(\{\omega\}),$$

and observe

$$\sum_{\omega\in\Omega} P^{**}(\{\omega\}) = 1 + \sum_{\omega\in\Omega} \xi(\omega)\frac{P^*(\{\omega\})}{2\|\xi\|^\vee} = 1 + \frac{E^*(\xi)}{2\|\xi\|^\vee} = 1,$$

since we claim that $E^*(\xi) = 0$. To check this last claim, note that $1 \in \mathcal{H}$ and $\langle 1, \xi \rangle = 0$ (since $\xi \in \mathcal{H}^\perp$) means $E^*(\xi 1) = E^*(\xi) = 0$. Also observe that

$$\left|\frac{\xi}{2\|\xi\|^\vee}\right| \leq \frac{1}{2},$$

so that

$$1 + \frac{\xi}{2\|\xi\|^\vee} > 0.$$

We conclude that P^{**} is a probability measure and that $P^{**} \equiv P^* \equiv P$. Furthermore, for any predictable process $\{(\phi_n^{(1)}, \ldots, \phi_n^{(d)}), 0 \leq n \leq N\}$, we may add the predictable component $\{\phi_n^{(0)}, 0 \leq n \leq N\}$ making the full family self-financing with $V_0 = 0$. So

$$\overline{V}_n(\phi) = \sum_{n=1}^{n} (\phi_n, \overline{\mathbf{d}}_n)$$

and $\overline{V}_N(\phi) \in \mathcal{H}$. Since $\xi \in \mathcal{H}^\perp$, we have

$$0 = \langle \overline{V}_N(\phi), \xi \rangle = E^*(\overline{V}_N(\phi)\xi) = \sum_{\omega\in\Omega} V_N(\phi)(\omega)\xi(\omega)P^*(\{\omega\}).$$

Therefore, using the martingale property and orthogonality,

$$\begin{aligned}
E^{**}\overline{V}_N(\phi) &= \sum_{\omega\in\Omega} \overline{V}_N(\phi)(\omega)\left(1 + \frac{\xi(\omega)}{2\|\xi\|^\vee}\right) P^*(\{\omega\}) \\
&= E^*\overline{V}_N(\phi) + E^*\left(\frac{\overline{V}_N(\phi)\xi}{2\|\xi\|^\vee}\right) \\
&= 0 + 0.
\end{aligned}$$

Since the predictable process is arbitrary, we apply Lemma 10.5.1 to conclude that $\{(\overline{S}_n, \mathcal{B}_n), 0 \le n \le N\}$ is a P^{**}-martingale. There is more than one equivalent martingale measure, a conclusion made possible by supposing the market was not complete. □

10.16.5 Option Pricing

Suppose X is a contingent claim which is attainable by using admissible strategy ϕ so that $X = V_N(\phi)$. We call $V_0(\phi)$ the *initial price* of the contingent claim.

If an investor sells a contingent claim X at time 0 and gains $V_0(\phi)$ dollars, the investor can invest the $V_0(\phi)$ dollars using strategy ϕ so that at time N, the investor has $V_N(\phi) = X$ dollars. So even if the investor has to pay to the buyer X dollars at time N, the investor is perfectly hedged.

Can we determine $V_0(\phi)$ without knowing ϕ? Suppose P^* is an equivalent martingale measure. Then we know $\{(\overline{V}_n(\phi), \mathcal{B}_n), 0 \le n \le N\}$ is a P^*-martingale so

$$E^*(\overline{V}_N(\phi)) = E^*(\overline{V}_0(\phi)) = \overline{V}_0(\phi) = V_0(\phi),$$

and

$$E^*(X/S_N^{(0)}) = V_0(\phi).$$

So the price to be paid is $E^*(X/S_N^{(0)})$. This does not require knowledge of ϕ, but merely knowledge of the contingent claim, the risk free asset and the equivalent martingale measure.

Furthermore, by the martingale property,

$$E^*(\overline{V}_N(\phi)|\mathcal{B}_n) = \overline{V}_n(\phi), \quad 0 \le n \le N;$$

that is,

$$V_n(\phi) = S_n^{(0)} E^*(V_N(\phi)/S_N^{(0)}|\mathcal{B}_n) = S_n^{(0)} E^*(X/S_N^{(0)}|\mathcal{B}_n).$$

This has the interpretation that if an investor sells the contingent claim at time n (with information $\mathcal{B}_n$ at his disposal), the appropriate sale price is $V_n(\phi)$ dollars because then there is a strategy ϕ which will make the $V_n(\phi)$ dollars grow to X dollars by time N and the investor is perfectly hedged. The price $V_n(\phi)$ can be determined as $S_n^{(0)} E^*(X/S_N^{(0)}|\mathcal{B}_n)$, so pricing at time n does not need knowledge of ϕ but only knowledge of the P^*, X and the risk free asset.

Is there more than one initial price that generates wealth X at time N? Put another way, is there more than one strategy ϕ whose initial values $V_0(\phi)$ are different but each generates the contingent claim X? Essentially, the answer is negative, provided the market is viable and complete. See Exercise 56.

10.17 Exercises

1. (a) If λ and μ are finite measures on $(\Omega, \mathcal{B})$ and if $\mu \ll \lambda$, then

$$\int f d\mu = \int f \frac{d\mu}{d\lambda} d\lambda$$

for any μ-integrable function f.

(b) If λ, μ and ν are finite measures on $(\Omega, \mathcal{B})$ such that $\nu \ll \mu$ and $\mu \ll \lambda$, then

$$\frac{d\nu}{d\lambda} = \frac{d\nu}{d\mu}\frac{d\mu}{d\lambda},$$

a.e. λ.

(c) If μ and ν are finite measures which are equivalent in the sense that each is AC with respect to the other, then

$$\frac{d\nu}{d\mu} = \left(\frac{d\mu}{d\nu}\right)^{-1},$$

a.e. μ, ν.

(d) If $\mu_k, k = 1, 2, \ldots$ and μ are finite measures on $(\Omega, \mathcal{B})$ such that

$$\sum_{i=1}^{\infty} \mu_k(A) = \mu(A)$$

for all $A \in \mathcal{B}$, and if the μ_k are AC with respect to a σ-finite measure λ, then $\mu \ll \lambda$ and

$$\frac{d\sum_{i=1}^{n} \mu_i}{d\lambda} = \sum_{i=1}^{n} \frac{d\mu_i}{d\lambda}, \quad \lim_{n\to\infty} \frac{d\sum_{i=1}^{n} \mu_i}{d\lambda} = \frac{d\mu}{d\lambda},$$

a.e. λ.

Hints: (a) The equation holds for f an indicator function, hence when f is simple and therefore for all integrable f. (b) Apply (a) with $f = d\nu/d\mu$.

2. Let $(\Omega, \mathcal{B}, P)$ be a probability space and let ξ be a positive random variable with $E(\xi) = 1$. Define $P^*(A) = \int_A \xi dP$. Show

(a) $E^*(X) = E(X\xi)$ for any random variable $X \geq 0$.

(b) Prove for any $X \geq 0$ that

$$E^*(X|Y_1, \ldots, Y_n) = \frac{E(X\xi|Y_1, \ldots, Y_n)}{E(\xi|Y_1, \ldots, Y_n)}$$

for any random variables $Y_1, \ldots, Y_n$.

3. If $X \in L_1, Y \in L_\infty, \mathcal{G} \subset \mathcal{B}$, then

$$E\big(YE(X|\mathcal{G})\big) = E(XE(Y|\mathcal{G})).$$

4. Suppose $Y \in L_1$ and $\mathbf{X}_1, \mathbf{X}_2$ are random vectors such that $\sigma(Y, \mathbf{X}_1)$ is independent of $\sigma(\mathbf{X}_2)$. Show almost surely

$$E(Y|\mathbf{X}_1, \mathbf{X}_2) = E(Y|\mathbf{X}_1).$$

(Hint: What is an expression for $\Lambda \in \sigma(\mathbf{X}_1, \mathbf{X}_2)$ in terms of $\mathbf{X}_1, \mathbf{X}_2$?

5. If $\{N(t), t \geq 0\}$ is a homogeneous Poisson process with rate λ, what is $E(N(s)|N(t))$ for $0 \leq s < t$.

6. If U is $U(0, 1)$ on $(\Omega, \mathcal{B}, P)$ define $Y = U(1 - U)$. For a positive random variable X, what is $E(X|Y)$.

7. Suppose X_1, X_2 are iid unit exponential random variables. What is

   (a) $E(X_1|X_1 + X_2)$?

   (b) $P[X_1 < 3|X_1 + X_2]$?

   (c) $E(X_1|X_1 \wedge t)$?

   (d) $E(X_1|X_1 \vee t)$?

8. Suppose X, Y are random variables with finite second moments such that for some decreasing function f we have $E(X|Y) = f(Y)$. Show that $\text{Cov}(X, Y) \leq 0$.

9. *Weak L_1 convergence.* We say X_n converges weakly in L_1 to X iff for any bounded random variable Y, we have $E(X_nY) \to E(XY)$. Show this implies $E(X_n|\mathcal{G})$ converges weakly in L_1 to $E(X|\mathcal{G})$ for any σ-field $\mathcal{G} \subset \mathcal{B}$.

10. Let X be an integrable random variable defined on the space $(\Omega, \mathcal{B}, P)$ and suppose $\mathcal{G} \subset \mathcal{B}$ is a sub-σ-field. Prove that on every non-null atom Λ of $\mathcal{G}$, the conditional expectation $E(X|\mathcal{G})$ is constant and

$$E(X|\mathcal{G})(\omega) = \int_\Lambda X dP/P(\Lambda), \quad \omega \in \Lambda.$$

Recall that Λ is a non-null atom of $\mathcal{G}$ if $P(\Lambda) > 0$ and Λ contains no subsets belonging to $\mathcal{G}$ other than $\emptyset$ and Ω.

11. Let $\phi : \mathbb{R} \mapsto \mathbb{R}$. Let X, Y be independent random variables. For each $x \in \mathbb{R}$ define

$$Q(x, A) := P[\phi(x, Y) \in A].$$

Show

$$P[\phi(X, Y) \in A|X] = Q(X, A)$$

almost surely.

Now assume ϕ is either bounded or non-negative. If $h(x := E(\phi(x, Y))$, then

$$E(\phi(X, Y)|X) = h(X),$$

almost surely.

12. (a) For $0 \leq X \in L_1$ and $\mathcal{G} \subset \mathcal{B}$, show almost surely

$$E(X|\mathcal{G}) = \int_0^\infty P[X > t|\mathcal{G}]dt.$$

(b) Show

$$P[|X| \geq t|\mathcal{G}] \leq t^{-k}E(|X|^k|\mathcal{G}).$$

13. If $\mathcal{B}_1 \subset \mathcal{B}_2 \subset \mathcal{B}$ and $E(X^2) < \infty$, then

$$E\left((X - E(X|\mathcal{B}_2))^2\right) \leq E\left((X - E(X|\mathcal{B}_1))^2\right).$$

14. Let $\{Y_n, n \geq 0\}$ be iid, $E(Y_n) = 0$, $E(Y_n^2) = \sigma^2 < \infty$. Set $X_0 = 0$ and show from first principles that

$$X_n = (\sum_{i=1}^{n} Y_k)^2 - n\sigma^2$$

is a martingale.

15. Suppose $\{(X_n, \mathcal{B}_n), n \geq 0\}$ is a martingale which is predictable. Show $X_n = X_0$ almost surely.

16. Suppose $\{(X_n, \mathcal{B}_n), n \geq 0\}$ and $\{(Y_n, \mathcal{B}_n), n \geq 0\}$ are submartingales. Show $\{(X_n \vee Y_n, \mathcal{B}_n), n \geq 0\}$ is a submartingale and that $\{(X_n + Y_n, \mathcal{B}_n), n \geq 0\}$ is as well.

17. **Polya urn.** (a) An urn contains b black and r red balls. A ball is drawn at random. It is replaced and moreover c balls of the color drawn are added. Let $X_0 = b/(b + r)$ and let X_n be the proportion of black balls attained at stage n; that is, just after the nth draw and replacement. Show $\{X_n\}$ is a martingale.

(b) For this Polya urn model, show X_n converges to a limit almost surely and in L_p for $p \geq 1$.

18. Suppose $(\Omega, \mathcal{B}, P) = ([0, 1), \mathcal{B}([0, 1)), \lambda)$ where λ is Lebesgue measure. Let $\mathcal{B}_n = \sigma([k2^{-n}, (k + 1)2^{-n}),\ 0 \leq k < 2^n)$. Suppose f is a Lebesgue integrable function on $[0, 1)$.

(a) Verify that the conditional expectation $E(f|\mathcal{B}_n)$ is a step function converging in L_1 to f. Use this to show the Mean Approximation Lemma (see

Exercise 37 of Chapter 5): If $\epsilon > 0$, there is a continuous function g defined on $[0, 1)$ such that

$$\int_{[0,1)} |f(x) - g(x)|dx < \epsilon.$$

(b) Now suppose that f is Lipschitz continuous meaning for some $K > 0$

$$|f(t) - f(s)| \leq K|t - s|, \quad 0 \leq s < t < 1.$$

Define

$$f_n(x) = \frac{\left(f((k+1)2^{-n}) - f(k2^{-n})\right)}{2^{-n}} 1_{\left[[k2^{-n},(k+1)2^{-n})\right]}(x), \quad x \in [0, 1)$$

and show that $\{(f_n, \mathcal{B}_n), n \geq 0\}$ is a martingale, that there is a limit f_∞ such that $f_n \to f_\infty$ almost surely and in L_1, and

$$f(b) - f(a) = \int_a^b f_\infty(s)ds, \quad 0 \leq a < b < 1.$$

19. Supppose $\{(X_n, \mathcal{B}_n), n \geq 0\}$ is a martingale such that for all $n \geq 0$ we have $X_{n+1}/X_n \in L_1$. Prove $E(X_{n+1}/X_n) = 1$ and show for any $n \geq 1$ that X_{n+1}/X_n and X_n/X_{n-1} are uncorrelated.

20. (a) Suppose $\{X_n, n \geq 0\}$ are non-negative random variables with the property that $X_n = 0$ implies $X_m = 0$ for all $m \geq n$. Define $D = \cup_{n=0}^{\infty}[X_n = 0]$ and assume

$$P[D|X_0, \ldots, X_n] \geq \delta(x) > 0 \text{ almost surely on } [X_n \leq x].$$

Prove

$$P\{D \cup [\lim_{n\to\infty} X_n = \infty]\} = 1.$$

(b) For a simple branching process $\{Z_n, n \geq 0\}$ with offspring distribution $\{p_k\}$ satisfying $p_1 < 1$, show

$$P[\lim_{n\to\infty} Z_n = 0 \text{ or } \infty] = 1.$$

21. Consider a Markov chain $\{X_n, n \geq 0\}$ on the state space $\{0, \ldots, N\}$ and having transition probabilities

$$p_{ij} = \binom{N}{j}\left(\frac{i}{N}\right)^j\left(1 - \frac{i}{N}\right)^{N-j}.$$

Show $\{X_n\}$ and

$$\{V_n = \frac{X_n(N - X_n)}{(1 - N^{-1})^n}, n \geq 0\}$$

are martingales.

22. Consider a Markov chain $\{X_n, n \geq 0\}$ on the state space $\{0, 1, \dots\}$ with transition probabilities

$$p_{ij} = \frac{e^{-i} i^j}{j!}, \quad j \geq 0,\ i \geq 0$$

and $p_{00} = 1$. Show $\{X_n\}$ is a martingale and that

$$P[\vee_{n=0}^{\infty} X_n \geq x | X_0 = i] \leq i/x.$$

23. Consider a game of repeatedly tossing a fair coin where the result ξ_k at round k has $P[\xi_k = 1] = P[\xi_k = -1] = 1/2$. Suppose a player stakes one unit in the first round and doubles the stake each time he loses and returns to the unit stake each time he wins. Assume the player has unlimited funds. Let X_n be the players net gain after the nth round. Show $\{X_n\}$ is a martingale.

24. Suppose $\{(X_n, \mathcal{B}_n), n \geq 0\}$ is a positive supermartingale and ν is a stopping time satisfying $X_\nu \geq X_{\nu-1}$ on $[0 < \nu < \infty]$. Show

$$M_n := \begin{cases} X_{(\nu-1)\wedge n}, & \text{if } \nu \geq 1, \\ 0, & \text{if } \nu = 0 \end{cases}$$

is a new positve supermartingale. In particular, this applies to the stopping time $\nu_a = \inf\{n \geq 0 : X_n > a\}$ and the induced sequence $\{M_n\}$ satisfies $0 \leq M_n \leq a$.

25. The Haar functions on $[0, 1]$ are defined by

$$H_1(t) \equiv 1, \quad H_2(t) = \begin{cases} 1, & \text{if } 0 \leq t < 1/2, \\ -1, & \text{if } 1/2 \leq t < 1, \end{cases}$$

$$H_{2^n+1}(t) = \begin{cases} 2^{n/2}, & \text{if } 0 \leq t < 2^{-(n+1)}, \\ -2^{n/2}, & \text{if } 2^{-(n+1)} \leq t < 2^{-n}, \\ 0, & \text{otherwise,} \end{cases}$$

$$H_{2^n+j}(t) = H_{2^n+1}\left(t - \frac{(j-1)}{2^n}\right), \quad j = 1, \dots, 2^n.$$

Plot the first 5 functions.

Let f be measurable with domain $[0, 1]$ and satisfying $\int_0^1 |f(s)| ds$. Define

$$A_k := \int_0^1 f(t) H_k(t) dt.$$

Let Z be a uniform random variable on $[0, 1]$. Show

$$f(Z) = \lim_{n \to \infty} \sum_{k=1}^{n} a_k H_k(Z)$$

almost surely and

$$\lim_{n\to\infty} \int_0^1 |f(s) - \sum_{k=1}^{n} a_k H_k(s)| ds = 0.$$

Hints: Define $\mathcal{B}_n = \sigma(H_i(Z), i \leq n)$, and show that $E(f(Z)|\mathcal{B}_n) = \sum_{k=1}^n a_k H_k(Z)$. Check that $EH_i(Z)H_k(Z) = 0$ for $i \neq k$.

26. Let $\{Y_n\}$ be random variables with $E(|Y_n|) < \infty$. Suppose for $n \geq 0$

$$E(Y_{n+1}|Y_0, \ldots, Y_n) = a_n + b_n Y_n,$$

where $b_n \neq 0$. Let

$$l_{n+1}(z) = a_n + b_n z, \quad l_{n+1}^{\leftarrow}(y) = \frac{y - a_n}{b_n}$$

and set

$$L_n(y) = l_1^{\leftarrow}(l_2^{\leftarrow}(\ldots(l_n^{\leftarrow}(y)\ldots)$$

(functional composition). Show for any k that

$$\{(X_n = kL_n(Y_n), \sigma(Y_0, \ldots, Y_n), n \geq 0\}$$

is a martingale.

Special cases include

(a) The Polya urn.

(b) The simple branching process.

(c) Supppse Y_0 is uniformly distributed on $[0, 1]$; given Y_n, we suppose Y_{n+1} is uniform on $[Y_n, 1]$. Then $X_n = 2^n(1 - Y_n)$ is a martingale.

27. If ν is a stopping time with respect to $\{\mathcal{B}_n, n \geq 0\}$, show that a random variable ξ is $\mathcal{B}_\nu$-measurable iff $\xi 1_{[\nu=n]} \in \mathcal{B}_n$ for $n \in \mathbb{N}$.

28. Suppose that $\{Y_n, n \geq 1\}$ are independent, positive random variables with $E(Y_n) = 1$. Put $X_n = \prod_{i=1}^n Y_i$.

(a) Show $\{X_n\}$ is an integrable martingale which converges a.s. to an integrable X.

(b) Suppose specifically that Y_n assumes the values $1/2$ and $3/2$ with probability $1/2$ each. Show that $P[X = 0] = 1$. This gives an example where

$$E\left(\prod_{i=1}^{\infty} Y_i\right) \neq \prod_{i=1}^{\infty} E(Y_i),$$

for independent, positive random variables. Show, however, that

$$E\left(\prod_{i=1}^{\infty} Y_i\right) \leq \prod_{i=1}^{\infty} E(Y_i)$$

always holds.

29. If $\{X_n\}$ is a martingale and it is bounded either above or below, then it is L_1-bounded.

30. Let $X_n, n \geq 0$ be a Markov chain with countable state space which we can take to be the integers and transition matrix $P = (p_{ij})$. A function ϕ on the state space is called excessive or superharmonic if

$$\phi(i) \geq \sum_j p_{ij}\phi(j).$$

Show using martingale theory that $\phi(X_n)$ converges with probability 1 if ϕ is bounded and excessive. Deduce from this that if the chain is irreducible and persistent, then ϕ must be constant.

31. Use martingale theory to prove the Kolmogorov convergence criterion: Suppose $\{Y_n\}$ is independent, $EY_n = 0$, $EY_n^2 < \infty$. Then, if $\sum_k EY_k^2 < \infty$, we have $\sum_k Y_k$ converges almost surely. Extend this result to the case where $\{Y_n\}$ is a martingale difference sequence.

32. Let $\{Z_0 = 1, Z_1, Z_2, \dots\}$ be a branching process with immigration. This process evolves in such a way that

$$Z_{n+1} = Z_n^{(1)} + \cdots + Z_n^{(Z_n)} + I_{n+1}$$

where the $\{Z_n^{(i)}, i \geq 1\}$ are iid discrete integer valued random variables, each with the offspring distribution $\{p_j\}$ and also independent of Z_n. Also $\{I_j, j \geq 1\}$ are iid with an immigration distribution (concentrating on the non-negative integers) and I_{n+1} is independent of Z_n for each n. Suppose $EZ_1 = m > 1$ and that $EI_1 = \lambda > 0$.

(a) What is $E(Z_{n+1}|Z_n)$?

(b) Use a martingale argument to prove that Z_n/m^n converges a.s. to a finite random variable.

33. Let $\{X_n, n \geq 1\}$ be independent, $E|X_n|^p < \infty$ for all n with $p \geq 1$. Prove

$$f(n) = E|\sum_{i=1}^{n}(X_i - E(X_i))|^p$$

is non-decreasing in n.

34. Let $\{Y_j\}$ be independent with

$$P[Y_j = 2^{2j}] = \frac{1}{2}, \quad P[Y_j = -2^{2j}] = \frac{1}{2}.$$

Define $X_0 = 0$, $X_n = \sum_{i=1}^{n} Y_i, n \geq 1$ and $\nu = \inf\{n : X_n > 0\}$. Then ν is not regular for $\{X_n\}$ even though

$$E\theta^\nu < \infty, \quad 0 \leq \theta < 2,$$

which implies $E\nu^n < \infty$ for $n \geq 1$. (Check that $EX_\nu = \infty$.)

35. Let $\{\xi_n\}$ be non-negative random variables satisfying

$$E(\xi_n|\xi_1,\dots,\xi_{n-1}) \le \delta_{n-1} + \xi_{n-1}$$

where $\delta_n \ge 0$ are constants and $\sum_n \delta_n < \infty$. Show $\xi_n \to \xi$ a.s. and ξ is finite a.s.

36. Let ν be a stopping time relative to the increasing sequence $\{\mathcal{B}_n, n \in \mathbb{N}\}$ of sub-σ-fields of $\mathcal{B}$ in the probability space $(\Omega, \mathcal{B}, P)$. For all $n \in \mathbb{N}$, denote by $\phi(n)$, the smallest integer p such that $[\nu = n] \in \mathcal{B}_p$. Show that $\phi(\nu)$ is a stopping time dominated by ν.

37. Let $\{X_n, n \in \mathbb{N}\}$, $\{\beta_n, n \in \mathbb{N}\}$, and $\{Y_n, n \in \mathbb{N}\}$ be 3 adapted sequences of finite positive random variables defined on the same probability space such that

$$E(X_{n+1}|\mathcal{B}_n) \le (1+\beta_n)X_n + Y_n, \quad n \in \mathbb{N}.$$

This relation expresses the fact that $\{X_n\}$ is almost a supermartingale. Show that the limit $\lim_{n\to\infty} X_n$ exists and is finite a.s. on the event

$$A = [\sum_n \beta_n < \infty, \sum_n Y_n < \infty].$$

(Hint: Consider the sequence $\{U_n, n \in \mathbb{N}\}$ defined by

$$U_n = X'_n - \sum_{m<n} Y'_n$$

where

$$X'_n = X_n/(1+\beta_1)\dots(1+\beta_{n-1}), \quad Y'_n = Y_n/(1+\beta_1)\dots(1+\beta_{n-1}),$$

and also the stopping times

$$\nu_a = \min\{n : \sum_{m\le n} Y_m/(1+\beta_1)\dots(1+\beta_{m-1}) > a\}.$$

Then observe that $(a + U_{\nu_a\wedge n}, n \in \mathbb{N}\}$ is a finite positive supermartingale.)

38. Suppose $\{\xi_n, n \ge 0\}$ is a sequence adapted to $\{\mathcal{B}_n, n \ge 0\}$; that is, $\xi_n \in \mathcal{B}_n$. Suppose the crystal ball condition $E(\sup_{n\ge 0} |\xi_n|) < \infty$ holds and that ν is an almost surely finite stopping time. Define

$$\nu^* := \inf\{n \ge 0 : \xi_n \ge E(\xi_\nu|\mathcal{B}_n)\}.$$

Show ν^* is an almost surely finite stopping time such that $\nu^* \le \nu$ and that for any $n \ge 0$

$$\xi_n < E(\xi_{\nu^*}|\mathcal{B}_n), \quad \text{on } [\nu^* > n].$$

39. Suppose $\{(X_n, \mathcal{B}_n), n \geq 0\}$ is a positive supermartingale and ν is a stopping time. Define
$$X'_n := E(X_{\nu \wedge n} | \mathcal{B}_n), \quad n \geq 0.$$
Show $\{(X'_n, \mathcal{B}_n), n \geq 0\}$ is again a positive supermartingale. (Use the pasting lemma or proceed from first principles.)

40. Let $\{X_n = \sum_{i=1}^n Y_i, n \geq 0\}$ be a sequence of partial sums of a sequence of mean 0 independent integrable random variables. Show that if the martingale converges almost surely and if its limit is integrable, then the martingale is regular. Thus for this particular type of martingale, L_1-boundedness, $\sup_n E(|X_n|) < \infty$, implies regularity.

(Hint: First show that $E(X_\infty - X_n | \mathcal{B}_n)$ is constant if $\mathcal{B}_n = \sigma(Y_1, \dots, Y_n)$ and $X_\infty = \lim_{n\to\infty} X_n$ almost surely.)

41. An integrable martingale $\{X_n, n \geq 0\}$ cannot converge in L_1 without also converging almost surely. On the other hand, an integrable martingale may converge in probability while being almost surely divergent.

Let $\{Y_n, n \geq 1\}$ be a sequence of independent random variables each taking the values ± 1 with probability 1/2. Let $\mathcal{B}_n = \sigma(Y_1, \dots, Y_n), n \geq 0$ and let $B_n \in \mathcal{B}_n$ be a sequence of events adapted to $\{\mathcal{B}_n, n \geq 0\}$ such that
$$\lim_{n\to\infty} P(B_n) = 0 \text{ and } P[\limsup_{n\to\infty} B_n] = 1.$$

Then the formulas
$$X_0 = 0, \quad X_{n+1} = X_n(1 + Y_{n+1}) + 1_{B_n} Y_{n+1}, \quad n \geq 0,$$
define an integrable martingale such that
$$\lim_{n\to\infty} P[X_n = 0] = 1, \quad P[\{X_n\} \text{ converges}] = 0.$$

(Note that $P[X_{n+1} \neq 0] \leq (1/2)P[X_n \neq 0] + P(B_n)$ and that on the set $[\{X_n\} \text{ converges}]$, the limit $\lim_{n\to\infty} 1_{B_n}$ exists.)

42. Suppose $\{(X_n, \mathcal{B}_n), n \geq 0\}$ is an L_1-bounded martingale. If there exists an integrable random variable Y such that $X_n \leq E(Y|\mathcal{B}_n)$ then $X_n \leq E(X_\infty|\mathcal{B}_n)$ for all $n \geq 0$ where $X_\infty = \lim_{n\to\infty} X_n$ almost surely.

43. (a) Suppose $\{\xi_n, n \geq 0\}$ are iid and $g : \mathbb{R} \mapsto \mathbb{R}_+$ satisfies $E(g(\xi)) = 1$. Show $X_n := \prod_{i=0}^n g(\xi_i)$ is a positive martingale converging to 0 provided $P[g(\xi_0) = 1) \neq 1$.

(b) Define $\{X_n\}$ inductively as follows: $X_0 = 1$ and X_n is uniformly distributed on $(0, X_{n-1})$ for $n \geq 1$. Show $\{2^n X_n, n \geq 0\}$ is a martingale which is almost surely convergent to 0.

44. Consider a random walk on the integer lattice of the positive quadrant in two dimensions. If at any step the process is at (m, n), it moves at the next step to $(m+1, n)$ or $(m, n+1)$ with probability 1/2 each. Start the process at $(0, 0)$. Let Γ be any curve connecting neighboring lattice points extending from the y-axis to the x-axis in the first quadrant. Show $E(Y_1) = E(Y_2)$, where Y_1, Y_2 denote the number of steps to the right and up respectively before hitting the boundary Γ. (Note (Y_1, Y_2) is the hitting point on the boundary.)

45. (a) Suppose that $E(|X|) < \infty$ and $E(|Y|) < \infty$ and that $E(Y|X) = X$ and $E(X|Y) = Y$. Show $X = Y$ almost surely.

Hint: Consider

$$\int_{[Y>c,X<c]} (Y - X)dP.$$

(b) If the sequence $\{X_n, -\infty < n < \infty\}$ is a martingale in both forward time and reverse time, then for any $m \neq n$, we have $X_n = X_m$ almost surely.

46. Suppose $\{B_n, n \geq 0\}$ is a sequence of events with $B_n \in \mathcal{B}_n$. What is the Doob decomposition of $X_n = \sum_{i=0}^n 1_{B_n}$?

47. **U-statistics.** Let $\{\xi_n, n \geq 1\}$ be iid and suppose $\phi : \mathbb{R}^m \mapsto \mathbb{R}$ is a symmetric function of m variables satisfying

$$E\big(|\phi(\xi_1, \dots, \xi_m)|\big) < \infty.$$

Define $\{U_{m,n}, n \geq m\}$ by

$$U_{m,n} := \sum_{1 \leq i_1 < \cdots < i_m \leq n} \phi(\xi_{i_1}, \dots, \xi_{i_m}) / \binom{n}{m}$$

and set $\mathcal{B}_n := \sigma(U_{m,n}, j \geq n\}$.

Some special cases of interest are

$$m = 1, \quad \phi(x) = x,$$
$$m = 2, \quad \phi(x_1, x_2) = (x_1 - x_2)^2/2.$$

(a) Show $\{U_{m,n}, n \geq m\}$ is a reversed martingale.

(b) Prove

$$\lim_{n\to\infty} U_{m,n} = E\big(\phi(\xi_1, \dots, \xi_m)\big),$$

almost surely and in L_1.

48. Let $\{(X_n, \mathcal{B}_n), n \geq 0\}$ be a submartingale such that $\vee_{n \geq 0} X_n < \infty$. If $E(\sup_j d_j^+) < \infty$, then $\{X_n\}$ is almost surely convergent.

49. **Ballot problem.** In voting, r votes are cast for the incumbent, s votes for the rival. Votes are cast successively in random order with $s > r$. The probability that the winner was ahead at every stage of the voting is $(s - r)/(s + r)$.

More generally, suppose $\{X_j, 1 \le j \le n\}$ are non-negative integer valued, integrable and iid random variables. Set $S_j = \sum_{i=1}^{j} X_i$. Then

$$P[S_j < j,\ 1 \le j \le n | S_n) = (1 - \frac{S_n}{n})^+.$$

Hint: Look at $S_n < n$ and consider $\{S_j/j\}$.

50. **Two genetics models.** Let $\{X_n, n \ge 0\}$ be a Markov chain on the state space $\{0, 1, \dots, M\}$ with transition matrix $\{p_{ij}, 0 \le i, j \le M\}$. For the following two models, compute

$$\psi_i := P[\text{ absorbtion at M} | X_0 = i]$$

and show ψ_i is the same for either model.

(a) Wright model:

$$p_{ij} = \binom{M}{j} \left(\frac{i}{M}\right)^j \left(1 - \frac{i}{M}\right)^{M-j},$$

for $i = 0, \dots, M;\ j = 0, \dots, M$.

(b) Moran model: Define

$$p_i = \frac{i(M - i)}{M^2}, \quad i = 0, \dots, M$$

and set

$$\begin{aligned} p_{ij} &= p_i, \quad j = i-1, i+1, i \ne 0 \text{ or } M, \\ p_{0j} &= \delta_{0j},\ p_{Mj} = \delta_{Mj}, \\ p_{ij} &= 0, \quad \text{otherwise.} \end{aligned}$$

51. (a) Suppose $\{(X_n, \mathcal{B}_n), n \ge 0\}$ is a positive supermartingale and ν is a stopping time. Show

$$E(X_0) \ge E(X_\nu 1_{[\nu<\infty]}).$$

(b) Suppose X_n represents an insurance company's assets at the start of year n and suppose we have the recursion $X_{n+1} = X_n + b - Y_n$, where b is a positive constant representing influx of premiums per year and Y_n, the claims in year n, has $N(\mu, \sigma^2)$ distribution where $\mu < b$. Assume that $\{Y_n, n \geq 1\}$ are iid and that $X_0 = 1$. Define the event

$$[\text{Ruin}] = \bigcup_{n=1}^{\infty} [X_n < 0].$$

Show

$$P[\text{Ruin}] \leq e^{-2(b-\mu)/\sigma^2}.$$

(Hint: Check that $\{\exp\{-2(b - \mu)\sigma^{-2} X_n, n \geq 0\}$ is a supermartingale. Save come computation by noting what is the moment generating function of a normal density.)

52. Suppose $\mathcal{B}_n \uparrow \mathcal{B}_\infty$ and $\{Y_n, n \in \overline{\mathbb{N}}\}$ is a sequence of random variables such that $Y_n \to Y_\infty$.

(a) If $|Y_n| \leq Z \in L_1$, then show almost surely that

$$E(Y_n|\mathcal{B}_n) \to E(Y_\infty|\mathcal{B}_\infty).$$

(b) If $Y_n \stackrel{L_1}{\to} Y_\infty$, then in L_1

$$E(Y_n|\mathcal{B}_n) \to E(Y_\infty|\mathcal{B}_\infty).$$

(c) Backwards analogue: Suppose now that $\mathcal{B}_n \downarrow \mathcal{B}_{-\infty}$ and $|Y_n| \leq Z \in L_1$ and $Y_n \to Y_{-\infty}$ almost surely. Show

$$E(Y_n|\mathcal{B}_n) \to E(Y_{-\infty}|\mathcal{B}_{-\infty})$$

almost surely.

53. A *potential* is a non-negative supermartingale $\{(X_n, \mathcal{B}_n), n \geq 0\}$ such that $E(X_n) \to 0$. Suppose the Doob decomposition (see Theorem 10.6.1) is $X_n = M_n - A_n$. Show

$$X_n = E(A_\infty|\mathcal{B}_n) - A_n.$$

54. Suppose f is a bounded continuous function on $\mathbb{R}$ and X is a random variable with distribution F. Assume for all $x \in \mathbb{R}$

$$f(x) = \int_{\mathbb{R}} f(x + y)F(dy) = E\big(f(x + X)\big).$$

Show using martingale theory that $f(x + s) = f(x)$ for each s in the support of F. In particular, if F has a density bounded away from 0, then f is constant. (Hint: Let $\{X_n\}$ be iid with common distribuion F and define an appropriate martingale.)

55. Suppose $\{X_j, j \geq 1\}$ are iid with common distribution F and let $\widehat{F}_n$ be the empirical distribution based on $X_1, \ldots, X_n$. Show

$$Y_n := \sup_{x \in \mathbb{R}} |\widehat{F}_n(x) - F(x)|$$

is a reversed submartingale.

Hint: Consider first $\{\widehat{F}_n(x) - F(x), n \geq 1\}$, then take absolute values, and then take the supremum over a countable set.)

56. Refer to Subsection 10.16.5. A *price system* is a mapping Π from the set of all contingent claims $\mathcal{X}$ to $[0, \infty)$ such that

$$\Pi(X) = 0 \text{ iff } X = 0, \quad \forall X \in \mathcal{X},$$
$$\Pi(aX + bX') = a\Pi(X) + b\Pi(X'),$$

for all $a \geq 0,\ b \geq 0,\ X, X' \in \mathcal{X}$. The price system Π is consistent with the market model if

$$\Pi(V_N(\phi)) = \Pi(V_0(\phi)),$$

for all admissible strategies ϕ.

(i) If P^* is an equivalent martingale measure, show

$$\Pi(X) := E^*(X/S_N^{(0)}), \quad \forall X \in \mathcal{X},$$

defines a price system that is consistent.

(ii) If Π is a consistent price system, show that P^* defined by

$$P^*(A) = \Pi(S_N^{(0)} 1_A), \quad \forall A \in \mathcal{B},$$

is an equivalent martingale measure.

(iii) If the market is complete, there is a unique initial price for a contingent claim.

References

[BD91] P. J. Brockwell and R. A. Davis. *Time Series: Theory and Methods (Second Edition)*. Springer: NY, 1991.

[Bil68] Patrick Billingsley. *Convergence of Probability Measures*. John Wiley & Sons Inc., New York, 1968.

[Bil95] Patrick Billingsley. *Probability and Measure (Third Edition)*. John Wiley & Sons Inc., New York, 1995.

[Bre92] Leo Breiman. *Probability*. SIAM: PA, 1992.

[Chu74] Kai Lai Chung. *A Course in Probability Theory*. Academic Press, New York-London, second edition, 1974. Probability and Mathematical Statistics, Vol. 21.

[Dur91] R. Durrett. *Probability: Theory and Examples*. Brooks Cole:CA, 1991.

[Fel68] William Feller. *An Introduction to Probability Theory and its Applications. Vol. I.* John Wiley & Sons Inc., New York, third edition, 1968.

[Fel71] William Feller. *An Introduction to Probability Theory and its Applications. Vol. II.* John Wiley & Sons Inc., New York, second edition, 1971.

[FG97] Bert Fristedt and Lawrence Gray. *A Modern Approach to Probability Theory*. Probability and its Applications. Birkhäuser Boston, Boston, MA, 1997.

[HK79] J. Michael Harrison and David M. Kreps. Martingales and arbitrage in multiperiod securities markets. *J. Econom. Theory*, 20(3):381–408, 1979.

[HP81] J. Michael Harrison and Stanley R. Pliska. Martingales and stochastic integrals in the theory of continuous trading. *StocProc*, 11:215–260, 1981.

[LL96] Damien Lamberton and Bernard Lapeyre. *Introduction to Stochastic Calculus Applied to Finance*. Chapman & Hall, London, 1996. Translated from the 1991 French original by Nicolas Rabeau and François Mantion.

[Loe77] M. Loève. *Probability Theory One (4th Ed)*. Springer: NY, 1977.

[Nev65] Jacques Neveu. *Mathematical foundations of the calculus of probability*. Holden-Day Inc., San Francisco, Calif., 1965. Translated by Amiel Feinstein.

[Nev75] J. Neveu. *Discrete-Parameter Martingales*. North-Holland Publishing Co., Amsterdam, revised edition, 1975. Translated from the French by T. P. Speed, North-Holland Mathematical Library, Vol. 10.

[Por94] Sidney C. Port. *Theoretical Probability for Applications*. Wiley Series in Probability and Mathematical Statistics: Probability and Mathematical Statistics. John Wiley & Sons Inc., New York, 1994. A Wiley-Interscience Publication.

[Res92] Sidney I. Resnick. *Adventures in Stochastic Processes*. Birkhäuser: Boston, MA, 1992.

[Rud74] Walter Rudin. *Real and Complex Analysis*. McGraw-Hill Book Co., New York, second edition, 1974.

Index